WORKBOOK FOR USE WITH

ANATOMY, PHYSIOLOGY, & DISEASE

Foundations for the Health Professions

Deborah Roiger, M.Ed.
St. Cloud Technical & Community College
St. Cloud, Minnesota

Nia Joyner Bullock, Ph.D.
Miller-Motte College
Cary, North Carolina

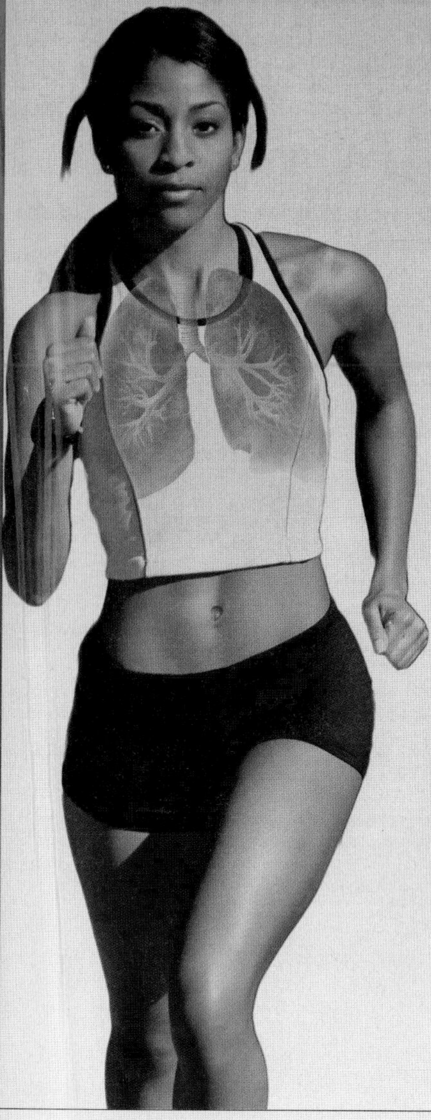

Mc Graw Hill
Connect Learn Succeed™

The McGraw·Hill Companies

WORKBOOK FOR USE WITH
ANATOMY, PHYSIOLOGY, & DISEASE: FOUNDATIONS FOR THE HEALTH PROFESSIONS
DEBORAH ROIGER AND NIA JOYNER BULLOCK

Published by McGraw-Hill, a business unit of The McGraw-Hill Companies, Inc., 1221 Avenue of the Americas, New York, NY 10020.

1 2 3 4 5 6 7 8 9 0 DOW/DOW 1 0 9 8 7 6 5 4 3

ISBN 978-0-07-740151-1
MHID 0-07-740151-4

www.mhhe.com

brief contents

preface

To the Student

Anatomy, Physiology, and Pathology is the foundational course for a wide variety of health careers. A firm understanding of the body's structures and normal or abnormal functions will be essential to your success in whichever career you choose. This workbook is directly tied to the main text, and it provides additional practice on every lesson. The following workbook features are designed to help you master this content:

- **Learning outcomes.** As with the textbook, each chapter in this workbook is driven by the same learning outcomes that are stated at the beginning of the textbook chapters.

learning outcomes

This chapter of the workbook is designed to help you learn the anatomy and physiology of the endocrine system. After completing this chapter in the text and this workbook, you should be able to:

9.1 Use medical terminology related to the endocrine system.

9.2 Compare and contrast the endocrine and nervous systems in terms of type, specificity, speed, and duration of communication.

9.3 Define *gland, hormone,* and *target tissue.*

9.4 List the major hormones, along with their target tissues and functions, of each of the endocrine system glands.

- **Word Roots & Combining Forms.** This section includes relevant word roots and combining forms to help you with medical terminology for each system.

word **roots** & combining **forms**

agglutin/o: clumping

blast/o: primitive cell

coagul/o: clot

cyt/o: cell

erythr/o: red

- **Coloring Book.** This section will help you locate the structures of the body and understand their relationship to each other. By coloring, you will become aware of the edges of each structure in the human anatomy in different views and of the way each structure relates to neighboring structures. Coloring or outlining will give you a broader understanding than would simply labeling structures on a diagram.

COLORING BOOK

Skeletal Muscles

Figures 6.1 to 6.7 show the skeletal muscles of the body by region. Color the box next to each term. Use the same color for the corresponding structures in the figures.

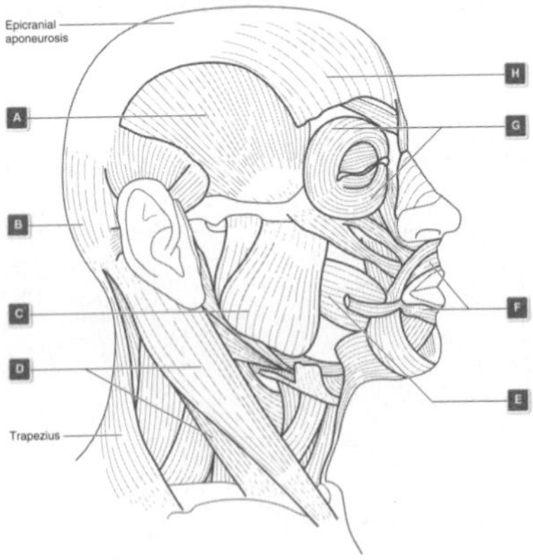

FIGURE 6.1 Muscles of the head and neck.

Muscles of the Head and Neck

- ☐ Temporalis(A)
- ☐ Occipitalis(B)
- ☐ Masseter(C)
- ☐ Sternocleidomastoid(D)
- ☐ Buccinator(E)
- ☐ Orbicularis oris(F)
- ☐ Orbicularis oculi(G)
- ☐ Frontalis(H)

- **Lab Exercises and Activities.** This section will help you understand the physiology of the human body. The exercises and activities will reinforce the concepts you have studied in the textbook.

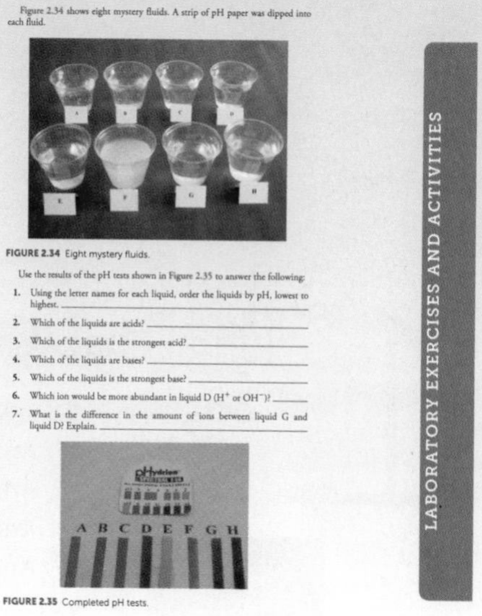
Figure 2.34 shows eight mystery fluids. A strip of pH paper was dipped into each fluid.

FIGURE 2.34 Eight mystery fluids.

Use the results of the pH tests shown in Figure 2.35 to answer the following:

1. Using the letter names for each liquid, order the liquids by pH, lowest to highest. ______
2. Which of the liquids are acids? ______
3. Which of the liquids is the strongest acid? ______
4. Which of the liquids are bases? ______
5. Which of the liquids is the strongest base? ______
6. Which ion would be more abundant in liquid D (H^+ or OH^-)? ______
7. What is the difference in the amount of ions between liquid G and liquid D? Explain. ______

FIGURE 2.35 Completed pH tests.

LABORATORY EXERCISES AND ACTIVITIES

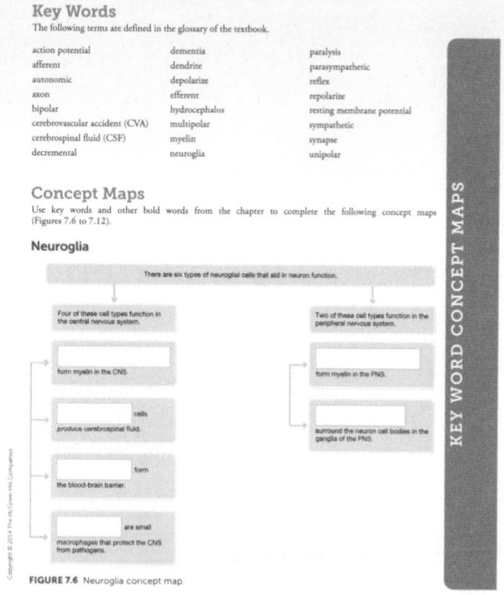
Key Words

The following terms are defined in the glossary of the textbook.

action potential	dementia	paralysis
afferent	dendrite	parasympathetic
autonomic	depolarize	reflex
axon	efferent	repolarize
bipolar	hydrocephalus	resting membrane potential
cerebrovascular accident (CVA)	multipolar	sympathetic
cerebrospinal fluid (CSF)	myelin	synapse
decremental	neuroglia	unipolar

Concept Maps

Use key words and other bold words from the chapter to complete the following concept maps (Figures 7.6 to 7.12).

Neuroglia

FIGURE 7.6 Neuroglia concept map.

KEY WORD CONCEPT MAPS

- **Key Word Concept Maps.** Each chapter of the workbook contains several concept maps that will help you understand the relationships between anatomy and physiology by linking concepts together. Making connections is a sure sign you are doing the critical thinking that will be invaluable in your new career.

- **Chapter Review Questions.** This section offers more practice to assess how much you have learned. The variety of exercises in this section includes **Word Deconstruction,** in which you will break down medical terms to their word roots, prefixes, and suffixes in addition to defining the term. Other exercise types include Multiple Select, Matching, Sentence Completion, Critical Thinking, and Case Study questions.

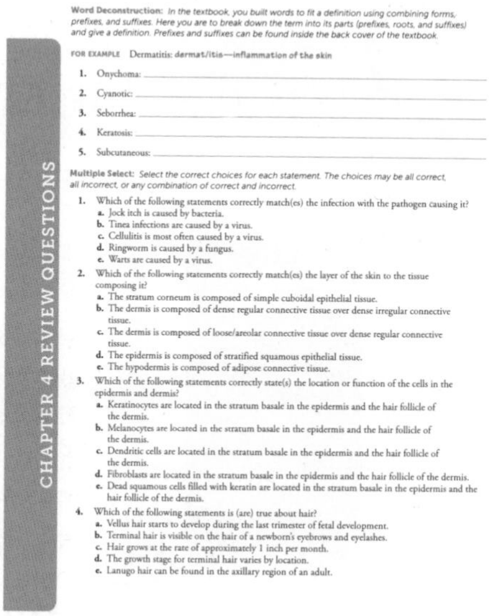
CHAPTER 4 REVIEW QUESTIONS

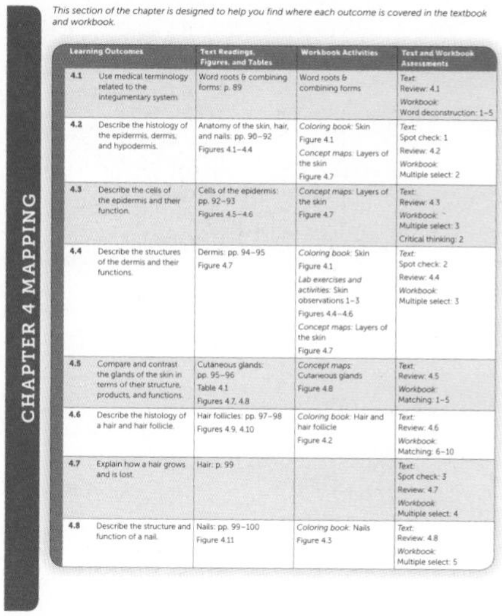
CHAPTER 4 MAPPING

- **Chapter Mapping.** Chapter Mapping at the end of each chapter quickly directs you to the location of all text, figures, tables, spot checks, and review questions in your textbook and workbook that are relevant to each learning outcome.

contents

3 Introduction to Pathology 60

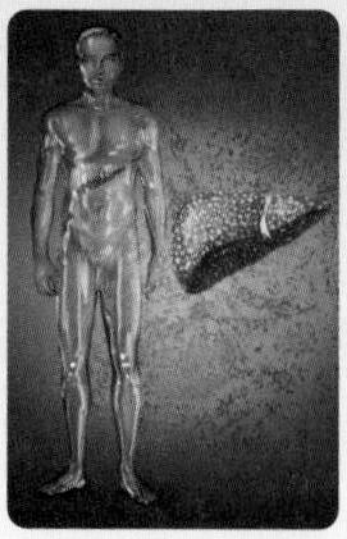

4 The Integumentary System 76

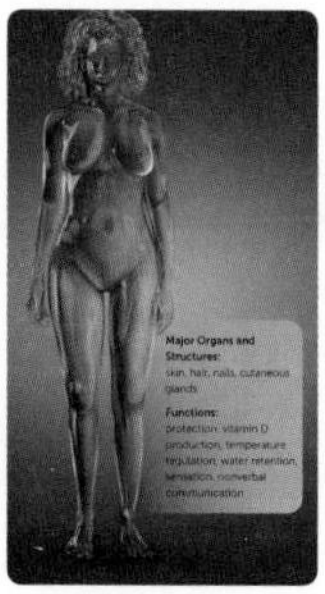

5 The Skeletal System 96

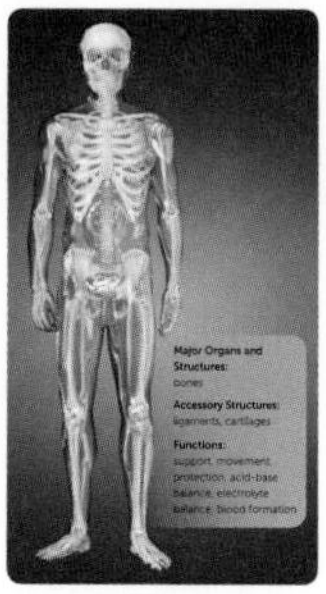

6 The Muscular System 126

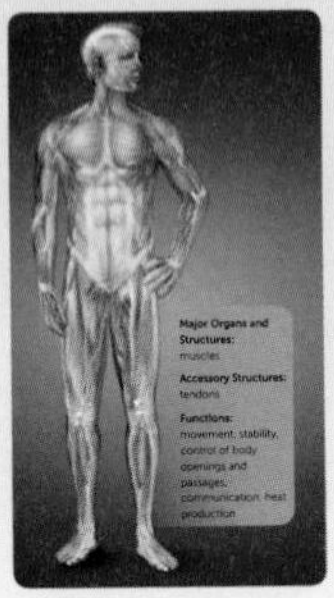

7 The Nervous System 152

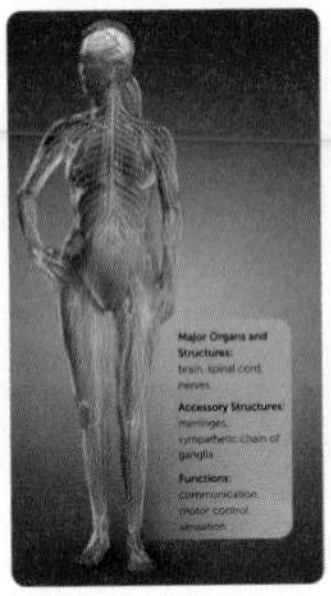

8 The Nervous System—Senses 174

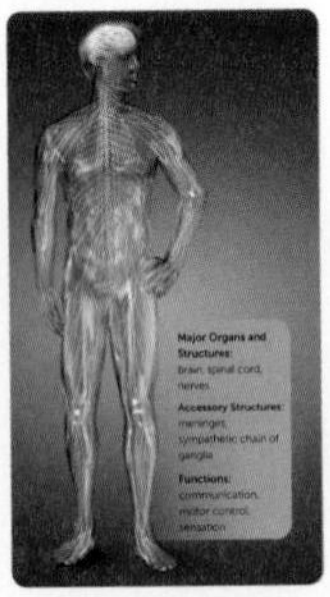

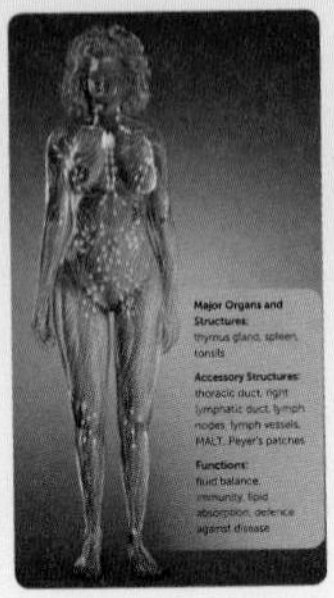

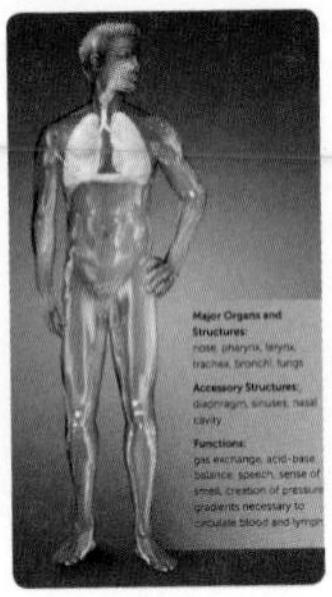

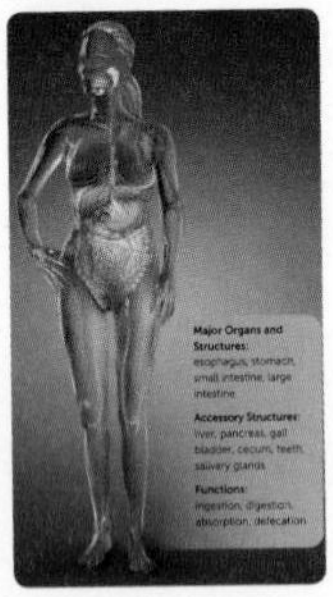

15 The Excretory/Urinary System 314

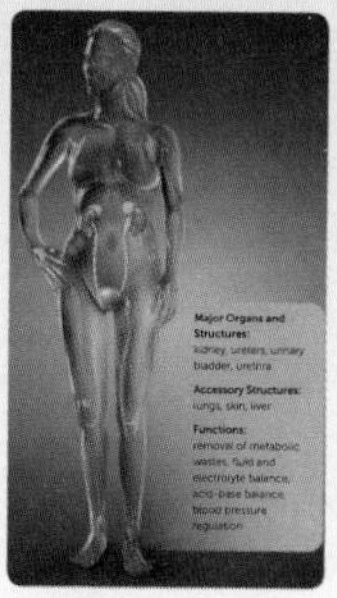

16 The Male Reproductive System 338

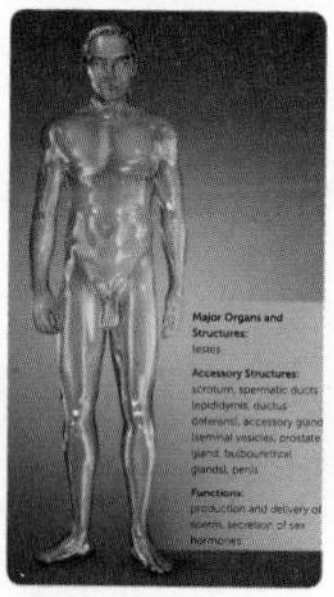

17 The Female Reproductive System 356

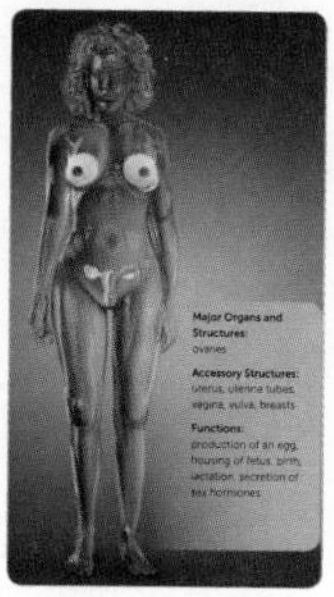

1 The Basics

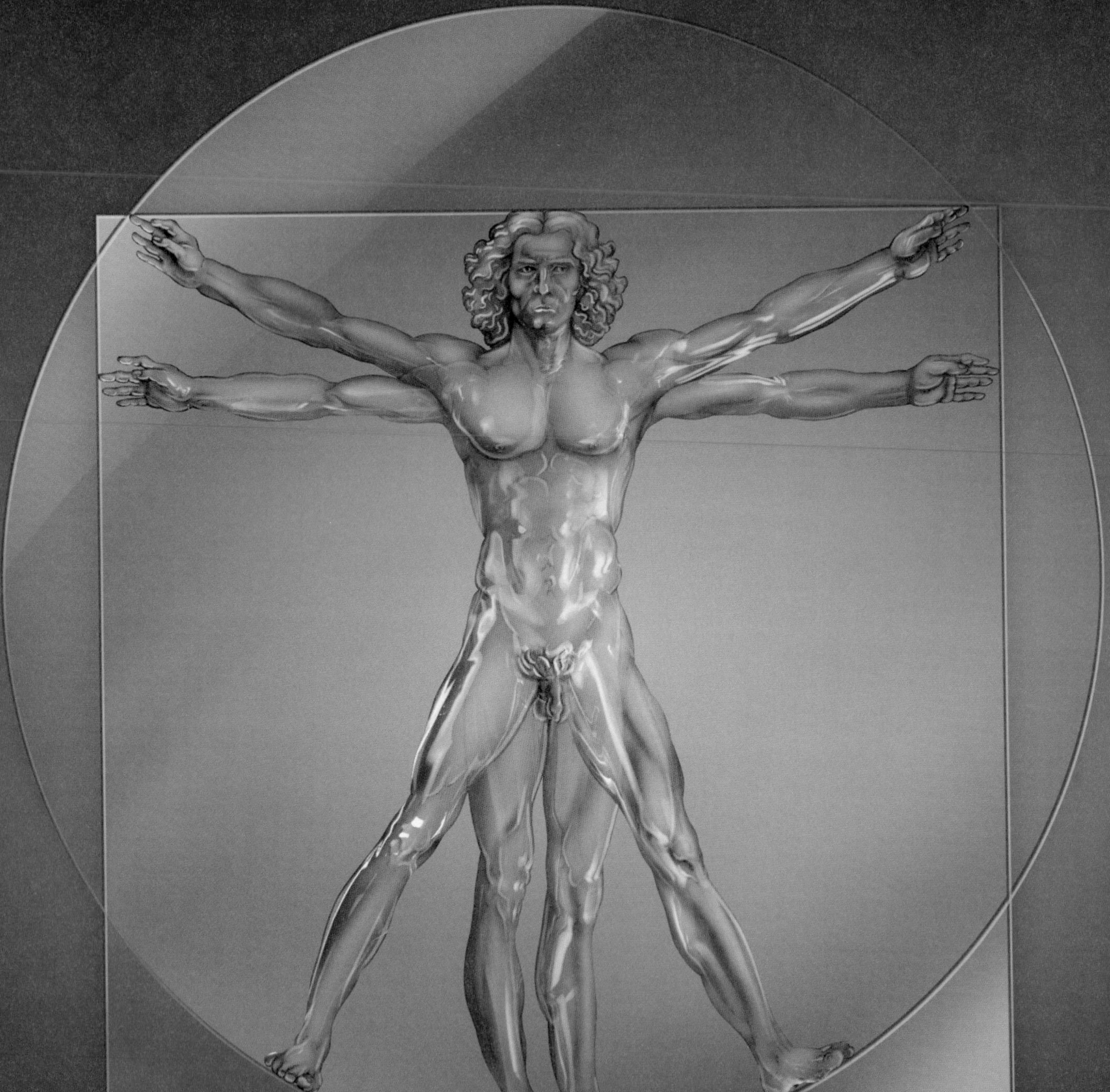

This chapter of the workbook is designed to help you learn the basic terms of anatomy and physiology of the human body. Like the following chapters in this workbook, this chapter is divided into five sections:

- **COLORING BOOK.** This section will help you locate the structures of the body and understand their relationship to each other.
- **LABORATORY EXERCISES AND ACTIVITIES.** This section will help you understand the physiology of the human body. The exercises and activities may include the use of common household items.
- **KEY WORD CONCEPT MAPS.** This section will help you understand the relationships between anatomy and physiology by linking concepts together.
- **CHAPTER REVIEW QUESTIONS.** This section is similar to the review questions in the text. It will give you more practice to assess how much you have learned.
- **CHAPTER MAPPING.** This section of the chapter helps you find where each learning outcome is addressed in the text and workbook.

learning outcomes

After completing this chapter in the text and this workbook, you should be able to:

1.1 Define *anatomy* and *physiology*.

1.2 Describe the location of structures in the human body using anatomical terms of direction, regions, planes, positions, and cavities.

1.3 Locate serous membranes by their individual names and relative location to organs.

1.4 Define *homeostasis* and explain why it is so important in human physiology.

1.5 Define *negative feedback* and *positive feedback* and explain their importance to homeostasis.

Anatomical Terms of Direction

Figure 1.1 shows the basic anatomical terms of direction. Color the box next to each term below. Use the same color for the corresponding arrow(s) in the drawing.

☐ **Cranial/superior$^{(A)}$.** These terms mean closer to the head end of the body or higher in the body than another structure. They are used for the axial region only.

☐ **Anterior/ventral$^{(B)}$.** These terms mean closer to the front or belly side of the body.

☐ **Posterior/dorsal$^{(C)}$.** These terms mean closer to the back side of the body.

☐ **Medial$^{(D)}$.** This term means closer to the midline of the body along the sagittal plane.

☐ **Lateral$^{(E)}$.** This term means farther from the midline of the body along the sagittal plane.

☐ **Proximal$^{(F)}$.** This term means closer to the attachment to the body. It is used for the appendicular region only.

☐ **Distal$^{(G)}$.** This term means farther from the attachment to the body. It is used for the appendicular region only.

☐ **Inferior$^{(H)}$.** This term means farther from the head end of the body or lower in the body than another structure. It is used for the axial region only.

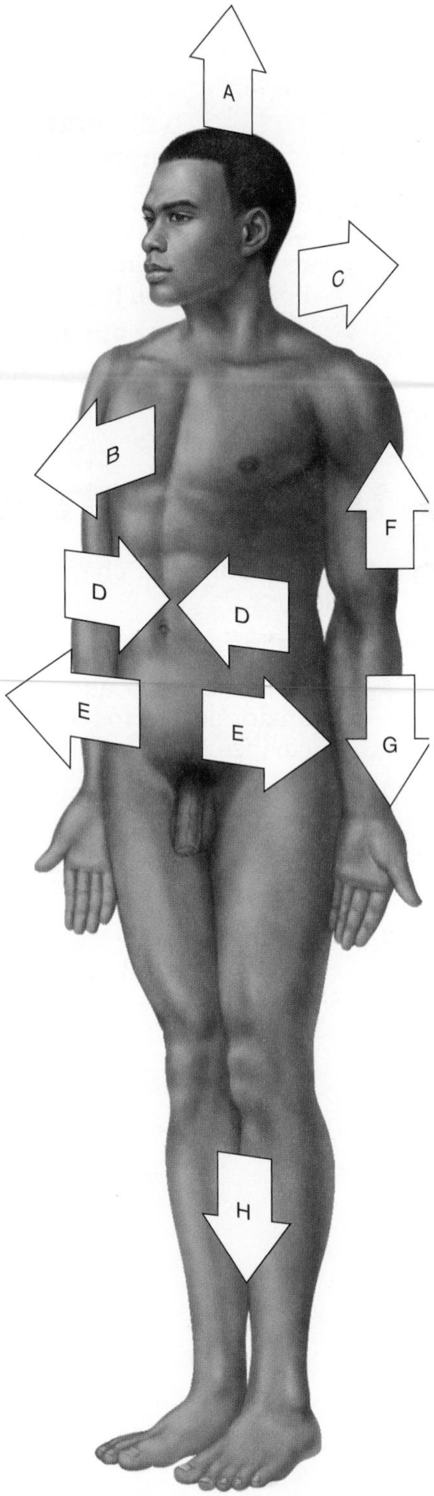

FIGURE 1.1 Anatomical terms of direction.

COLORING BOOK

Anatomical Regions

Figure 1.2 shows anatomical regions. Color the box next to each region name below. Use the same color for the arrow pointing to the corresponding region on the drawing.

Axial Region

- ☐ Abdominal region (A)
- ☐ Axillary region (B)
- ☐ Cranial/cephalic region (C)
- ☐ Cervical region (D)
- ☐ Facial region (E)
- ☐ Inguinal region (F)
- ☐ Pelvic region (G)
- ☐ Thoracic region (H)
- ☐ Umbilical region (I)

Appendicular Region

- ☐ Brachial region (J)
- ☐ Carpal region (K)
- ☐ Cubital region (L)
- ☐ Femoral region (M)
- ☐ Palmar region (N)
- ☐ Patellar region (O)
- ☐ Plantar region (P)
- ☐ Tarsal region (Q)

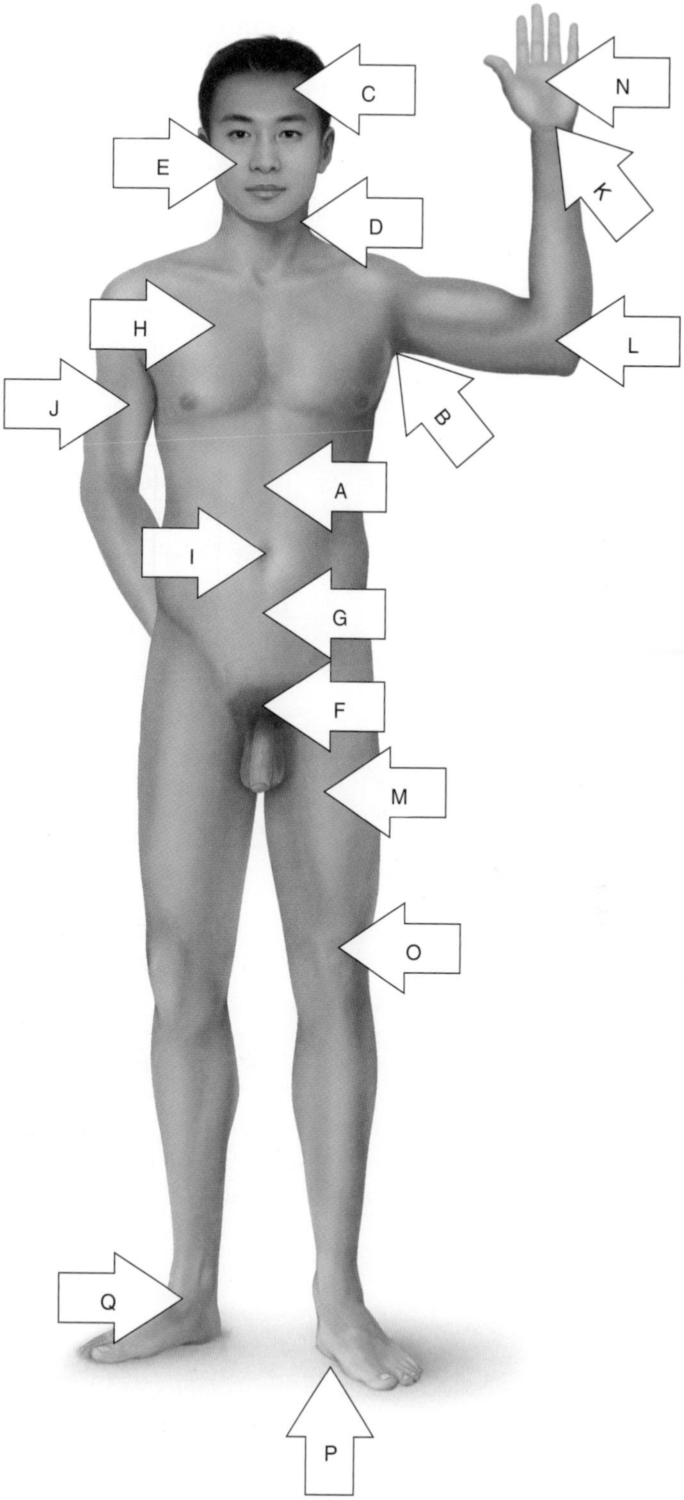

FIGURE 1.2 Anatomical regions.

COLORING BOOK

Anatomical Planes

Figure 1.3 shows the anatomical planes of the human body. Color the box next to the name of each plane below. Use the same color for the corresponding plane in the drawing.

☐ **Sagittal plane(B).** This plane separates right from left.

☐ **Midsagittal plane(A).** This is a sagittal plane down the midline of the body.

☐ **Frontal/coronal plane(C).** This plane separates front from back.

☐ **Transverse plane(D).** This plane separates top from bottom.

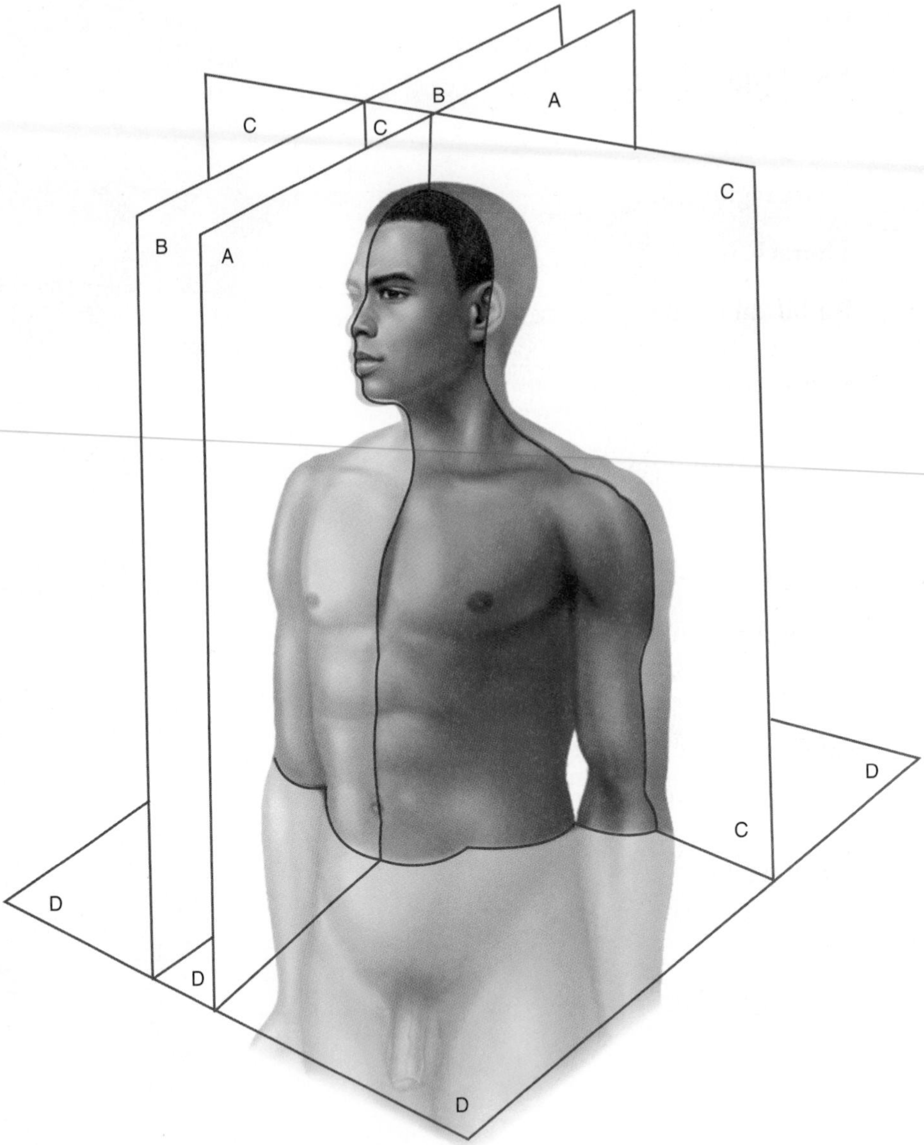

FIGURE 1.3 Anatomical planes.

COLORING BOOK

Anatomical Cavities

Figure 1.4 shows the cavities in the human body. Color the box next to the name of each cavity below. Use the same color for the corresponding cavity in the drawing.

- ☐ **Cranial cavity**[(A)]
- ☐ **Thoracic cavity**[(B)]
- ☐ **Abdominal cavity**[(C)]
- ☐ **Pelvic cavity**[(D)]
- ☐ **Vertebral cavity**[(E)]
- ☐ **Pleural cavity**[(F)]
- ☐ **Pericardial cavity**[(G)]

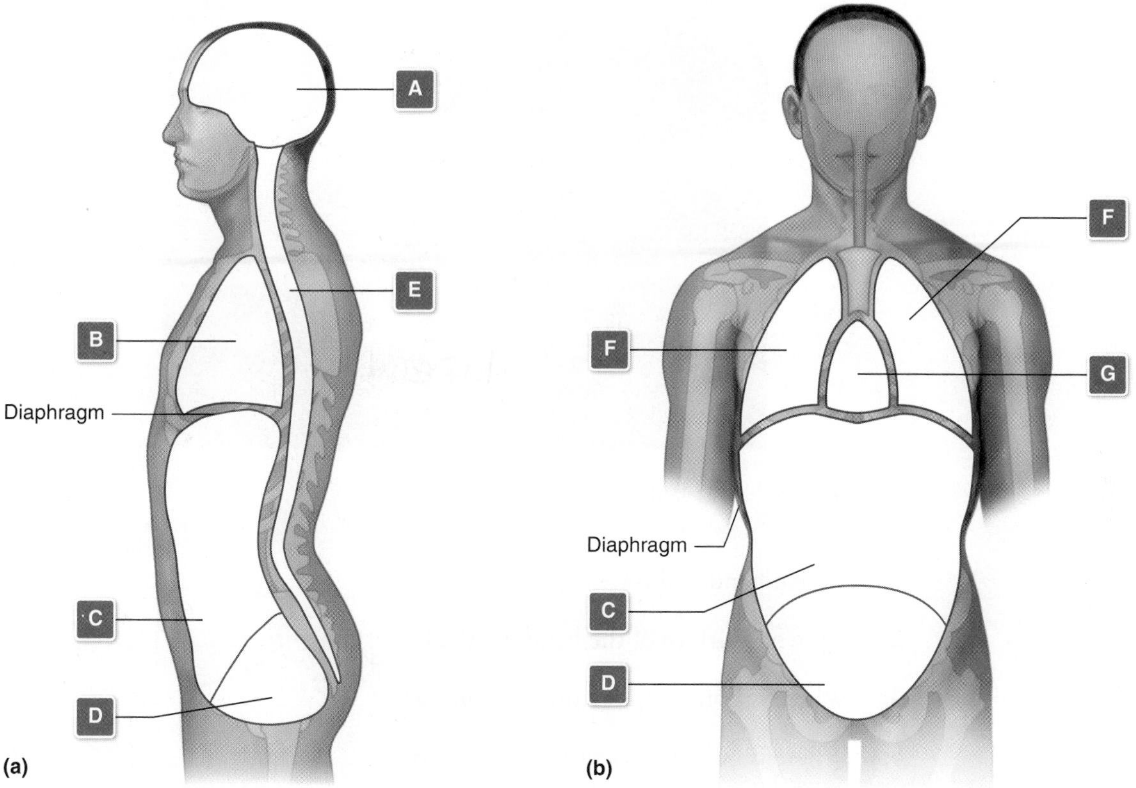

FIGURE 1.4 Cavities: (a) sagittal view, (b) anterior view.

COLORING BOOK

Hand

One of the advantages of studying human body anatomy is that you have all the structures of the human body with you every day. You can use your own body to practice your terms. **Figure 1.5** shows the bones of the hand. Use anatomical terms of direction to answer the following questions.

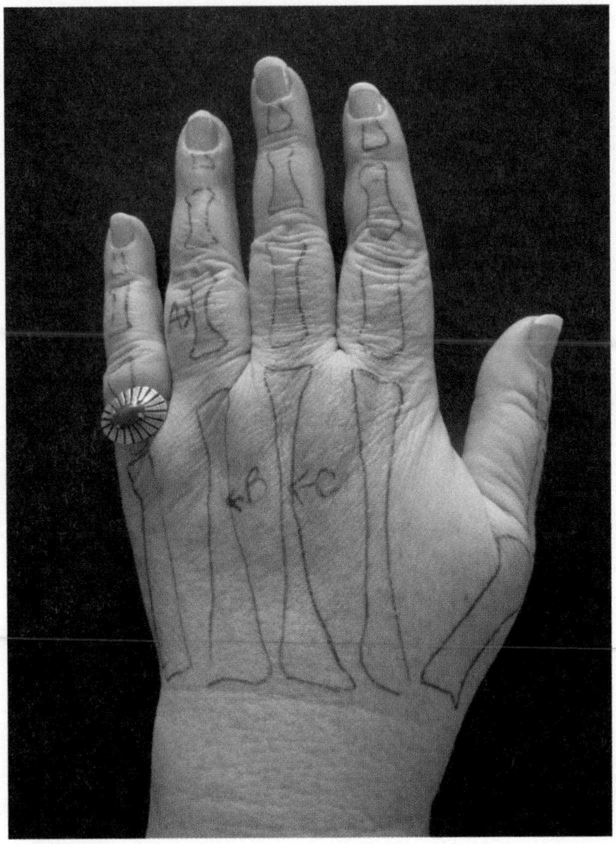

FIGURE 1.5 Hand with bones drawn on it.

1. Which hand is this? ____________________
2. On what surface of the hand are the bones drawn? ____________________
3. Where is bone A in relation to bone B? ____________________
4. Where is bone C in relation to bone B? ____________________

warning

Did you remember to put the hand in standard anatomical position before answering these questions? For which question(s) does standard anatomical position make a difference?

____

Heart

Figure 1.6 shows a heart surgery. Use the appropriate anatomical term(s) to answer the following questions.

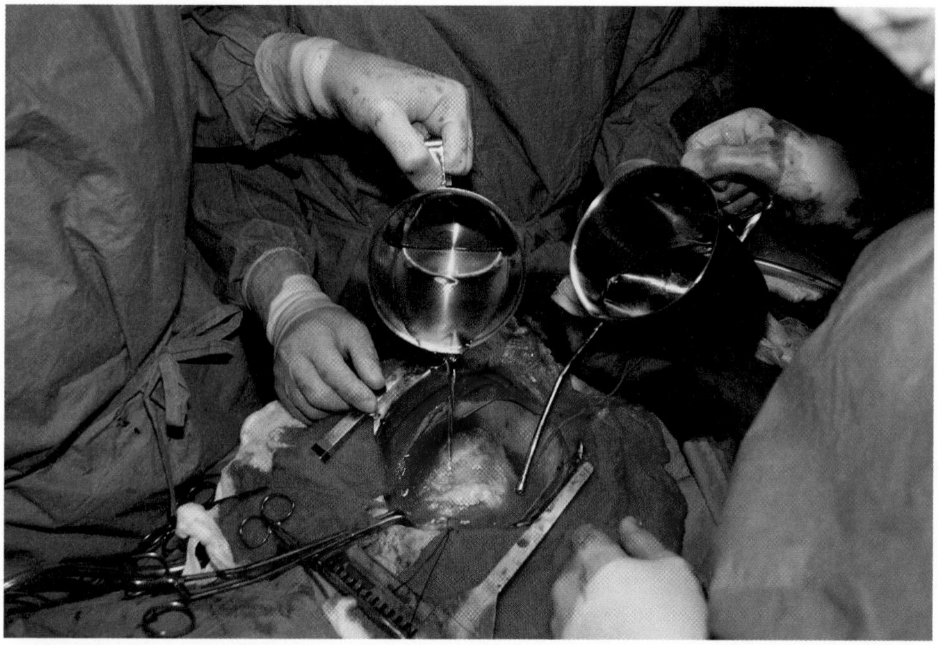

FIGURE 1.6 Heart surgery.

5. In what general cavity is the heart located? ____________________

6. What specific cavities are lateral to the heart? ____________________

7. What specific membrane would need to be cut in order to expose the heart? ____________________

8. The aorta is a major vessel exiting the heart. The trachea and esophagus run posterior to the heart. In what space in the chest are the aorta, trachea, and esophagus located? ____________________

Key Words

The following terms are defined in the glossary of the textbook.

abdominal	meninges	pleura
anatomy	mesentery	positive feedback
appendicular	negative feedback	proximal
axial	parietal	serous membrane
cranial	pelvic	thoracic
greater omentum	peritoneum	visceral
homeostasis	physiology	

Concept Maps

Complete the boxes in the following concept maps (**Figures 1.7** to **1.10**). Each map contains at least one key word.

Anatomical Terms

Anatomical terms of direction describe the location of one body part in relation to another.

Anterior or ventral refers to the belly side of the body, while posterior or Dorsal refers to the back side of the body.

Eyes are superior and lateral to the nose, while the nose is inferior and medial to the eyes.

Superior refers to closer to the top, while inferior refers to closer to the bottom.

The elbow is proximal to the wrist, while the ankle is distal to the knee.

The epidermis (outer layer of the skin) is superficial to the dermis, while the subcutaneous layer of the skin is deep to the dermis and epidermis.

FIGURE 1.7 Anatomical terms concept map.

Anatomical Regions

Anatomical regions describe various areas of the body.

The head, neck, and trunk are part of the Axial region.

The arms and legs are part of the Apendicular region.

This region can be further divided into:

- The Cephalic region, which is the head.
- The cervical region, which is the neck.
- The Abdominal region, which is the belly.
- The pelvic region, which is the lower end of the trunk.

FIGURE 1.8 Anatomical regions concept map.

Anatomical Cavities

Anatomical cavities describe divisions or compartments of the body.

The dorsal cavity contains the [cranial] cavity and the [vertebral] cavity.

These cavities are lined by the [meninges].

The two [Pleural] cavities, the [Pericardial] cavity and the mediastinum, are all located in the [thoracic] cavity.

The abdominopelvic cavity contains the [abdominal] cavity and the [pelvic] cavity.

These cavities are lined by double-layered [pleural] membranes with fluid between the layers.

The layer of the membrane in contact with an organ is the [visceral] layer, while the layer of the membrane that does not come in contact with an organ is the [parietal] layer.

The [pleura] is the doubled-layered membrane that surrounds each lung in the thoracic cavity.

The [pericardium] is the double-layered membrane that surrounds the heart.

The [peritoneum] is the double-layered membrane that surrounds the organs in the abdominopelvic cavity.

FIGURE 1.9 Anatomical cavities concept map.

Homeostasis

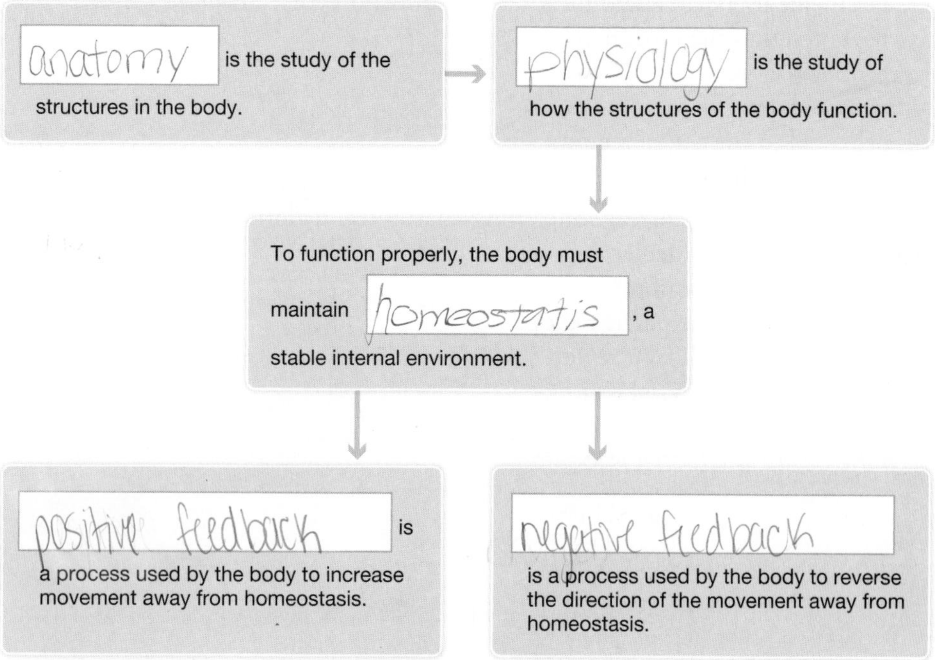
anatomy is the study of the structures in the body.
physiology is the study of how the structures of the body function.
To function properly, the body must maintain homeostatis, a stable internal environment.
positive feedback is a process used by the body to increase movement away from homeostasis.
negative feedback is a process used by the body to reverse the direction of the movement away from homeostasis.

FIGURE 1.10 Homeostasis concept map.

Multiple Select: *Select the correct choices for each statement. The choices may be all correct, all incorrect, or any combination of correct and incorrect.*

1. Which of the following statements use(s) an anatomical term of direction correctly?
   - **a.** The anterior surface of the arm has the most hair.
   - **b.** The liver is distal to the diaphragm.
   - **c.** The elbow is superior to the wrist.
   - **d.** The cubital region is proximal to the carpal region.
   - **e.** The esophagus is posterior to the heart.
2. Which of the following statements is (are) correct concerning anatomical regions?
   - **a.** The plantar region is part of the appendicular region.
   - **b.** The palmar region is part of the axial region.
   - **c.** The abdominal region is superior to the pelvic region.
   - **d.** The pelvic region is part of the appendicular region.
   - **e.** The cranial region is superior to the cervical region.

Matching: *Match the organ to the cavity in which it is located.*

_______	**1.** Pleural cavity	**a.** Uterus
_______	**2.** Pericardial cavity	**b.** Brain
_______	**3.** Abdominal cavity	**c.** Lung
_______	**4.** Pelvic cavity	**d.** Heart
_______	**5.** Cranial cavity	**e.** Liver

Completion: *Fill in the blanks to complete the following statements concerning the cadaver images in Figures 1.11 to 1.13. Use a term for an anatomical plane in statements 1 to 3.*

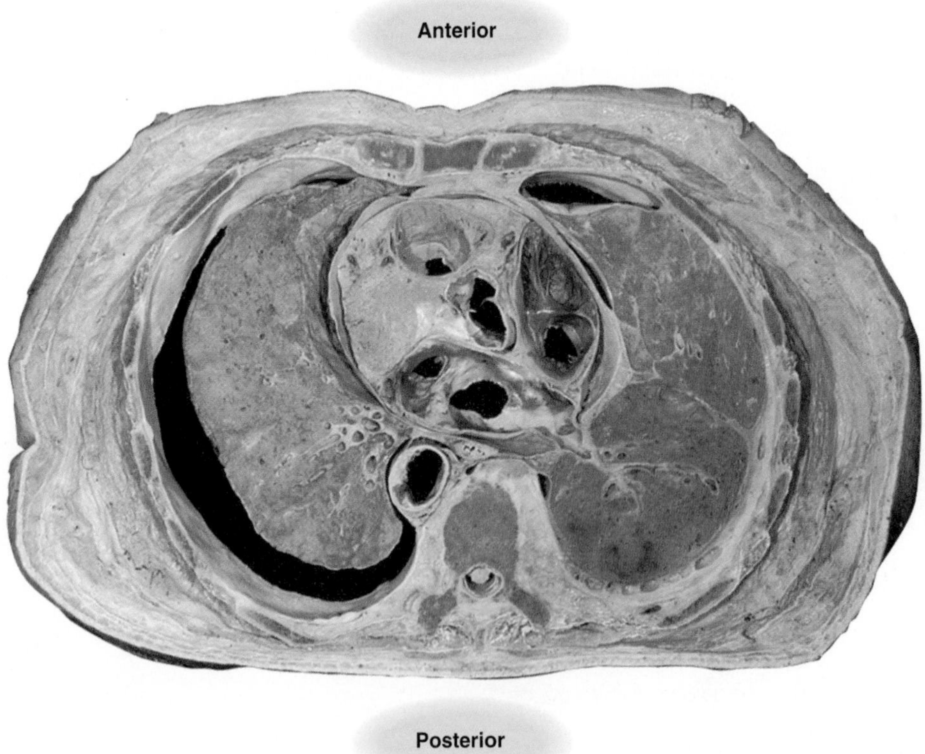

FIGURE 1.11

CHAPTER 1 REVIEW QUESTIONS

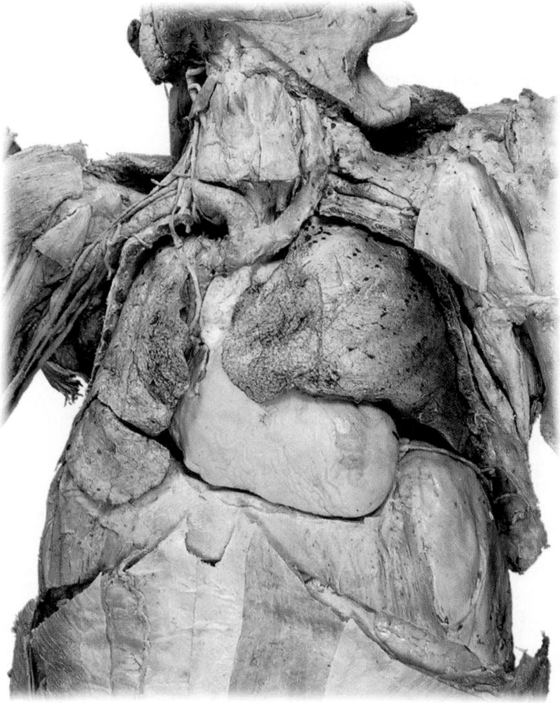

FIGURE 1.12

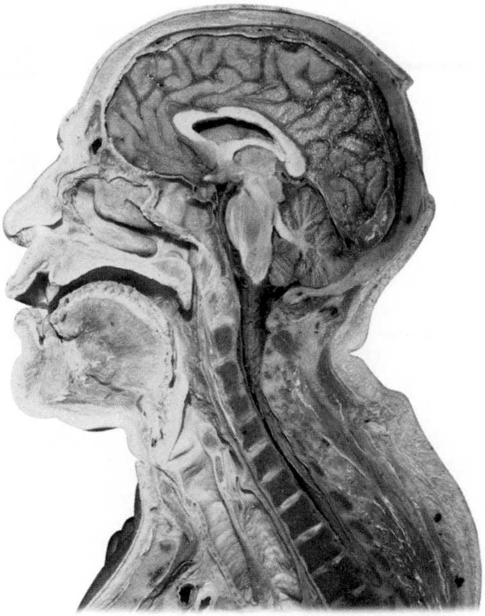

FIGURE 1.13

1. The body in **Figure 1.11** was cut along a ____________________ plane.
2. The body in **Figure 1.12** was cut along a ____________________ plane.
3. The body in **Figure 1.13** was cut along a ____________________ plane.
4. The definition of *homeostasis* is ____________________.
5. The definition of *physiology* is ____________________.

Critical Thinking

1. An arrow entered the left axillary region of a hunter and lodged in his heart. List in order the layers of serous membrane the arrow would have pierced. (**Hint:** You may want to sketch the arrow entering the thorax and the relative serous membranes.)

2. What anatomical plane of the body would be used to illustrate the relative position of the liver, stomach, and spinal column? ____________________

   What anatomical plane of the body would be used to illustrate the relative position of the urinary bladder, the stomach, and the liver? ____________________

Case Study

1. A patient comes to the emergency room with an elevated heart rate and a low blood pressure that continues to fall. He is diagnosed as having a bleeding ulcer. The bleeding is stopped with medical intervention and he is given a blood transfusion. His heart rate and blood pressure return to normal. What homeostatic feedback mechanisms, if any, are evident in this situation?

   ____

   ____

   ____

   ____

   ____

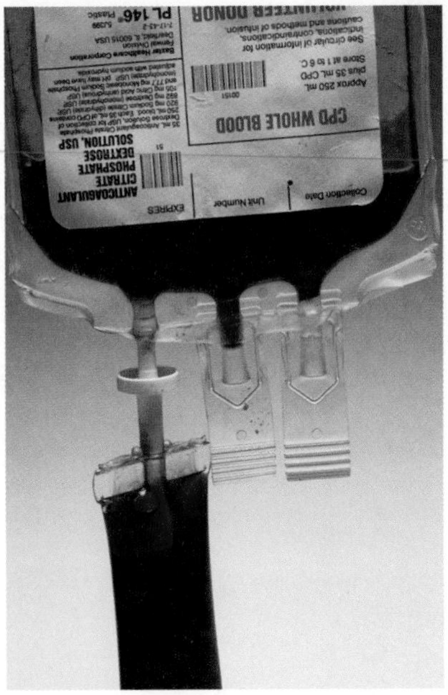

CHAPTER 1 REVIEW QUESTIONS

This section of the chapter is designed to help you find where each outcome is covered in the textbook and workbook.

Learning Outcomes		Text Readings, Figures, and Tables	Workbook Activities	Text and Workbook Assessments
1.1	Define *anatomy* and *physiology.*	Overview: p. 4	*Concept maps:* Homeostasis Figure 1.10	*Text:* Review: 1.1 *Workbook:* Completion: 5
1.2	Describe the location of structures in the human body using anatomical terms of direction, regions, planes, positions, and cavities.	Anatomical terms: pp. 4–11 Tables 1.1–1.5 Figures 1.1–1.8	*Coloring book:* Anatomical terms of direction; Anatomical regions; Anatomical planes; Anatomical cavities Figures 1.1–1.4 *Lab exercises and activities:* Hand; Heart Figures 1.5, 1.6 *Concept maps:* Anatomical terms; Anatomical regions; Anatomical cavities Figures 1.7–1.9	*Text:* Spot Check: 1, 2 Review: 1.2 *Workbook:* Multiple Select: 1, 2 Matching: 1–5 Completion: 1–3
1.3	Locate serous membranes by their individual names and relative location to organs.	Serous membranes: pp. 11–14 Figures 1.9–1.13	*Concept maps:* Anatomical cavities Figure 1.9	*Text:* Spot Check: 3 Review: 1.3 *Workbook:* Critical Thinking: 1
1.4	Define *homeostasis* and explain why it is so important in human physiology.	Homeostasis: p. 14	*Concept maps:* Homeostasis Figure 1.10	*Text:* Review: 1.4 *Workbook:* Completion: 4
1.5	Define *negative feedback* and *positive feedback* and explain their importance to homeostasis.	Homeostasis: pp. 14–16 Figure 1.14	*Concept maps:* Homeostasis Figure 1.10	*Text:* Spot Check: 4 Review: 1.5 *Workbook:* Critical Thinking: 2 Case Study: 1

2 Levels of Organization of the Human Body

learning outcomes

This chapter of the workbook is designed to help you learn fundamentals of the human body's levels of organization. After completing this chapter in the text and this workbook, you should be able to:

2.1 List the levels of organization of the human body from simplest to most complex.

2.2 Define the terms *matter, element, atom,* and *isotope.*

2.3 Define *molecule* and describe two methods of bonding that may form molecules.

2.4 Summarize the five functions of water in the human body and give an explanation or example of each.

2.5 Compare solutions based on tonicity.

2.6 Determine whether a substance is an acid or a base and its relative strength if given its pH.

2.7 Describe the four types of organic molecules in the body by giving the elements present in each, their building blocks, an example of each, the location of each example in the body, and the function of each example.

2.8 Explain three factors governing the speed of chemical reactions.

2.9 Write the equation for cellular respiration using chemical symbols and describe it in words.

2.10 Explain the importance of ATP in terms of energy use in the cell.

2.11 Describe cell organelles and explain their functions.

2.12 Compare four methods of passive transport and active transport across a cell membrane in terms of materials moved, direction of movement, and the amount of energy required.

2.13 Describe bulk transport, including endocytosis and exocytosis.

2.14 Describe the processes of transcription and translation in protein synthesis in terms of location and the relevant nucleic acids involved.

2.15 Describe what happens to a protein after translation.

2.16 Explain the possible consequences of mistakes in protein synthesis.

2.17 Describe the process of mitosis, including a comparison of the chromosomes in a parent cell to the chromosomes in the daughter cells.

2.18 Explain the possible consequences of mistakes in replication.

2.19 Describe the effects of aging on cell division.

2.20 Describe the four classifications of tissues in the human body.

2.21 Describe the modes of tissue growth, change, shrinkage, and death.

2.22 Identify the human body systems and their major organs.

The Cell

Figure 2.1 is a drawing of a generic cell. It is a composite of the organelles possible in various cells. Each type of cell in the human body has its own unique set of organelles necessary to carry out its function. Color the box next to each term. Use the same color for the corresponding structure in **Figure 2.1**.

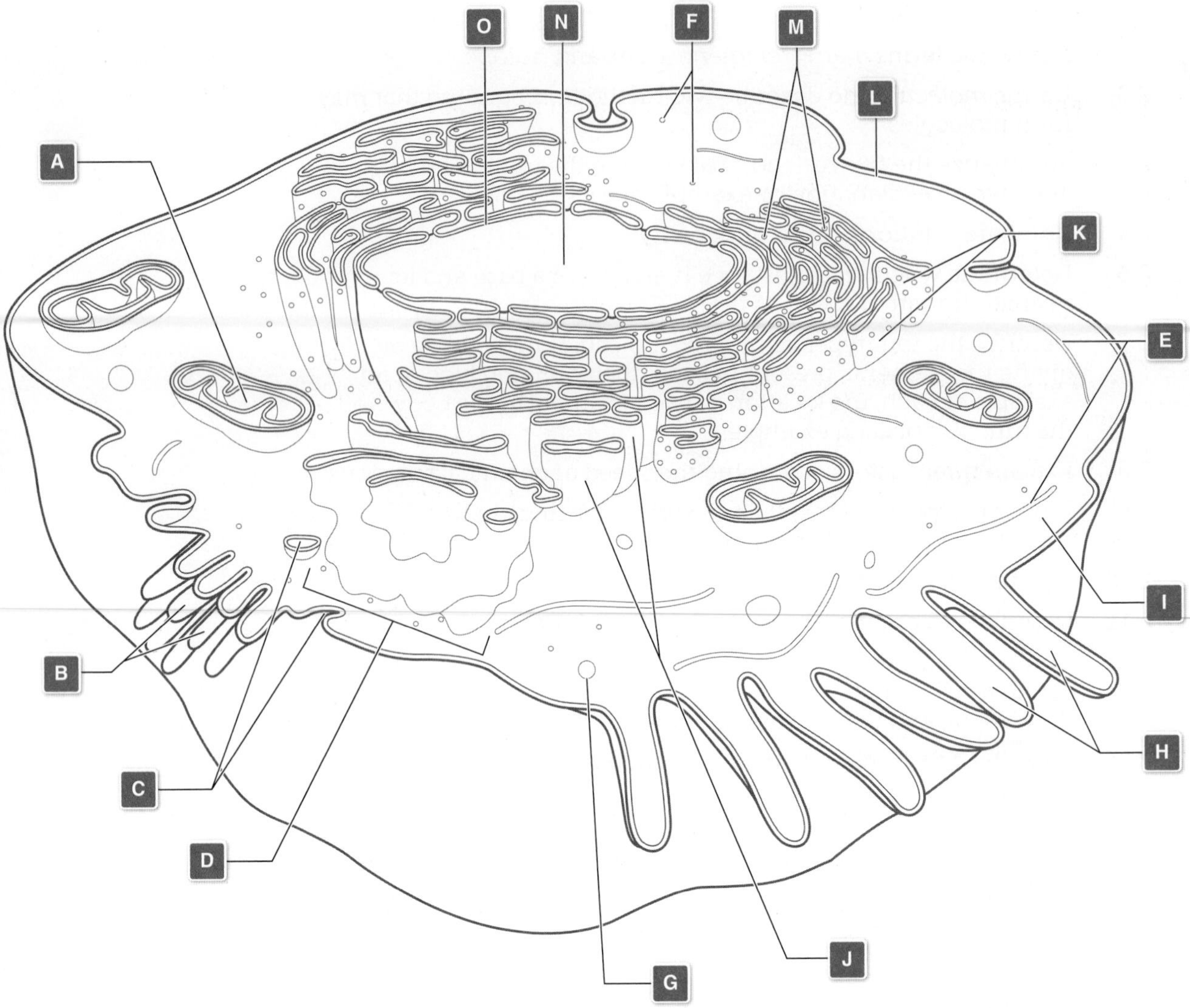

FIGURE 2.1 Generic cell.

☐ **Mitochondrion**(A). Rod-shaped organelle; carries out cellular respiration and uses the energy produced from it to create ATP.

☐ **Microvilli**(B). Hairlike extensions of the cell membrane that are used for extra surface area.

☐ **Secretory vesicles**(C). Membrane packages bubbled off the Golgi complex that are used to transport cell products.

☐ **Golgi complex**(D). Organelle of membrane folds close to the endoplasmic reticulum. It inspects and modifies the proteins and lipids produced in the cell.

☐ **Cytoskeleton**(E). Protein fibers that suspend the organelles in the cytoplasm.

☐ **Free ribosomes**(F). Organelles composed of large and small subunits that function together to assemble proteins.

☐ **Lysosome**(G). Membrane-bound package of digesting enzymes.

☐ **Cilia**(H). Hairlike extensions of the cell membrane that are used to move materials past the cell.

☐ **Cytoplasm**(I). Fluid within the cell. This solution must contain all the solutes the cell needs to function.

☐ **Smooth endoplasmic reticulum**(J). Extension of the nuclear membrane that has no ribosomes on its surface. It is the site of lipid production.

☐ **Rough endoplasmic reticulum**(K). Extension of the nuclear membrane that has ribosomes on its surface. It is the site of protein synthesis.

☐ **Cell membrane**(L). Phospholipid bilayer that gives definition to the cell and regulates what may go in or out of the cell.

☐ **Fixed ribosomes**(M). Organelles located on the rough endoplasmic reticulum. They are composed of large and small subunits that function together to assemble proteins.

☐ **Chromatin**(N). This is not an organelle. It is the DNA spread out loosely in the nucleus so that it can be used.

☐ **Nucleus**(O). Organelle enclosed by a membrane. It contains the DNA for the cell.

Tissues

Cells come together to function as tissues. There are four classes of tissues: epithelial, connective, muscle, and nervous. There may be different types of tissue in each class.

Epithelial Tissue

All of the types of tissue in this class either cover structure surfaces or line hollow structures. Therefore, there will always be an open edge with these tissues. To identify this type of tissue, first look for an open edge, then the tissue (as described below), and finally a basement membrane that separates the epithelial tissue from other tissues.

Epithelial tissues are named for whether they are ciliated, their cell shape, and the amount of layering of the cells. The cells may be squamous (flat), cuboidal (like a cube), or columnar (like a column). There may be a single layer (simple), multiple layers (stratified), or a false layering (pseudostratified) that looks as though it is layered but in which all cells touch the basement membrane. **Figure 2.2** shows cell shapes and layering. Color the box next to the term. Use the same color for the corresponding shape or layering in **Figure 2.2**.

- Squamous(A)
- Cuboidal(B)
- Columnar(C)
- Simple(D)
- Stratified(E)
- Pseudostratified(F)

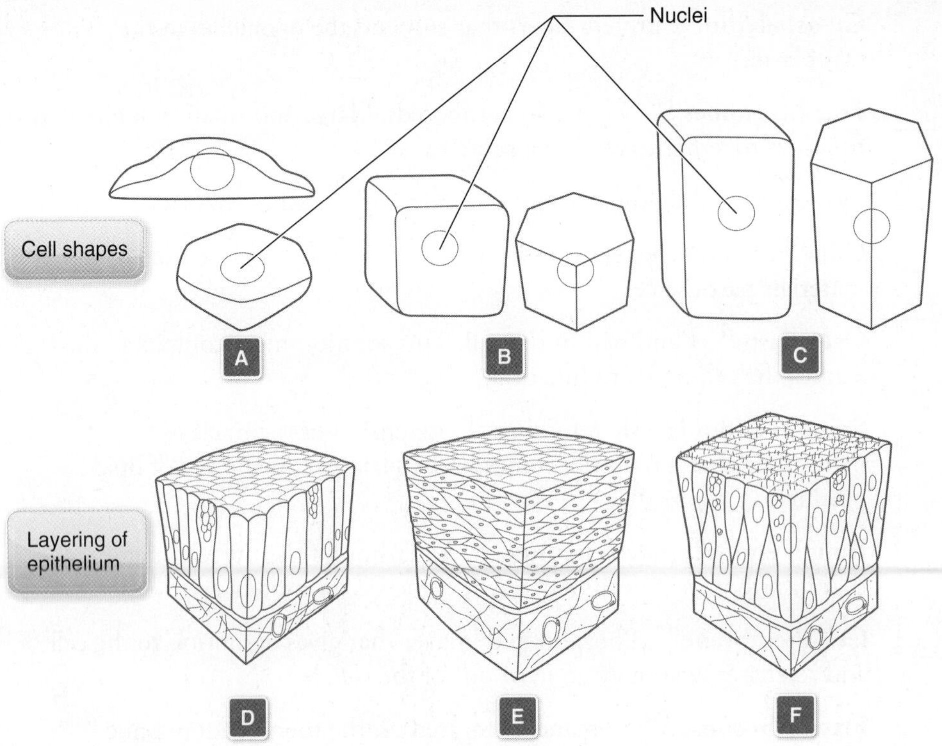

FIGURE 2.2 Cell shapes and layering.

Five types of epithelial tissue are shown with other tissues in **Figures 2.3** to **2.7**. Color the box next to the name of the epithelial tissue. Use the same color for just the epithelial tissue and its basement membrane in the corresponding figure.

Simple Squamous Epithelial Tissue This tissue can be found lining the air sacs of the lungs. See **Figure 2.3**.

- Squamous epithelial cells(A)
- Basement membrane(B)

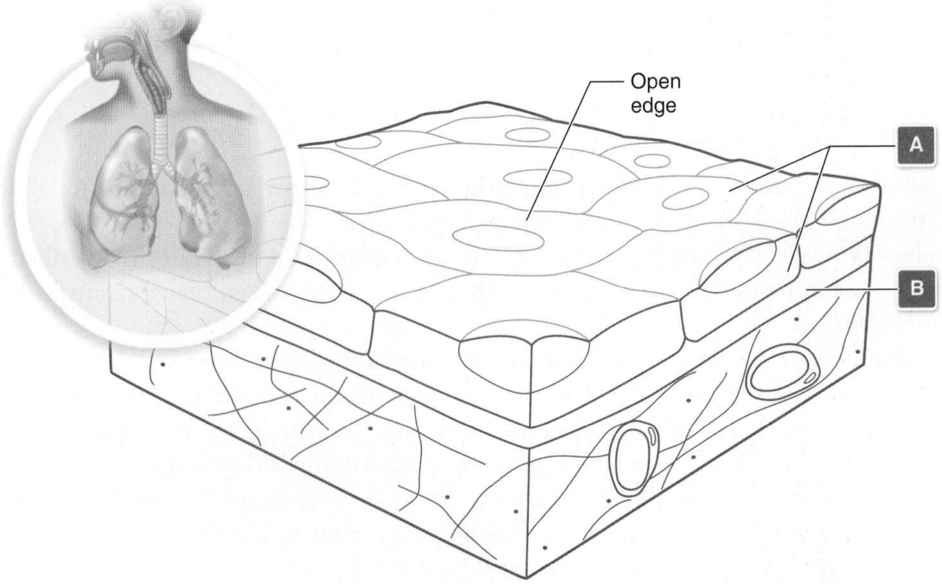

FIGURE 2.3 Simple squamous epithelial tissue.

Simple Columnar Epithelial Tissue This tissue can be found in the lining of the stomach and intestines. See **Figure 2.4**.

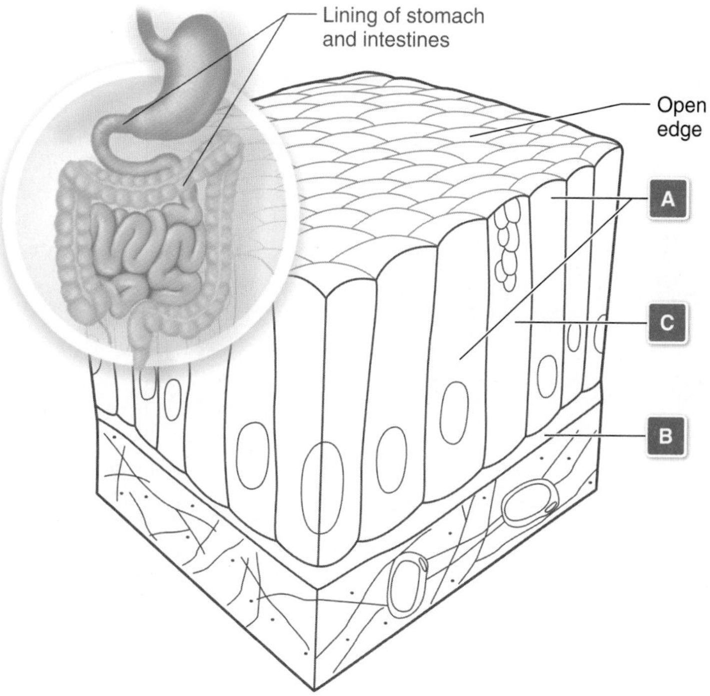

- [ ] Columnar cells(A)
- [ ] Basement membrane(B)
- [ ] Goblet cell containing mucus(C)

FIGURE 2.4 Simple columnar epithelial tissue.

Ciliated Pseudostratified Columnar Epithelial Tissue This tissue can be found lining parts of the respiratory tract. See **Figure 2.5**.

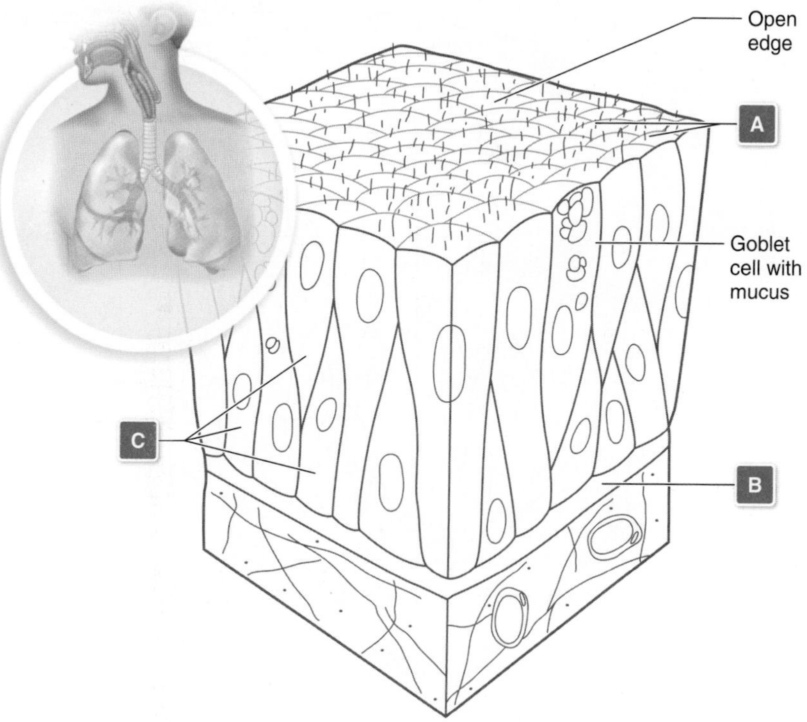

- [ ] Cilia(A)
- [ ] Basement membrane(B)
- [ ] Pseudostratified columnar cells(C)

FIGURE 2.5 Ciliated pseudostratified columnar epithelial tissue.

COLORING BOOK

Stratified Cuboidal Epithelial Tissue This tissue can be found in glandular ducts of the skin. Although this tissue is not mentioned in the text, this diagram will be good practice in naming a tissue new to you. See **Figure 2.6**.

- ☐ Cuboidal cells(A)
- ☐ Basement membrane(B)

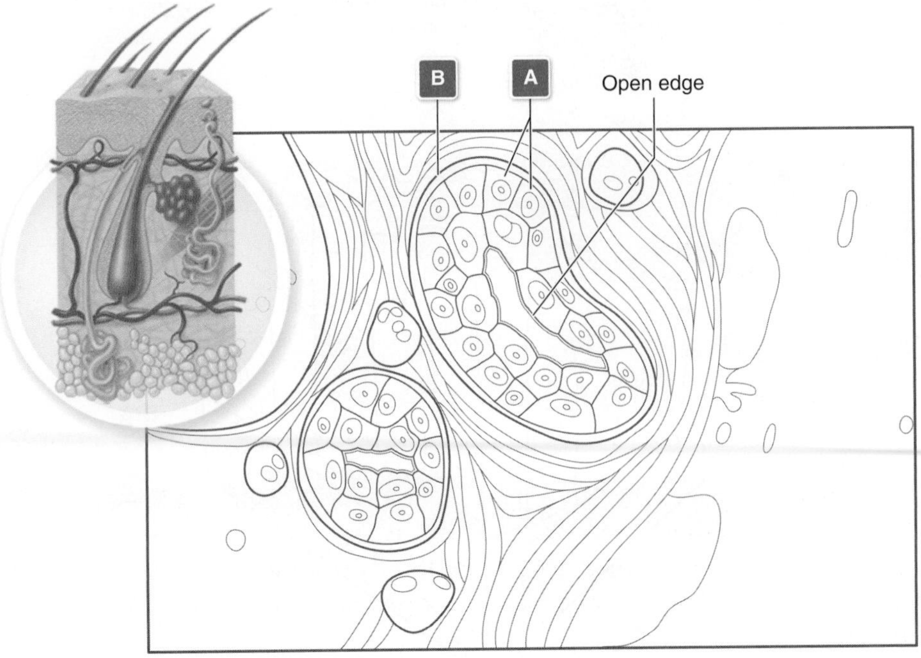

B
A
Open edge

FIGURE 2.6 Stratified cuboidal epithelial tissue.

Transitional Epithelial Tissue This tissue is stratified, but the shape of the cells is difficult to determine. This type of epithelial tissue is designed to be able to stretch. It can be found lining the urinary bladder. See **Figure 2.7**.

- ☐ Basement membrane(A)
- ☐ Epithelial cells(B)

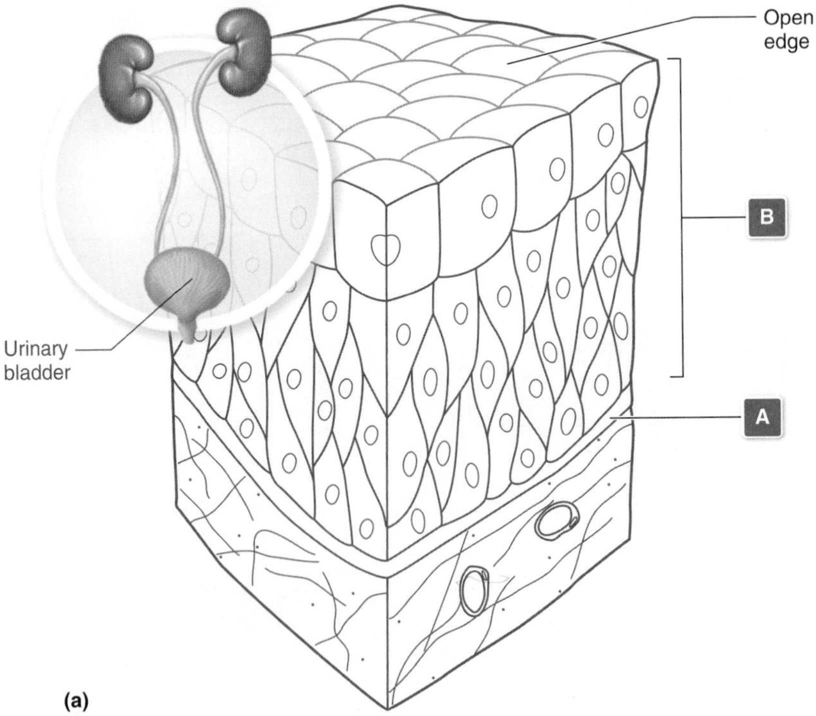

Open edge
B
A
Urinary bladder
(a)

FIGURE 2.7 Transitional epithelial tissue: (a) tissue not stretched, (b) tissue stretched.

COLORING BOOK

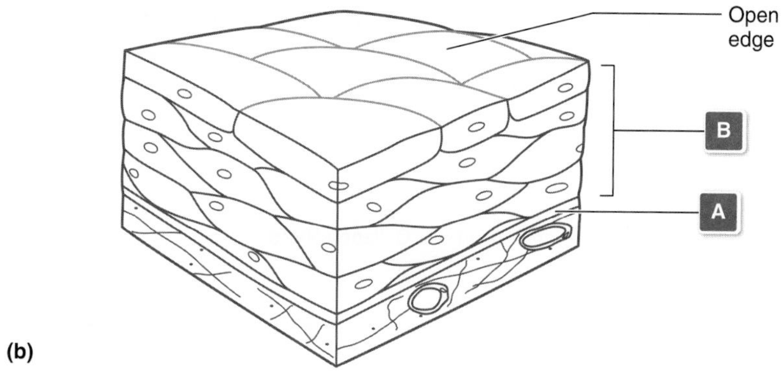
Open
edge
B
A
(b)

FIGURE 2.7 concluded

Connective Tissue

All types of tissues in this class have cells in a matrix (a background substance) and possible fibers. The matrix may be very fluid (as in blood) or as hard as concrete (as in bone).

Six types of connective tissue—loose/areolar, dense regular, adipose, blood, bone, and cartilage—are shown in **Figures 2.8** to **2.15**. Color the box next to the name of the connective tissue. Use the same color for the cells and fibers of the connective tissue in the corresponding figure unless additional structures are indicated.

Loose/Areolar Connective Tissue This tissue can be found in the dermis of the skin. See **Figure 2.8**.

☐ Fibroblast$^{(A)}$

☐ Fibers$^{(B)}$

☐ Matrix$^{(C)}$

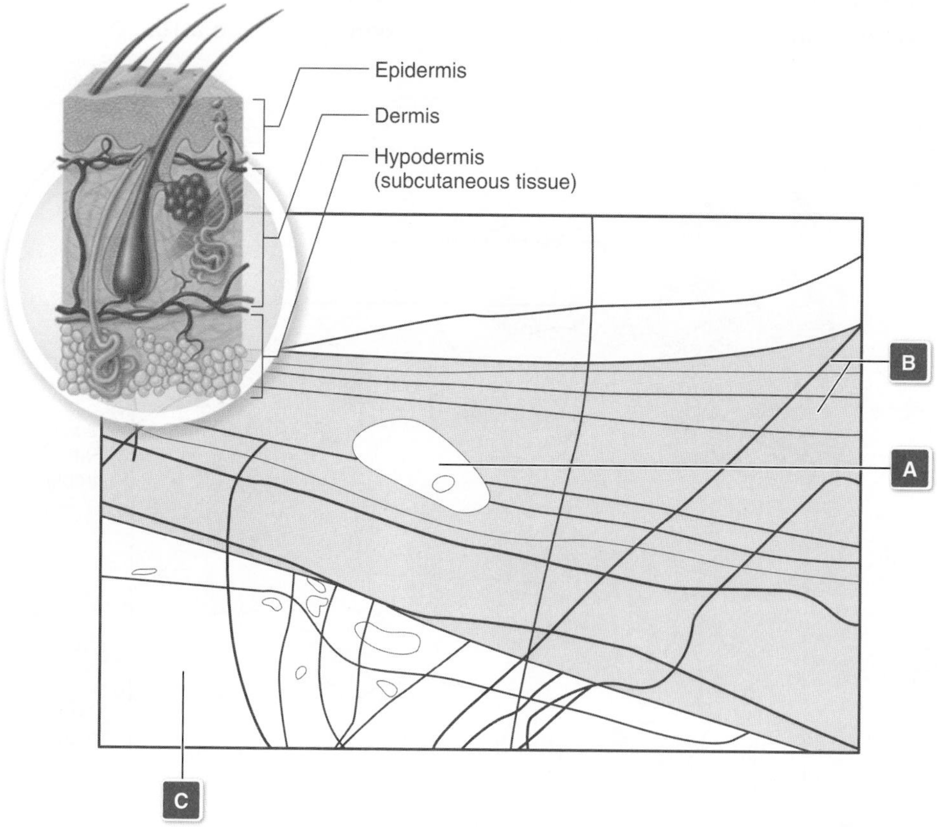
Epidermis
Dermis
Hypodermis
(subcutaneous tissue)
B
A
C

FIGURE 2.8 Loose/areolar connective tissue.

COLORING BOOK

Dense Regular Connective Tissue The collagen fibers of this tissue are tightly packed. Fiber-making cells are occasionally interspersed between fibers. This tissue can be found in tendons and ligaments. See **Figure 2.9**.

☐ Cells

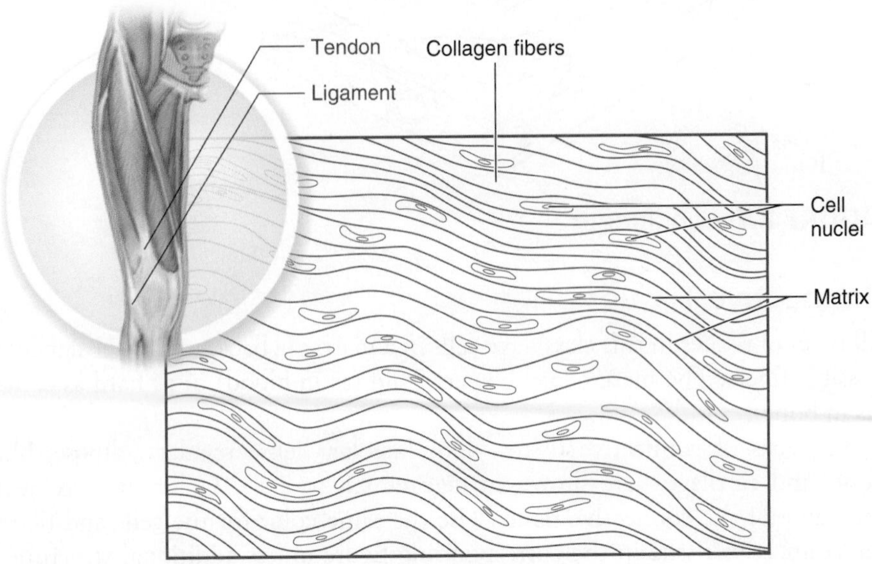

FIGURE 2.9 Dense regular connective tissue.

Adipose Connective Tissue These cells are tightly packed. Each cell is so filled with a fat droplet that the nucleus of the cell is pushed off to the side. Adipose connective tissue is found deep to the skin. See **Figure 2.10**.

☐ Adipose cells

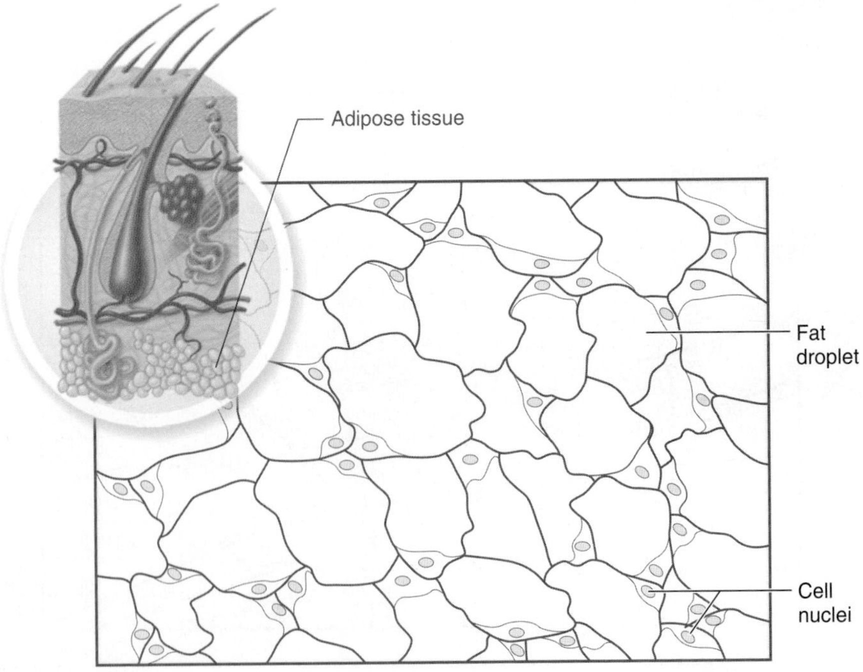

FIGURE 2.10 Adipose connective tissue.

Blood Connective Tissue This connective tissue has red blood cells (without nuclei), white blood cells (with nuclei), and platelets (cell fragments) in a very fluid matrix called *plasma.* It can be found traveling through the heart and all of the blood vessels of the body, such as the aorta. See **Figure 2.11**.

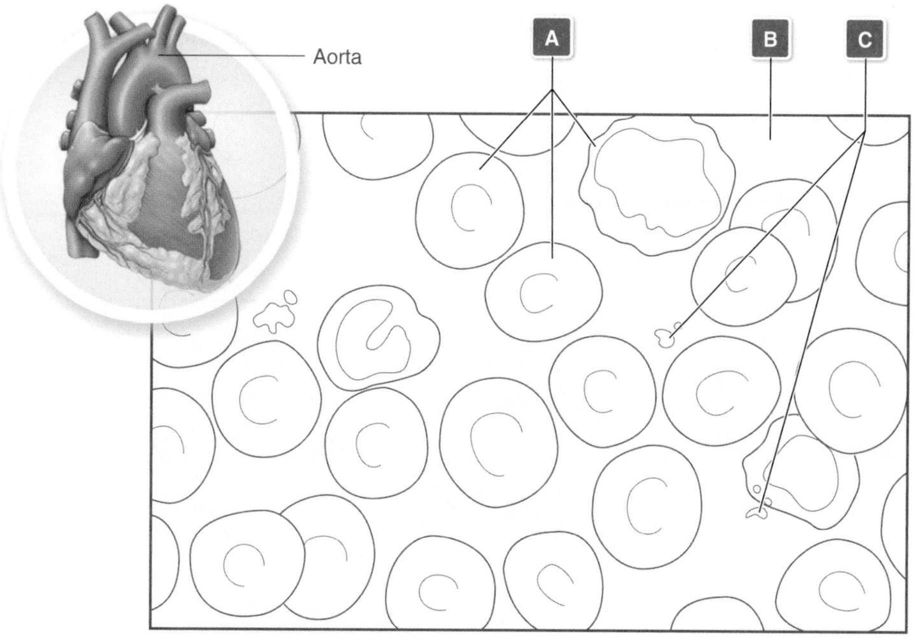

- ☐ Blood cells(A)
- ☐ Plasma (matrix)(B)
- ☐ Platelets(C)

FIGURE 2.11 Blood connective tissue.

Bone Connective Tissue This tissue has cells isolated in small spaces within a very hard matrix. It is found in all of the bones of the body. See **Figure 2.12**.

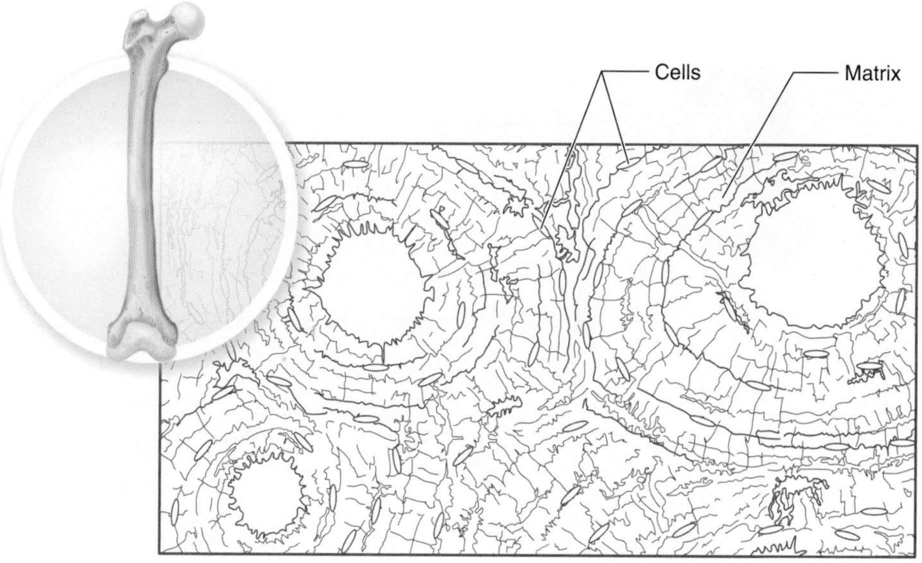

- ☐ Cells
- ☐ Matrix

FIGURE 2.12 Bone connective tissue.

Cartilage Connective Tissue Three types of cartilage connective tissue—hyaline, elastic, and fibrocartilage—are shown in **Figures 2.13** to **2.15**. In general, this connective tissue is characterized by a background matrix that looks like colored gelatin. The cells appear to be located within bubbles in the gelatin. Fibers help determine the type of cartilage. Color the box next to the name of the cartilage. Use the same color for the matrix of the cartilage.

Hyaline cartilage connective tissue. The fibers in this tissue are so fine that they may be hard to detect. Hyaline cartilage connective tissue can be found in the costal cartilages that connect the ribs to the sternum. See **Figure 2.13**.

☐ Hyaline cartilage matrix

☐ Cartilage cells

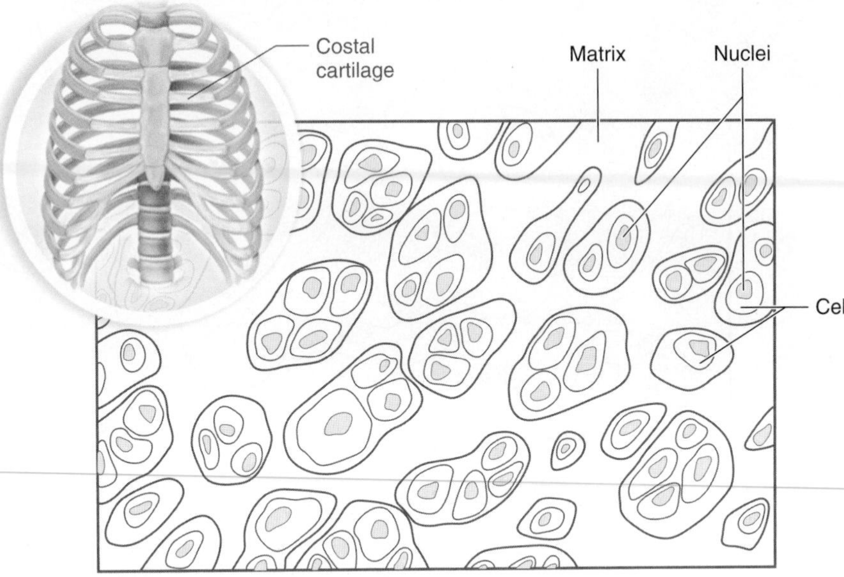

FIGURE 2.13 Hyaline cartilage connective tissue.

Elastic cartilage connective tissue. This tissue can be found in the outer ear. Its fibers go in many directions. See **Figure 2.14**.

☐ Elastic cartilage matrix

☐ Cartilage cells

☐ Fibers

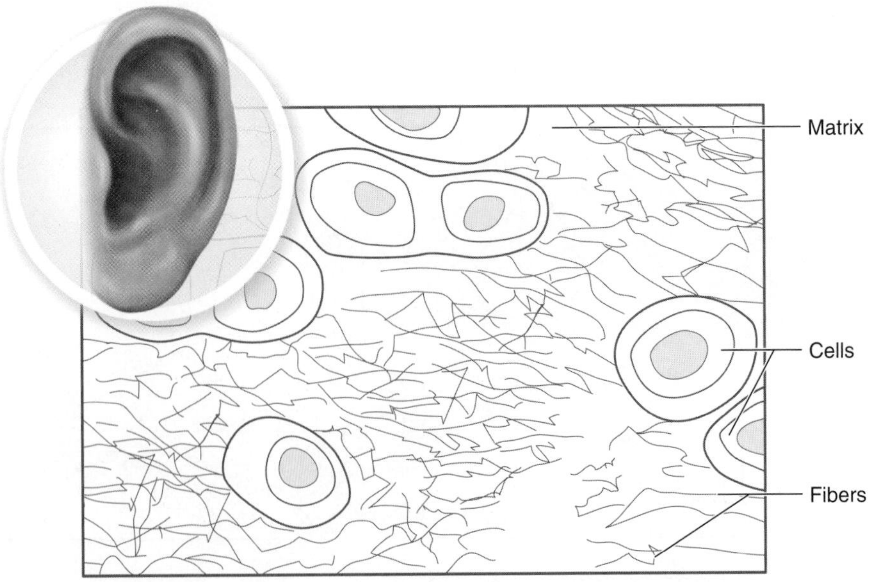

FIGURE 2.14 Elastic cartilage connective tissue.

Fibrocartilage connective tissue. This tissue can be found in intervertebral disks. The fibers go in one direction. See **Figure 2.15**.

☐ Fibrocartilage matrix

☐ Cartilage cells

☐ Fibers

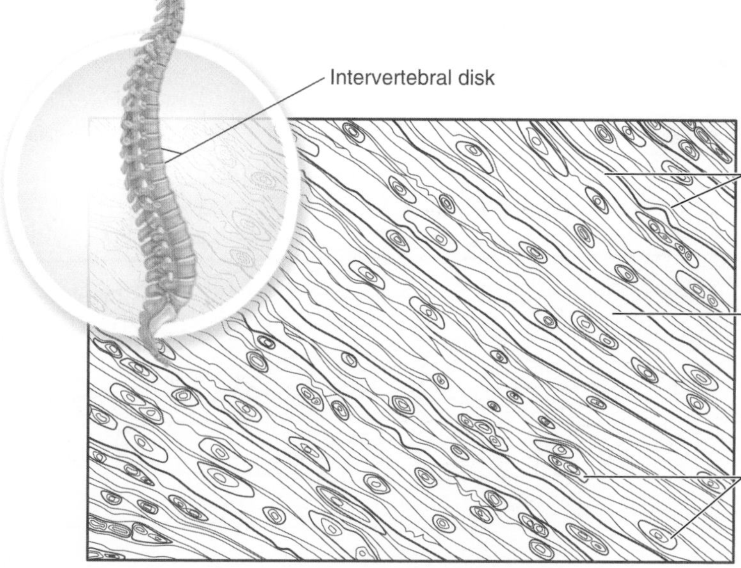

FIGURE 2.15 Fibrocartilage connective tissue.

Muscle Tissue

The three types of muscle tissue—skeletal muscle, smooth muscle, and cardiac muscle—all contain cells with a high concentration of proteins. The appearance of each type of muscle tissue differs on the basis of the arrangement of the proteins within the cells. Color the box next to the name of the muscle tissue. Use the same color for the cells of the muscle tissue in the corresponding figure.

Skeletal Muscle Tissue This tissue has striated (striped), cylindrical cells, each with many nuclei pushed off to the side. This tissue can be found in skeletal muscles that move the body, such as the biceps brachii. See **Figure 2.16**.

☐ Skeletal muscle cells

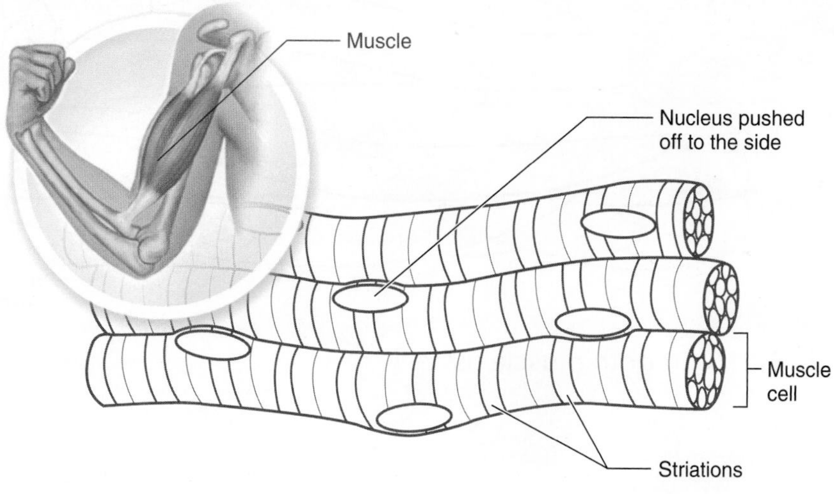

FIGURE 2.16 Skeletal muscle tissue.

COLORING BOOK

Cardiac Muscle Tissue This tissue has striated, branching cells with one nucleus per cell and specialized junctions, called *intercalated disks,* between cells. It can be found in the heart. See **Figure 2.17**.

☐ Cardiac muscle cells

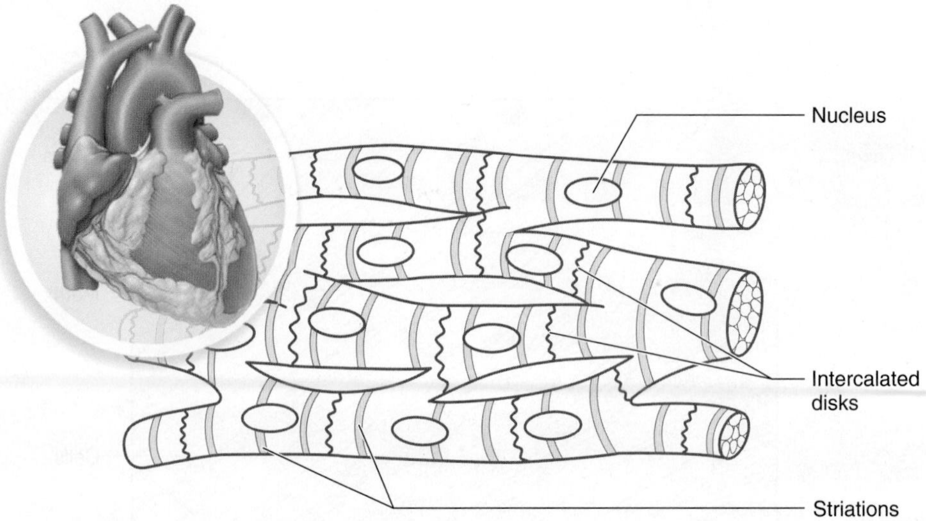

FIGURE 2.17 Cardiac muscle tissue.

Smooth Muscle Tissue This tissue has nonstriated cells that have one nucleus per cell and taper at the ends. This tissue can be found in the walls of hollow organs such as the stomach and intestines. See **Figure 2.18**.

☐ Smooth muscle cells

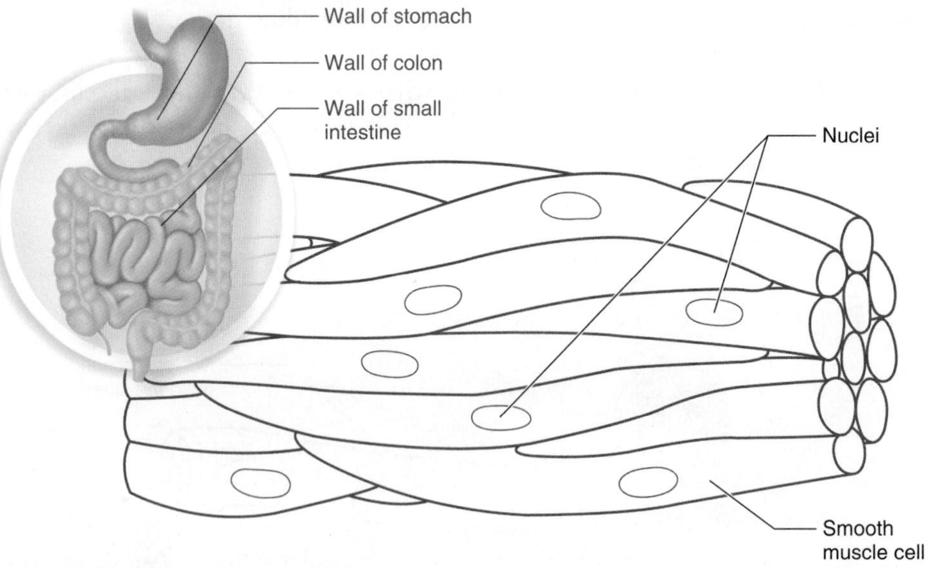

FIGURE 2.18 Smooth muscle tissue.

Nervous Tissue

This tissue is composed of nerve cells called *neurons* and many more support cells called *neuroglia* (not shown here) that protect and assist neurons in their function. Neurons can vary greatly in size and shape. They can be found in the brain, spinal cord, and nerves. See **Figure 2.19**.

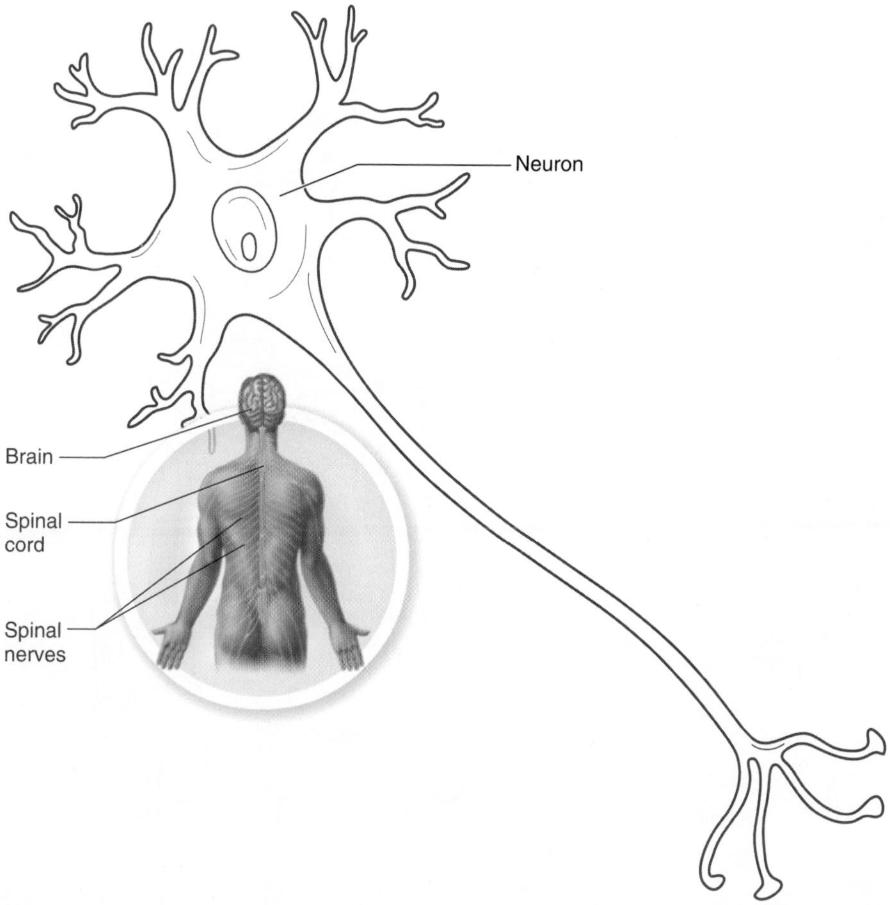
Neuron
Brain
Spinal cord
Spinal nerves

☐ Nerve cell (neuron)

FIGURE 2.19 Nervous tissue.

Systems

Organs and accessory structures come together to function as a human body system. Color the box next to the name of the system. Use the same color for the major organs and structures in the corresponding figure.

COLORING BOOK

Integumentary System

- Major organs and structures: skin, hair, nails, cutaneous glands.
- Functions: protection, vitamin D production, temperature regulation, water retention, sensation, and nonverbal communication.

See **Figure 2.20**.

☐ Integumentary system

Skeletal System

- Major organs and structures: bones.
- Accessory structures: ligaments and cartilages.
- Functions: support, movement, protection, acid-base balance, electrolyte balance, blood formation.

See **Figure 2.21**.

☐ Skeletal system

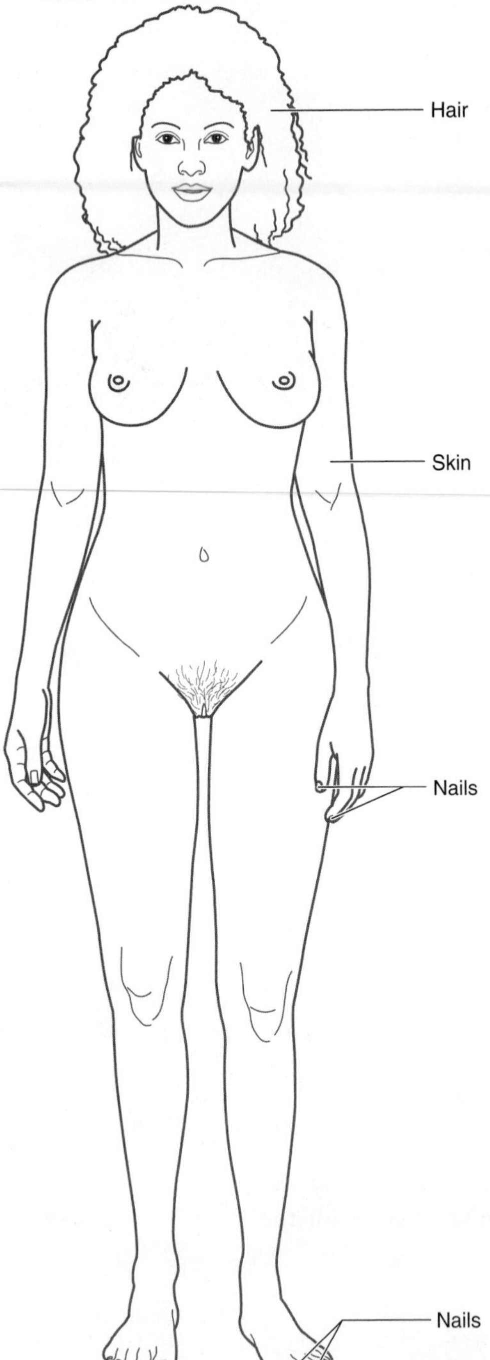

FIGURE 2.20 Integumentary system.

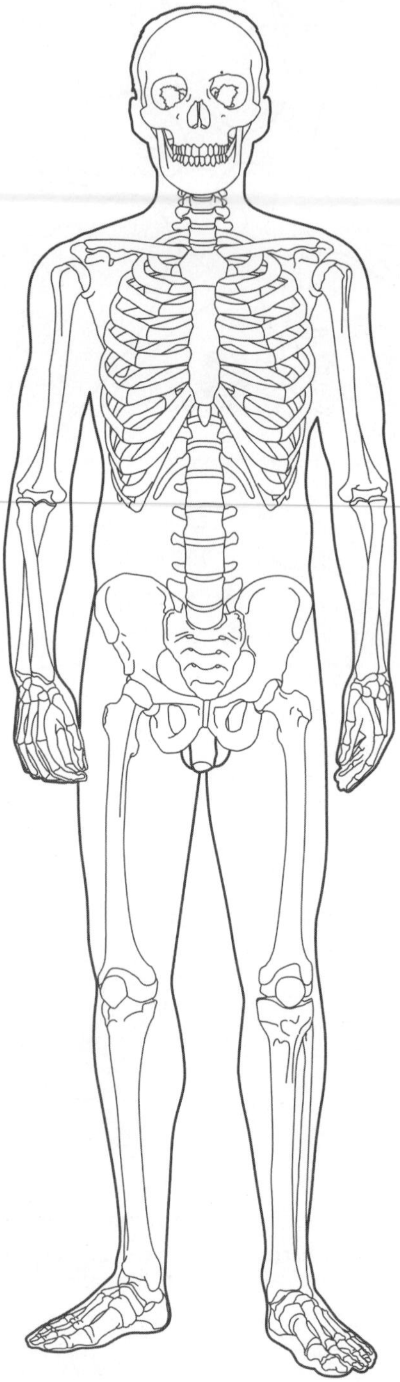

FIGURE 2.21 Skeletal system.

Muscular System

- Major organs and structures: muscles.
- Accessory structures: tendons.
- Functions: movement, stability, control of body openings and passages, communication, heat production.

See **Figure** 2.22.

☐ Muscular system

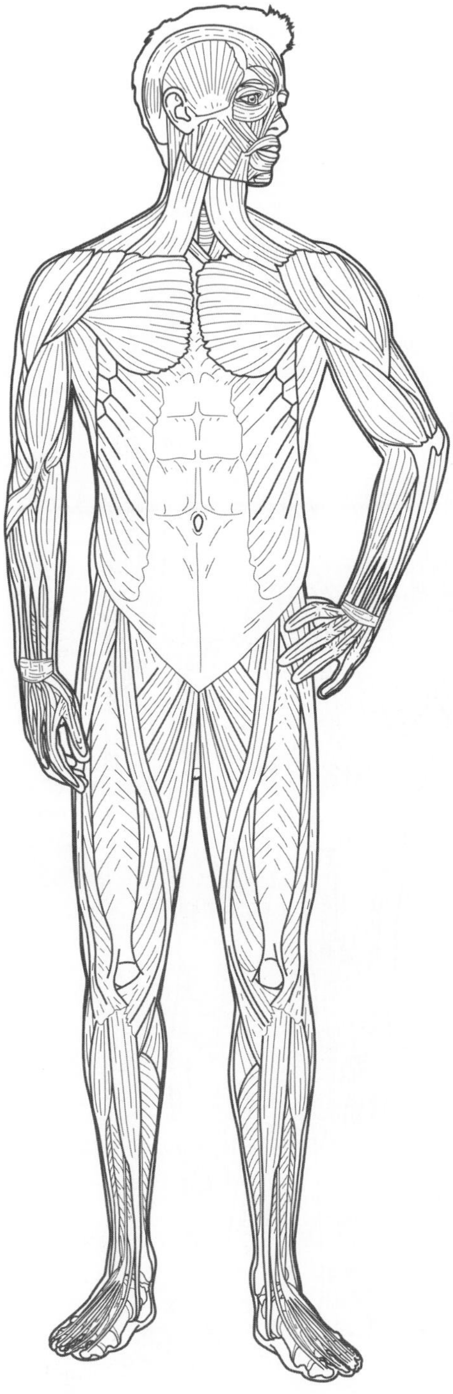

FIGURE 2.22 Muscular system.

Nervous System

- Major organs and structures: brain, spinal cord, nerves.
- Accessory structures: meninges, sympathetic chain of ganglia.
- Functions: communication, motor control, sensation.

See **Figure** 2.23.

☐ Nervous system

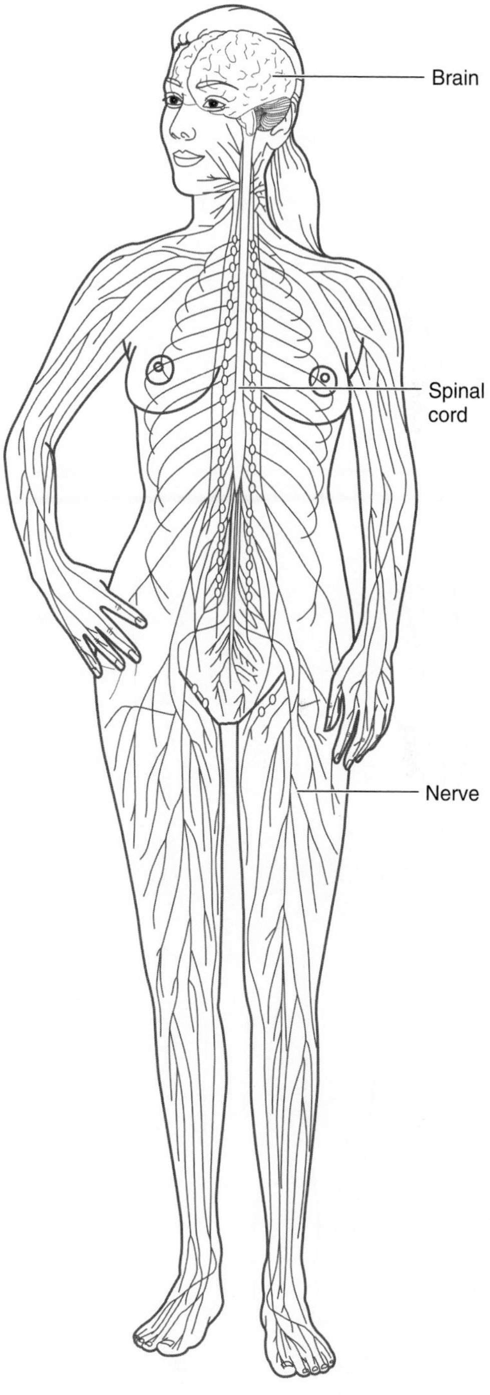

FIGURE 2.23 Nervous system.

COLORING BOOK

Endocrine System

- Major organs and structures: pineal gland, hypothalamus, pituitary gland, thyroid gland, adrenal glands, pancreas, testes, ovaries.
- Functions: communication, hormone production.

See **Figure 2.24**.

☐ Endocrine system

Cardiovascular System

- Major organs and structures: heart, aorta, superior and inferior venae cavae.
- Accessory structures: arteries, veins, capillaries.
- Functions: transportation, protection by fighting foreign invaders and clotting to prevent its own loss, acid-base balance, fluid and electrolyte balance, temperature regulation.

See **Figure 2.25**.

☐ Cardiovascular system

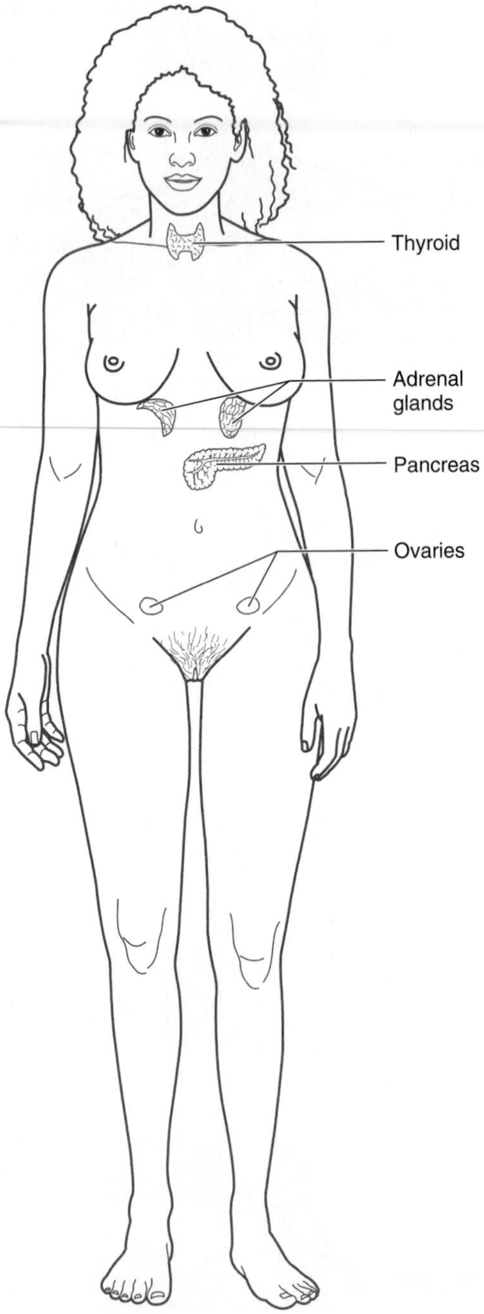

FIGURE 2.24 Endocrine system.

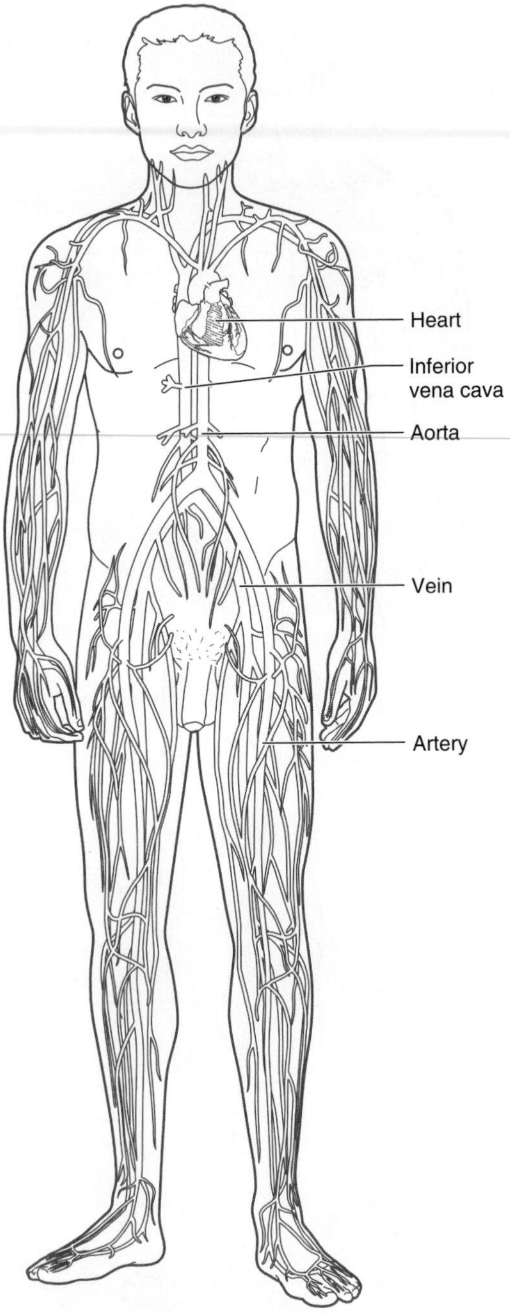

FIGURE 2.25 Cardiovascular system.

COLORING BOOK

Lymphatic System

- Major organs and structures: thymus gland, spleen, tonsils.
- Accessory structures: thoracic duct, right lymphatic duct, lymph nodes, lymph vessels, MALT, Peyer's patches.
- Functions: fluid balance, immunity, lipid absorption, defense against disease.

See **Figure 2.26**.

☐ Lymphatic system

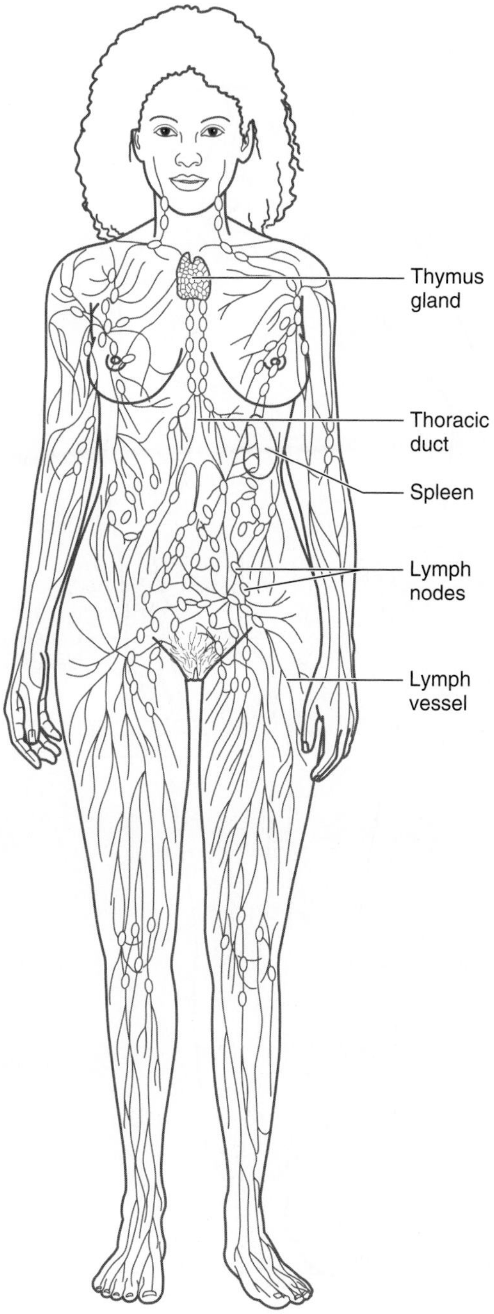

FIGURE 2.26 Lymphatic system.

Respiratory System

- Major organs and structures: nose, pharynx, larynx, trachea, bronchi, lungs.
- Accessory structures: diaphragm, sinuses, nasal cavity.
- Functions: gas exchange, acid-base balance, speech, sense of smell, creation of pressure gradients necessary to circulate blood and lymph.

See **Figure 2.27**.

☐ Respiratory system

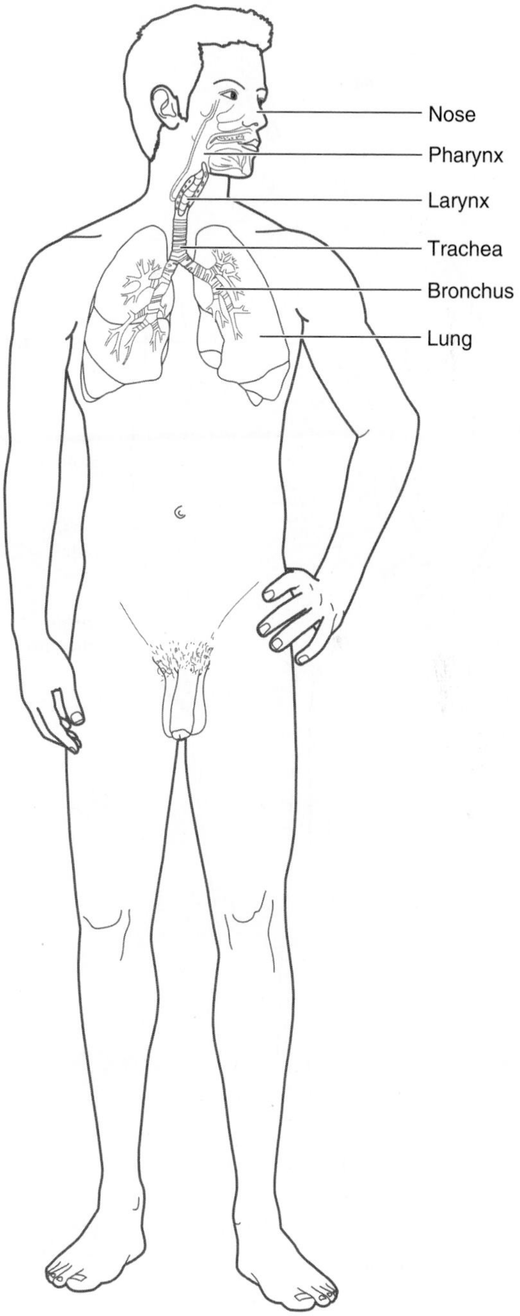

FIGURE 2.27 Respiratory system.

COLORING BOOK

Digestive System

- Major organs and structures: esophagus, stomach, small intestine, large intestine.
- Accessory structures: liver, pancreas, gallbladder, cecum, teeth, salivary glands.
- Functions: ingestion, digestion, absorption, defecation.

See **Figure 2.28.**

☐ Digestive system

Excretory/Urinary System

- Major organs and structures: kidneys, ureters, urinary bladder, urethra.
- Accessory structures: lungs, skin, liver.
- Functions: removal of metabolic wastes, fluid and electrolyte balance, acid-base balance, blood pressure regulation.

See **Figure 2.29.**

☐ Excretory/urinary system

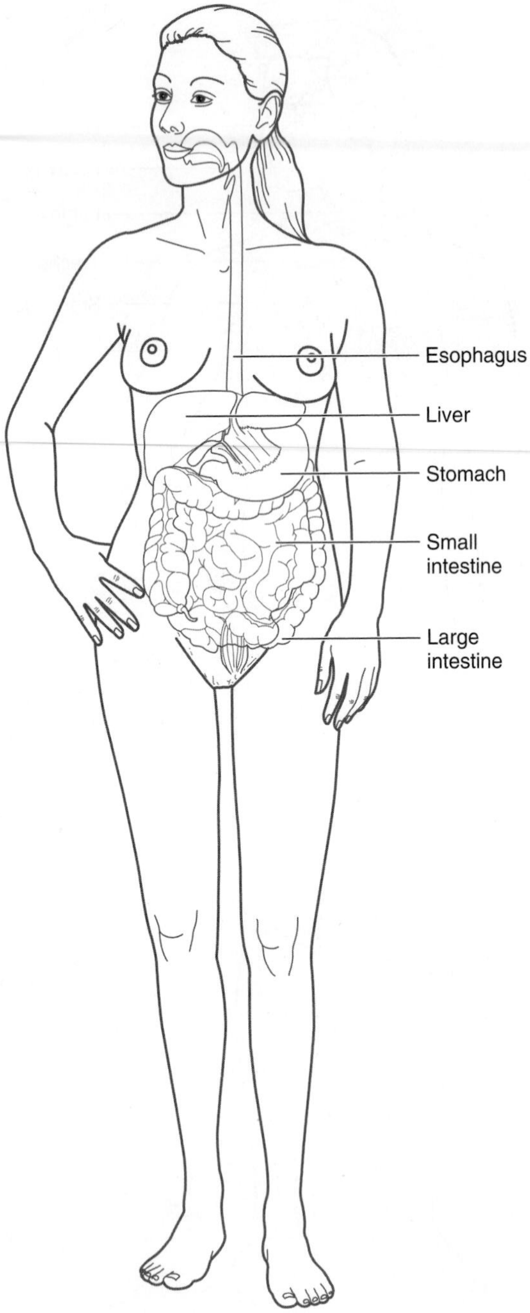

FIGURE 2.28 Digestive system.

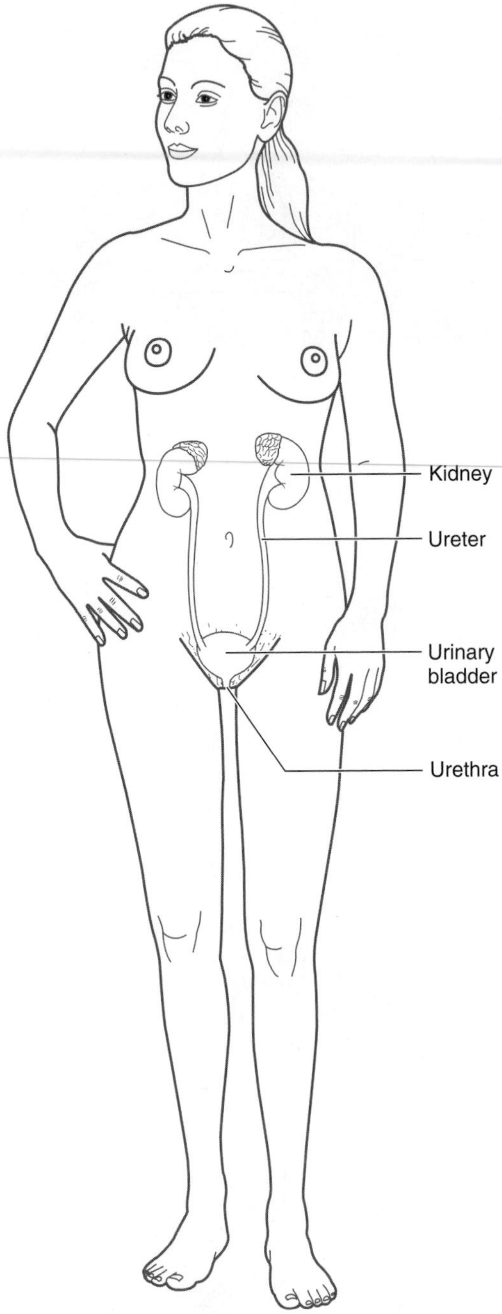

FIGURE 2.29 Excretory/urinary system.

COLORING BOOK

Male Reproductive System

- Major organs and structures: testes.
- Accessory structures: scrotum, spermatic ducts (epididymis, ductus deferens), accessory glands (seminal vesicles, prostate gland, bulbourethral glands), penis.
- Functions: production and delivery of sperm, secretion of sex hormones.

See **Figure 2.30**.

☐ Male reproductive system

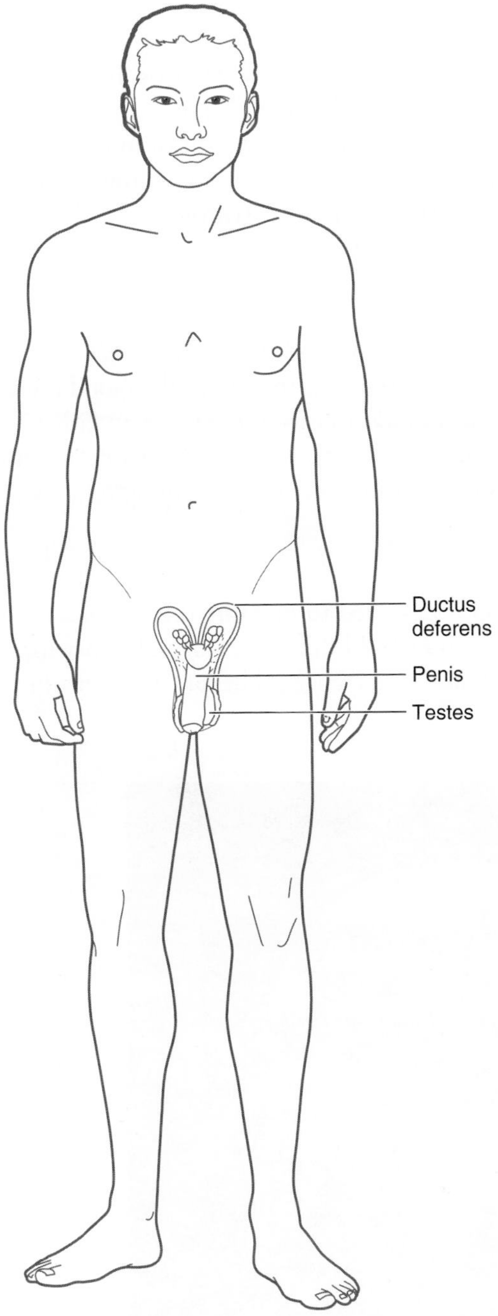

FIGURE 2.30 Male reproductive system.

Female Reproductive System

- Major organs and structures: ovaries.
- Accessory structures: uterus, uterine tubes, vagina, vulva, breasts
- Functions: production of an egg, housing of the fetus, birth, lactation, secretion of sex hormones.

See **Figure 2.31**.

☐ Female reproductive system

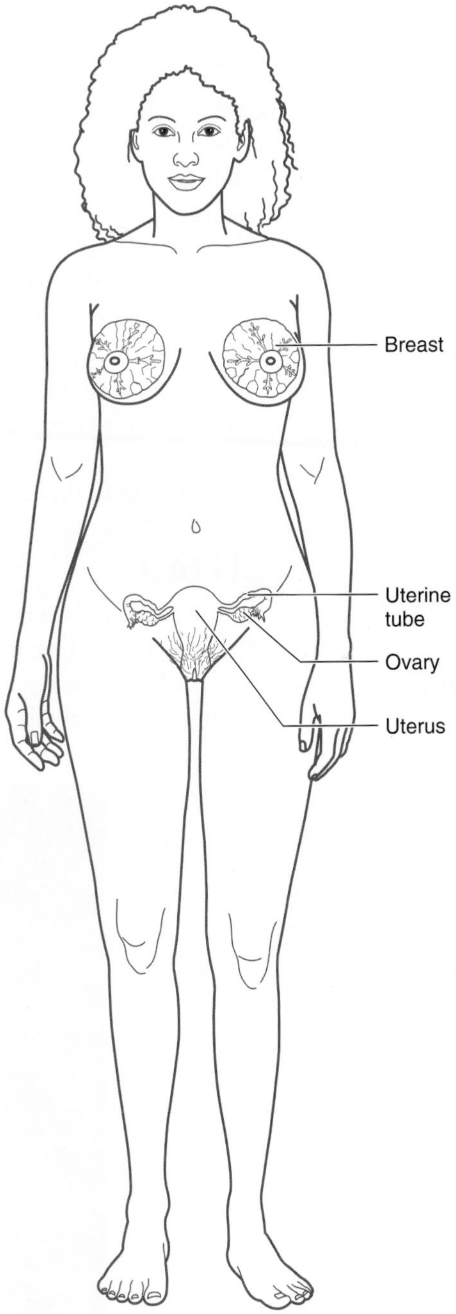

FIGURE 2.31 Female reproductive system.

pH

The pH scale is a means of measuring the acidity or alkalinity of a substance. It is basically a scale from 0 to 14. Substances with a pH closer to 0 are increasingly acidic. Substances with a pH closer to 14 are increasingly basic, or alkaline. Substances with a pH of 7 are neutral. See Figure 2.32.

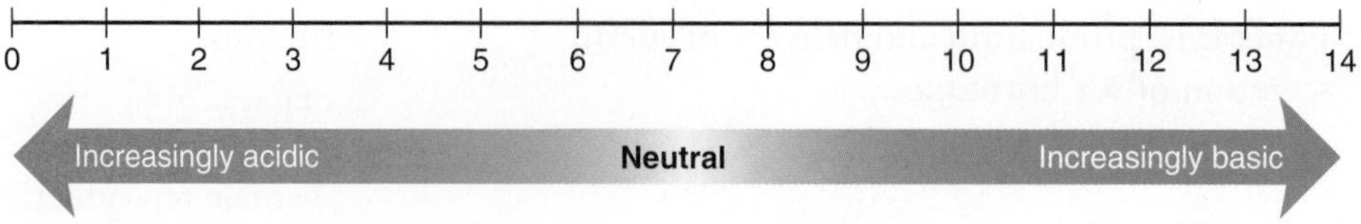

FIGURE 2.32 pH scale.

In general, what is being measured is hydrogen ions (H^+) on the acidic end of the scale and hydroxide ions (OH^-) on the basic end. For each one-number difference in pH, there is a 10-times difference in the amount of ions present. So a substance with a pH of 2 is acidic, and it has 10 times more hydrogen ions than a substance with a pH of 3, 100 times (10×10) more hydrogen ions than a substance with a pH of 4, and 1,000 ($10 \times 10 \times 10$) times more hydrogen ions than a substance with a pH of 5.

Blood pH is vitally important in the physiology of the human body. For example, the normal pH of blood is 7.35 to 7.45. A condition called *acidosis* results if the pH of the blood is below 7.35. *Alkalosis* results if the pH of the blood is greater than 7.45. The body attempts to return blood pH to homeostasis in either situation by changing the respiratory rate or urine composition. It is very important that you understand this concept, as it is relevant in many of the body system chapters.

pH test

The pH paper shown in Figure 2.33 is designed to measure the pH of a fluid by changing color. To perform the test, you would dip a strip of pH paper into the fluid and then compare it to a standard color chart. The pH is determined by finding a color match.

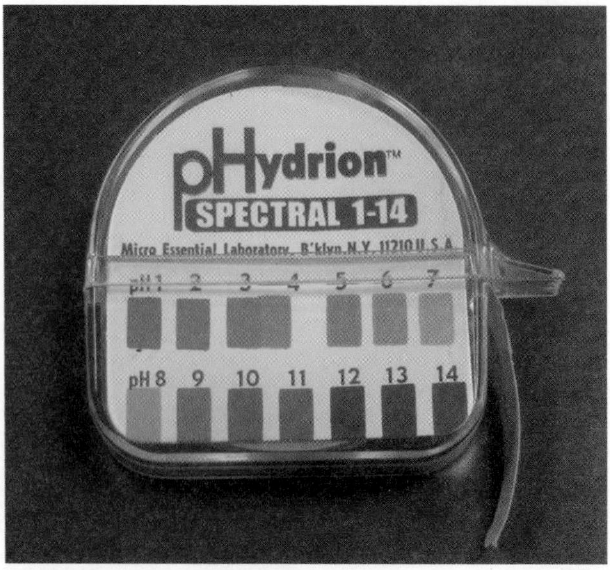

FIGURE 2.33 pH paper.

Figure 2.34 shows eight mystery fluids. A strip of pH paper was dipped into each fluid.

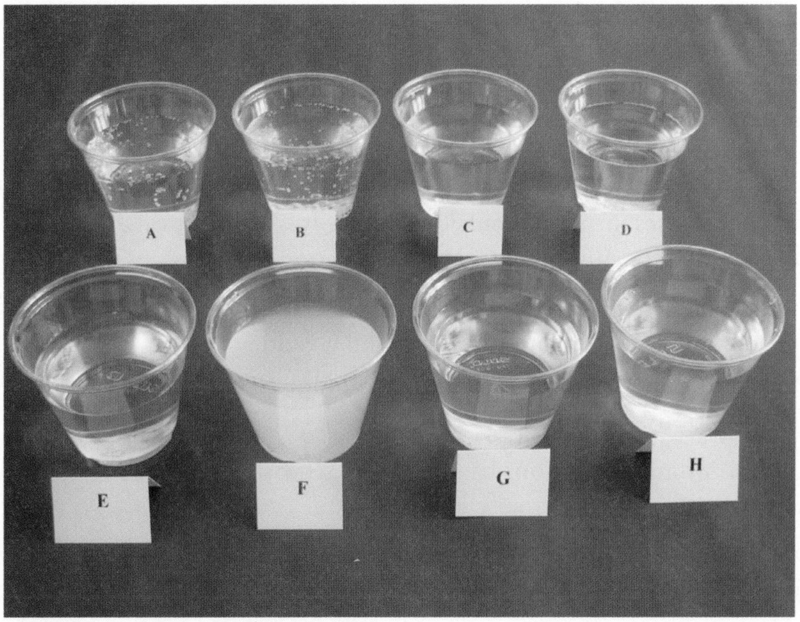

FIGURE 2.34 Eight mystery fluids.

Use the results of the pH tests shown in Figure 2.35 to answer the following:

1. Using the letter names for each liquid, order the liquids by pH, lowest to highest. ________________________________
2. Which of the liquids are acids? ________________________________
3. Which of the liquids is the strongest acid? ________________________
4. Which of the liquids are bases? ________________________________
5. Which of the liquids is the strongest base? ________________________
6. Which ion would be more abundant in liquid D (H^+ or OH^-)? ________
7. What is the difference in the amount of ions between liquid G and liquid D? Explain. ________________________________

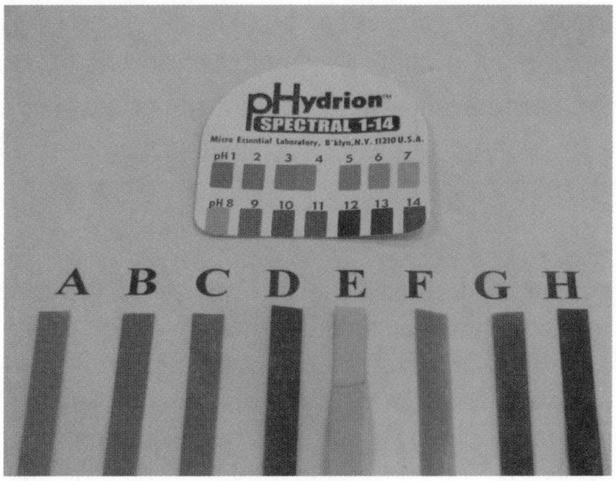

FIGURE 2.35 Completed pH tests.

Osmosis

A cell is defined by its cell membrane because the membrane separates the intracellular fluid—the cytoplasm—inside the cell from the extracellular fluid outside the cell. Both the cytoplasm and the extracellular fluid are solutions composed of solutes and a solvent (water). Cell membranes are selectively permeable, which means they regulate what can pass into or out of the cell. Not all solutes can cross a selectively permeable cell membrane.

Osmosis is the movement of water across a cell membrane to equalize the concentration of solutes inside and outside the cell. It is a passive method of transport, so it does not require any energy. Osmosis will occur if two conditions are met:

1. There is a concentration gradient (difference in concentration of solutes) between the cytoplasm and the extracellular fluid.
2. The solutes cannot move across the membrane to equalize the concentrations of the solutions.

Concentrations of solutions can be compared using the term *tonicity* in three ways. They are described in the following list:

- A hypertonic solution has a higher concentration of solutes than another solution.
- An isotonic solution has the same concentration of solutes as another solution.
- A hypotonic solution is less concentrated than another solution.

Egg Lab

Osmosis can be demonstrated in an egg lab as described below.

A raw egg can be used as a model for a cell. In preparing for the egg lab, the egg's shell must first be removed without damaging the delicate membrane underneath. As the egg is swirled in a container of hydrochloric acid and water, the acid reacts to dissolve the minerals in the eggshell without harming the delicate membrane deep to the shell.

clinical point

Hydrochloric acid is the same acid produced in your stomach. It is a strong acid with a pH of 0.8. In the egg lab, the pH of the hydrochloric acid rises as more and more of the shell is dissolved. This is similar to what happens when you take an antacid tablet. The hydrochloric acid in your stomach reacts with the antacid as it is dissolved to neutralize the acid. The stomach contents become less acidic, and the pH rises.

Once the egg is prepared, the egg lab can begin. First, the egg is weighed without the shell, and the weight is recorded. The egg weighs 54.6 g. The egg is then placed in a beaker of tap water. Because there may be small amounts of chemicals (chlorine or fluorine added by the city) and dissolved minerals (from the plumbing pipes delivering the water to the tap), tap water is not pure water. One thing is for sure: Tap water is a far less concentrated solution than the liquid found inside a raw egg.

Make a prediction: What do you think will happen to the weight of the raw egg after it has been placed in a container of tap water?

____

At 15-minute intervals, the egg is removed from the beaker of water. It is carefully dried using a paper towel, weighed, and then placed back in a beaker of tap water. The weights are given in Table 2.1.

TABLE 2.1 Weights of the egg

Starting weight of egg	54.6 g
15 min	58.4 g
30 min	60.0 g
45 min	60.8 g
60 min	61.6 g
75 min	62.4 g
90 min	63.0 g
105 min	63.6 g
120 min	64.0 g
135 min	64.3 g
150 min	64.6 g
165 min	64.8 g
180 min	65.0 g

Make a graph: Plot the weights on the student graph in Figure 2.36. The first three weights have been added for you.

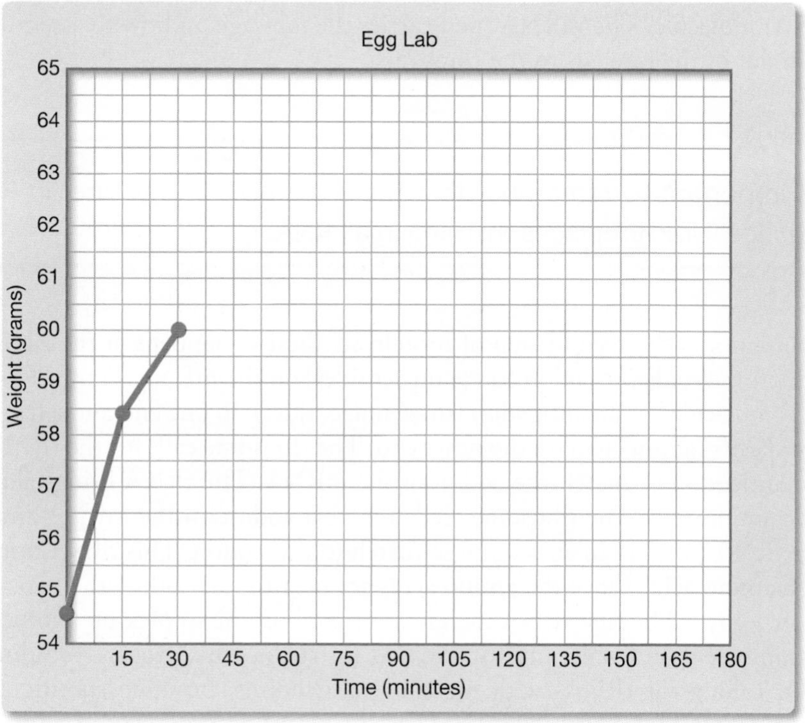

FIGURE 2.36 Graph of egg lab.

Analyze the graph:

1. Did the weight of the egg change the same amount every 15 minutes? __________

2. What happens to the amount of change in every 15-minute interval as time goes on? __________

3. What happens to the speed of the change? __________

4. Does the graph of the egg form a straight line? __________

5. What material(s) moved across the membrane? __________

6. By what process did the material(s) move? __________

7. In what type of solution was the egg placed (hypotonic, isotonic, or hypertonic)? __________

Protein Synthesis

Making proteins is a function of many cells. Protein synthesis is a two-step process—transcription and translation. The first step is transcription, and its purpose is to take the information on how to assemble the protein from the DNA in the nucleus to the ribosome on the rough endoplasmic reticulum. To do that, the DNA is first split between the nucleotide rungs of the twisted ladder. Free nucleotides inside the nucleus then bind to one side of the open DNA. C's match to G's and U's match to A's. This forms a single-stranded messenger RNA (mRNA) molecule. The mRNA then carries the message on how to assemble the protein out of the nucleus to the ribosome.

warning

It is important to remember that there are no T's (thymine) in RNA language. Uracil replaces thymine in all RNA.

Translation—the second step of protein synthesis—happens at the ribosome. The ribosome reads the codon (three nucleotides) on the mRNA. A transfer-RNA (tRNA) molecule has an anticodon (three nucleotides) on one end and a handle to carry a specific amino acid on the other end. The ribosome calls for the tRNA that has an anticodon to match the codon of the mRNA. The tRNA brings along its specific amino acid. The ribosome reads the next codon on the mRNA and calls for the tRNA that has an anticodon to match to this codon. This tRNA brings its specific amino acid. Once the amino acids are next to each other, the two amino acids are joined. The first tRNA then leaves the ribosome to pick up another specific amino acid and waits until it is called for again. This process—reading the mRNA, calling for tRNAs with matching anticodons, bringing specific amino acids, and joining the amino acids together—continues until all of the mRNA has been read and the string of amino acids has been formed. The string of amino acids then bends, folds, and pleats to create a unique shape that depends on the order of the amino acids. The protein has now been synthesized. From here it goes to the Golgi complex to be inspected and modified.

Protein Synthesis Example

In this example, you will follow a hypothetical DNA molecule through the process of protein synthesis. It may take thousands of amino acids to make a protein. To keep the example very simple, we use a small section of DNA that codes for a string of nine amino acids within a larger protein.

Given the following sequence of one side of a DNA molecule:

CTTTGGCGCGGGAATTCCTTTGTCCAC

1. What would be the rest of the sequence of nucleotides in the mRNA formed for this section of DNA?

   GAAAC . . . ______________________________

2. During what process is the mRNA formed? ____________________

3. Where is the mRNA formed? ____________________

Use the mRNA you just formed, and the following tRNAs with their corresponding amino acids (shown in **Figure 2.37**) to answer the rest of the questions.

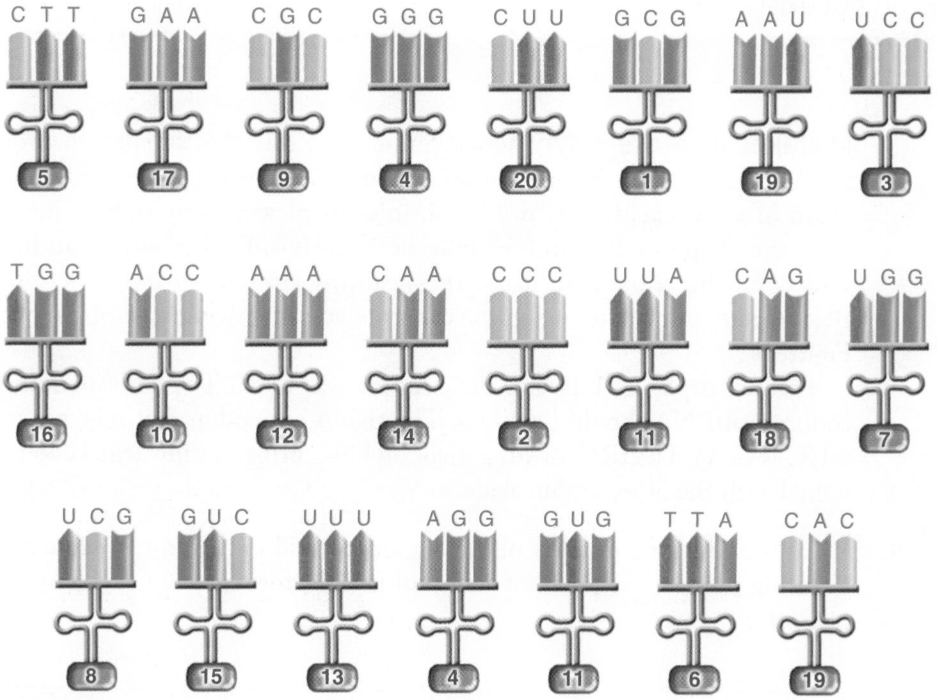

FIGURE 2.37 tRNA and amino acids. The anticodon is shown in red, and the different amino acids are listed by blue numbers instead of by name.

4. What is the sequence of anticodons that matches the mRNA you formed in question 1?

____

5. Each tRNA brings its specific amino acid. What is the amino acid sequence for the resulting protein? (List the numbers.)

____

6. During what process are the amino acids assembled into a protein?

____

7. Where does that process take place?

____

8. Where does the protein go from there?

____

warning

Did you notice that some of the tRNAs in Figure 2.37 were impossible? Did you remember that RNAs cannot contain thymine (T's)? Cross out any tRNAs with thymine in their anticodons because they do not exist.

Mistakes in protein synthesis: Sometimes mistakes happen in protein synthesis. If the wrong amino acid is brought during assembly, the chain of amino acids may not bend, fold, or pleat exactly right. Therefore, the shape of the protein may be slightly off. One wrong amino acid could be enough to cause the resulting protein to not function. Below are two mistakes using the example you just worked through. See **Figure 2.37**.

The last triplet of DNA in the example was CAC. The corresponding codon of mRNA should be GUG. The tRNA anticodon that matches to GUG is CAC. The tRNA with anticodon CAC brings amino acid 19 to be joined with the other amino acids.

9. What would be the number of the last amino acid in the protein sequence if during transcription the last codon of mRNA formed was GCG instead of GUG?

____

10. What would be the number of the last amino acid in the protein sequence if during transcription the last codon of mRNA formed was UUA instead of GUG? ____________________

____

11. Do the mistakes in transcription described in questions 9 and 10 result in an abnormal protein? Explain.

____

____

____

____

Key Words

The following terms are defined in the glossary of the textbook.

acid
atom
base
cellular respiration
chemical reaction
epithelial tissues
histology
integumentary system
membrane transport
metabolism
mitosis
molecule
mutation
organelles
organic molecules
osmosis
protein synthesis
reactants
replication
solution

Concept Maps

Complete the boxes in the following concept maps (**Figures 2.38** to **2.43**). Each map contains at least one key word.

Chemical Level

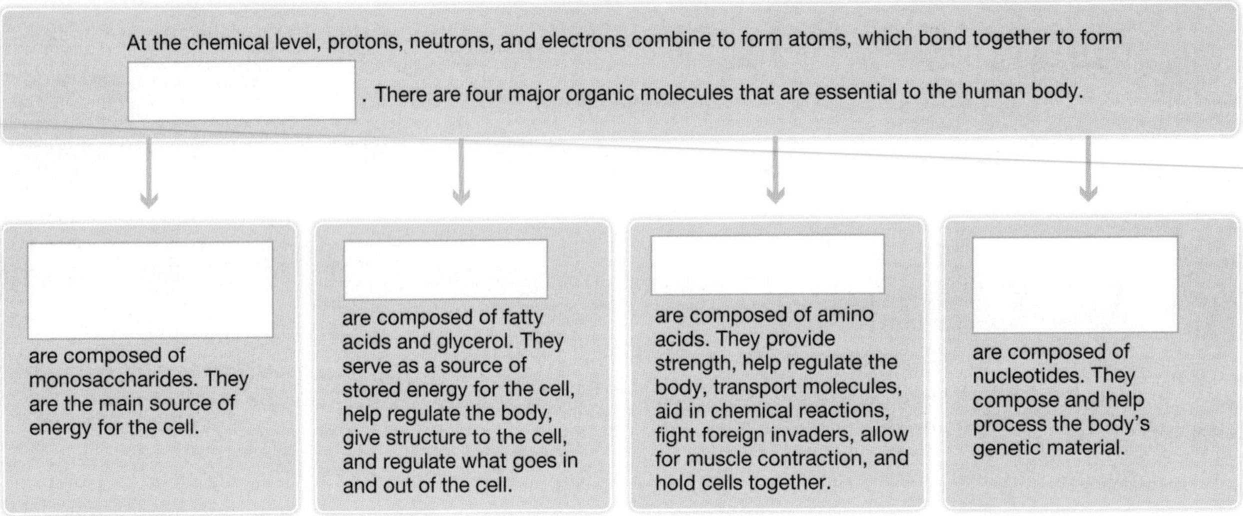
At the chemical level, protons, neutrons, and electrons combine to form atoms, which bond together to form
. There are four major organic molecules that are essential to the human body.
are composed of monosaccharides. They are the main source of energy for the cell.
are composed of fatty acids and glycerol. They serve as a source of stored energy for the cell, help regulate the body, give structure to the cell, and regulate what goes in and out of the cell.
are composed of amino acids. They provide strength, help regulate the body, transport molecules, aid in chemical reactions, fight foreign invaders, allow for muscle contraction, and hold cells together.
are composed of nucleotides. They compose and help process the body's genetic material.

FIGURE 2.38 Chemical level concept map.

Organelle Level

Organelles are small organlike structures found suspended within the cell's fluid or ______.

The ______ houses the ______ (genetic material of the cell).

Has a nuclear membrane from which the ______ extends.

______ has

______, which assemble proteins.

______ is the site of lipid synthesis.

______ perform cellular respiration and use the energy obtained from it to create ______—the usable form of energy for the cell.

______ modifies and packages proteins and lipids, which are carried to other parts of the cell by

______.

The ______ is a bilayer of phospholipids that controls what may enter or exit the cell.

FIGURE 2.39 Organelle level concept map.

Cellular Level

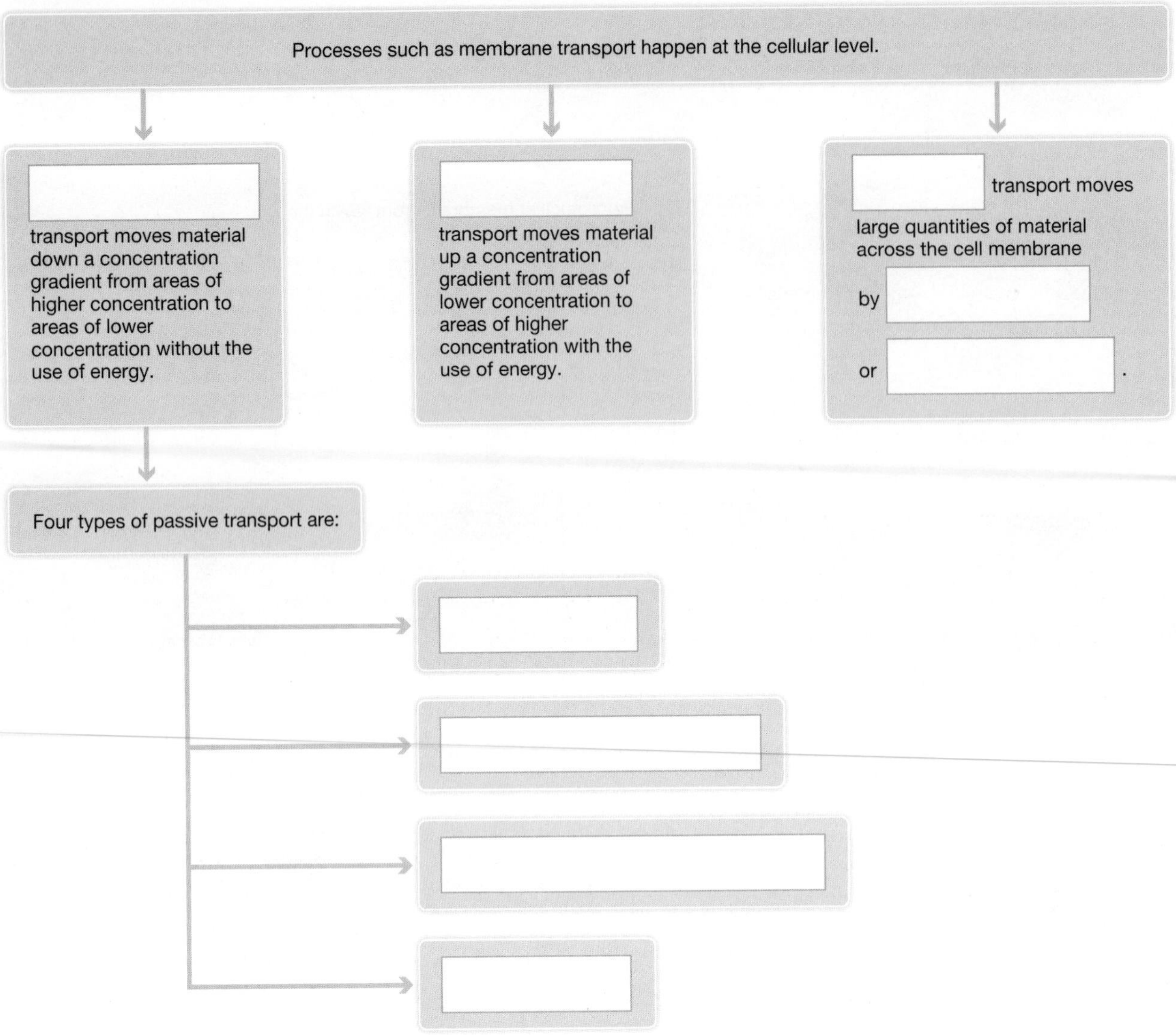
Processes such as membrane transport happen at the cellular level.
transport moves material down a concentration gradient from areas of higher concentration to areas of lower concentration without the use of energy.
transport moves material up a concentration gradient from areas of lower concentration to areas of higher concentration with the use of energy.
transport moves large quantities of material across the cell membrane
by
or .
Four types of passive transport are:

FIGURE 2.40 Cellular level concept map.

Cell Division

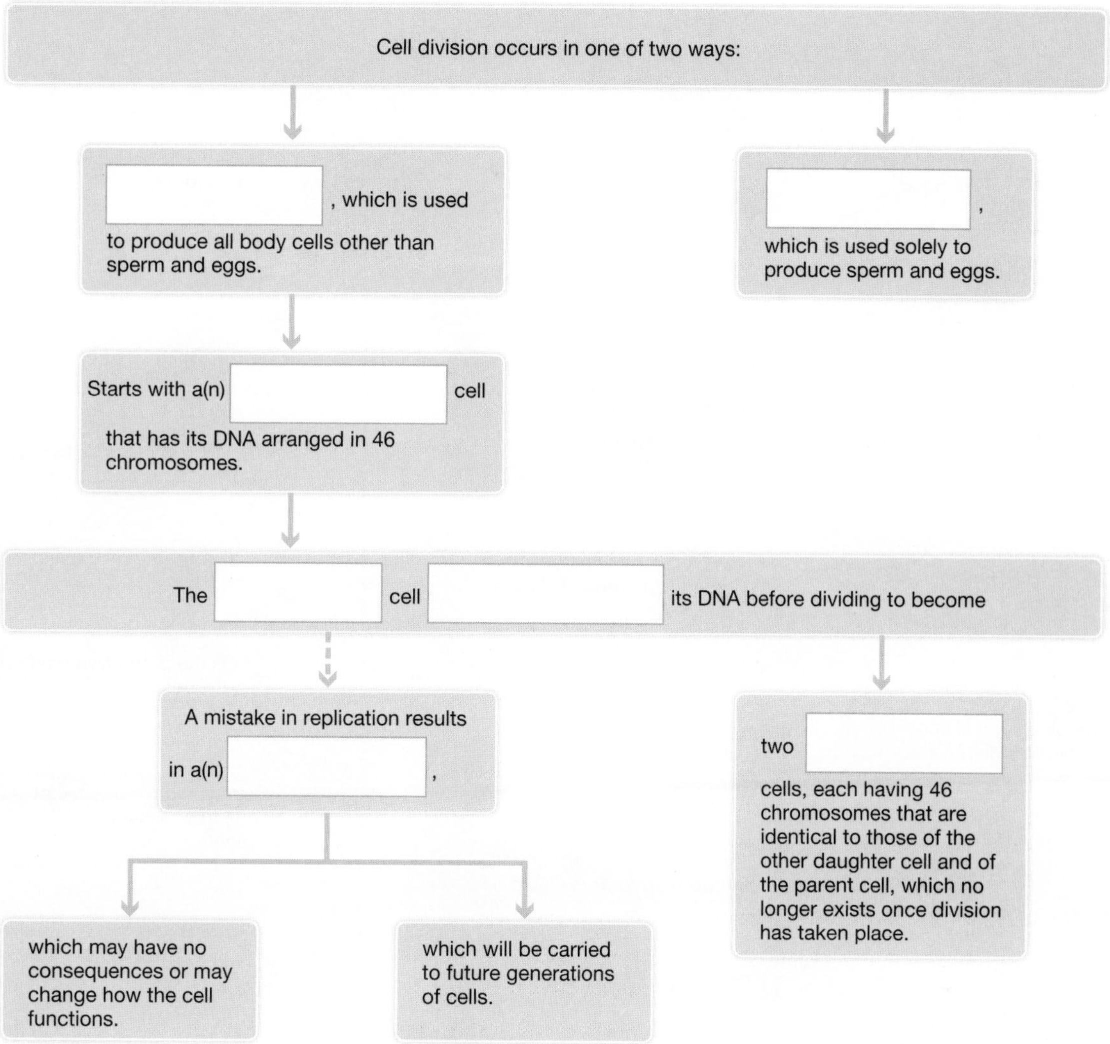
Cell division occurs in one of two ways:
, which is used to produce all body cells other than sperm and eggs.
, which is used solely to produce sperm and eggs.
Starts with a(n) cell that has its DNA arranged in 46 chromosomes.
The cell its DNA before dividing to become
A mistake in replication results in a(n) ,
two cells, each having 46 chromosomes that are identical to those of the other daughter cell and of the parent cell, which no longer exists once division has taken place.
which may have no consequences or may change how the cell functions.
which will be carried to future generations of cells.

FIGURE 2.41 Cell division concept map.

Tissue Level

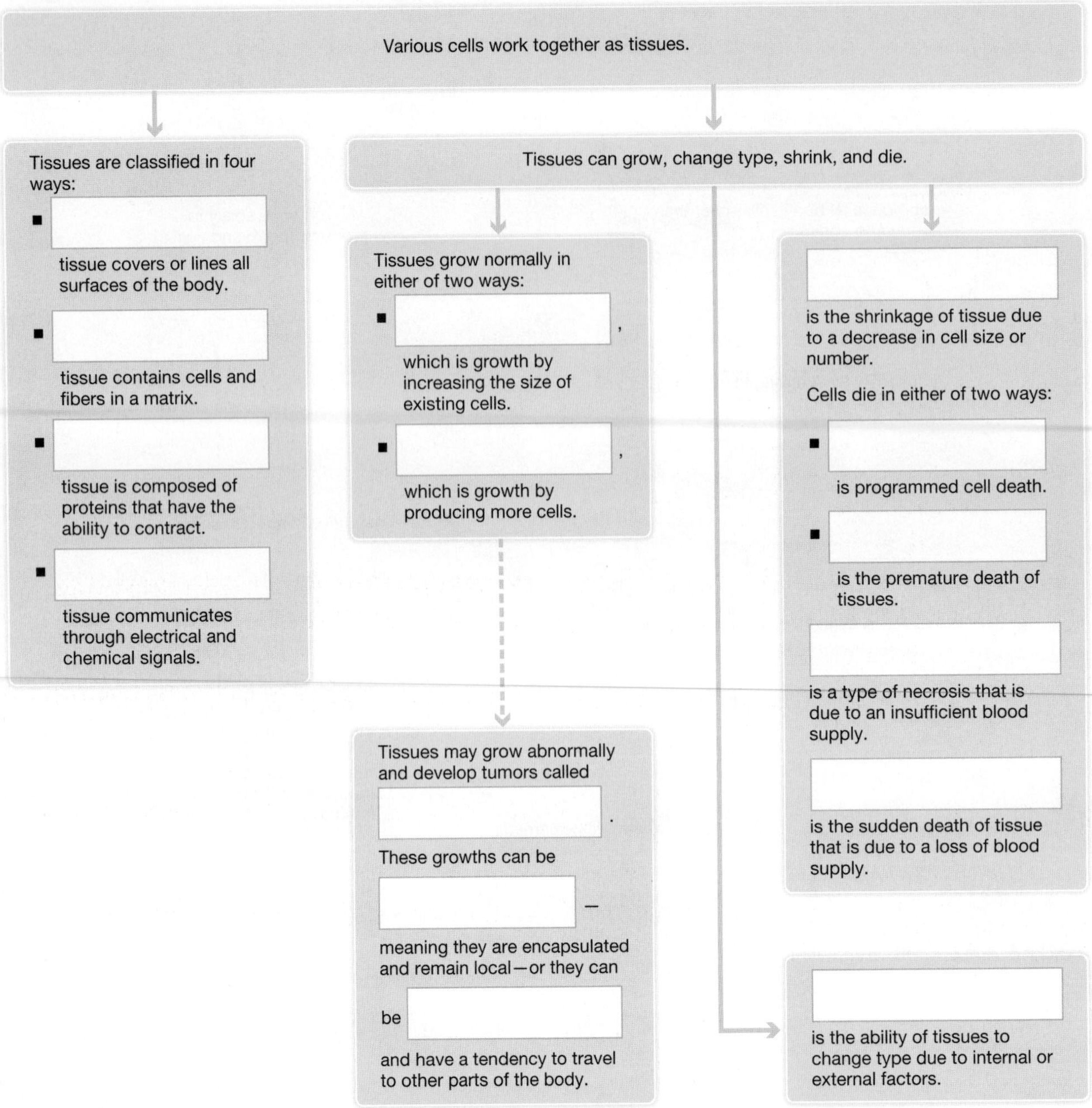
Various cells work together as tissues.
Tissues are classified in four ways:
tissue covers or lines all surfaces of the body.
tissue contains cells and fibers in a matrix.
tissue is composed of proteins that have the ability to contract.
tissue communicates through electrical and chemical signals.
Tissues can grow, change type, shrink, and die.
Tissues grow normally in either of two ways:
which is growth by increasing the size of existing cells.
which is growth by producing more cells.
Tissues may grow abnormally and develop tumors called
These growths can be
meaning they are encapsulated and remain local—or they can
be
and have a tendency to travel to other parts of the body.
is the shrinkage of tissue due to a decrease in cell size or number.
Cells die in either of two ways:
is programmed cell death.
is the premature death of tissues.
is a type of necrosis that is due to an insufficient blood supply.
is the sudden death of tissue that is due to a loss of blood supply.
is the ability of tissues to change type due to internal or external factors.

FIGURE 2.42 Tissue level concept map.

Organ and System Levels

Organs work together to function as systems that make up the human body.

- ______ system contains major organs and structures such as skin, hair, nails, and cutaneous glands.
- ______ system contains bones as its major organs.
- ______ system contains muscles as its major organs.
- ______ system contains major organs and structures such as the brain, spinal cord, and nerves.
- ______ system contains major organs and structures such as the pineal body, hypothalamus, pituitary gland, thyroid gland, adrenal glands, pancreas, testes, and ovaries.
- ______ system contains major organs and structures such as the heart, aorta, and the superior and inferior venae cavae.
- ______ system contains major organs and structures such as the thymus gland, spleen, and tonsils.
- ______ system contains major organs and structures such as the nose, pharynx, larynx, trachea, bronchi, and lungs.
- ______ system contains major organs and structures such as the esophagus, stomach, and small and large intestines.
- ______ or ______ system contains major organs and structures such as the kidneys, ureters, urinary bladder, and urethra.
- ______ system contains major organs and structures such as the uterus, uterine tubes, vagina, ovaries, and breasts.
- ______ system contains major organs and structures such as the testes, penis, ductus deferens, and urethra.

FIGURE 2.43 Organ and system levels concept map.

KEY WORD CONCEPT MAPS

Multiple Select: *Select the correct choices for each statement. The choices may be all correct, all incorrect, or any combination of correct and incorrect.*

1. What may increase the speed of chemical reactions?
  a. Enzymes that act as catalysts.
  b. Decreasing the amount of the reactants.
  c. Decreasing the temperature of the reactants.
  d. Increasing the amount of the reactants.
  e. Increasing the temperature of the reactants.

2. What is (are) the function(s) of water in the human body?
  a. Water acts as a lubricant.
  b. Water aids in chemical reactions.
  c. Water is used for transportation of wastes.
  d. Water separates ionically bonded molecules.
  e. Water is used for temperature regulation.

3. Which of the following statements is (are) true about molecules?
  a. Molecules are composed of two or more elements bonded together.
  b. In the molecule CO_2, there are two atoms of oxygen and one atom of carbon.
  c. Atoms bind together to form molecules to fill their outer shells with electrons.
  d. Water and carbon dioxide are organic molecules.
  e. Atoms share electrons in a covalent bond.

4. A urine pH test came back as normal with a pH of 6. What is true about this urine?
  a. It is a strong acid.
  b. It is a weak acid.
  c. It is a weak base.
  d. It is a strong base.
  e. It has more $H+$ ions than pure water.

5. What is true about osmosis?
  a. It requires ATP to move materials up a concentration gradient.
  b. It is a passive process.
  c. It is used for solutes that can cross the cell membrane.
  d. It will occur across a selectively permeable membrane if there is a concentration gradient.
  e. It speeds up as time goes by and concentrations become equal.

6. What is true about active transport?
  a. Active transport moves materials from low to high concentration across a membrane.
  b. Active transport moves materials from high to low concentration across a membrane.
  c. Active transport requires ATP.
  d. Active transport is a passive process.
  e. Active transport continues until the concentrations are equal.

7. What is true about the comparison of DNA and RNA?
  a. DNA is double-stranded; RNA is single-stranded.
  b. DNA is the genetic material of the cell; RNA processes it.
  c. DNA and RNA contain guanine (G), cytosine (C), and adenine (A).
  d. DNA and RNA can be found in the nucleus.
  e. DNA and RNA molecules are composed of the elements C, H, O, N, and P.

8. Which of the following statements is (are) true about protein synthesis?
  a. Anticodons match to identical codons in translation.
  b. Anticodons are present in mRNA.
  c. Mistakes in protein synthesis result in mutations.

d. Transfer RNA carries the message from the DNA in the nucleus to the ribosome.
e. All mistakes in translation result in proteins that do not function properly.

9. What may happen to tissues?
   a. They may grow by increasing the number of cells, a process called *hyperplasia*.
   b. They may die a programmed death, called *necrosis*.
   c. They may die a sudden death, called an *infarct*.
   d. They may change from one type to another, a process called *metaplasia*.
   e. They may shrink through disuse, a process called *apoptosis*.

10. What is true about the levels of organization in the human body?
   a. The organism level is the most complex.
   b. The chemical level is the simplest level.
   c. Organelles work together to form systems.
   d. There are four classifications of tissues in the human body.
   e. Tissues work together to function as organs.

Matching: *Match the type of organic molecule to the example. Choices may be used more than once.*

______	1. Phospholipid	a. Protein
______	2. Glycogen	b. Nucleic acid
______	3. Steroid	c. Carbohydrate
______	4. Fats	d. Lipids
______	5. RNA	

Matching: *Match the organelle to its function. Choices may be used more than once.*

______	6. Inspects and modifies proteins	a. Secretory vesicles
______	7. Is the site of lipid synthesis	b. Mitochondria
______	8. Produces ATP	c. Rough endoplasmic reticulum
______	9. Is the site of protein synthesis	d. Golgi complex
______	10. Are packages of materials for transport	e. Smooth endoplasmic reticulum

Completion: *Fill in the blanks in the following statements.*

1. Cellular respiration can be written as: Glucose + ______________________ yields carbon dioxide + water + ____________________.
2. The chemical formula for cellular respiration is: ____.
3. In a sugar solution, water is the ____________________ and sugar is the ____________________.
4. When placed in water, ionically bonded molecules become ions in solution called ____________________.
5. DNA is arranged as 46 ____________________ when it is about to divide, but it is arranged as ____________________ through most of its life cycle so that it can be used.

CHAPTER 2 REVIEW QUESTIONS

Critical Thinking

1. Consider the egg lab (p. 40). Draw a general graph of what you would expect to happen to the weight of an egg placed in each of the following solutions. See **Figures 2.44** to **2.46**.

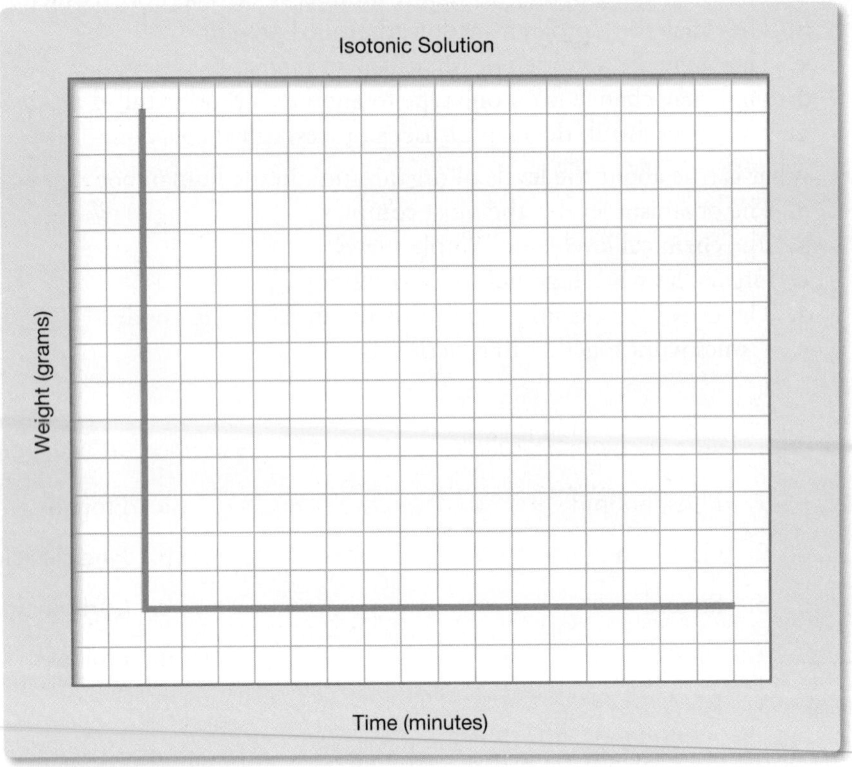

FIGURE 2.44 Isotonic solution.

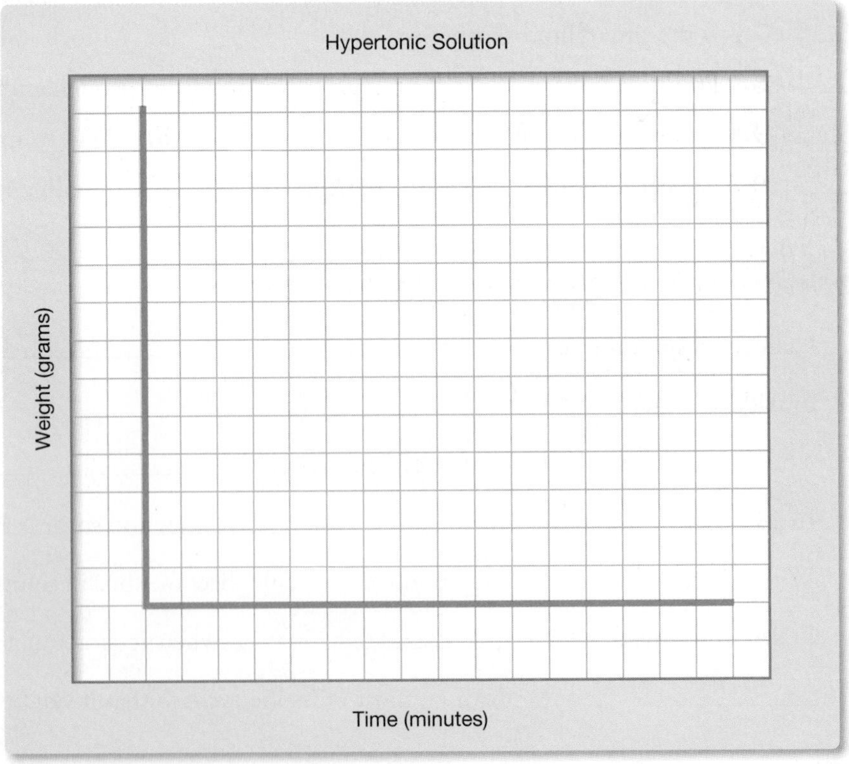

FIGURE 2.45 Hypertonic solution.

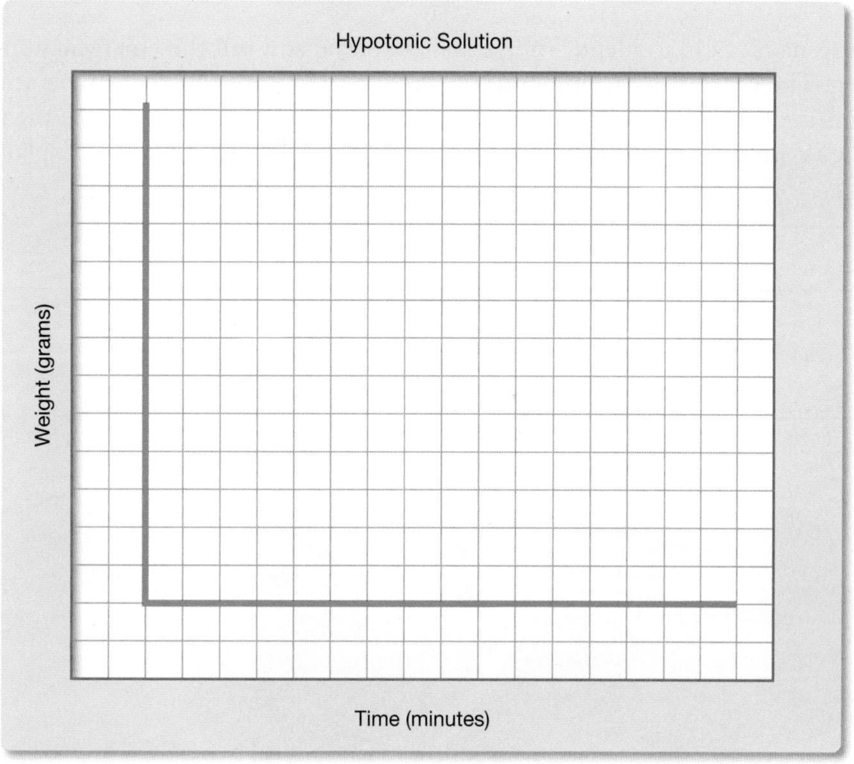

FIGURE 2.46 Hypotonic solution.

2. Cells in the ovary produce the hormone estrogen. Estrogen is a steroid. What would be the relative amount of organelles in these cells in order to carry out this function?

____

____

____

____

____

Case Study

1. Paramedics arrive on the scene of a car accident. They assess the scene and call the emergency room with the victim's condition. The ER doctor recommends starting an IV, not to treat the patient at this time, but to establish an intravenous line should drugs need to be quickly administered later on the way to the hospital. Should the IV fluids be hypotonic, isotonic, or hypertonic to blood plasma? Explain.

____

____

____

____

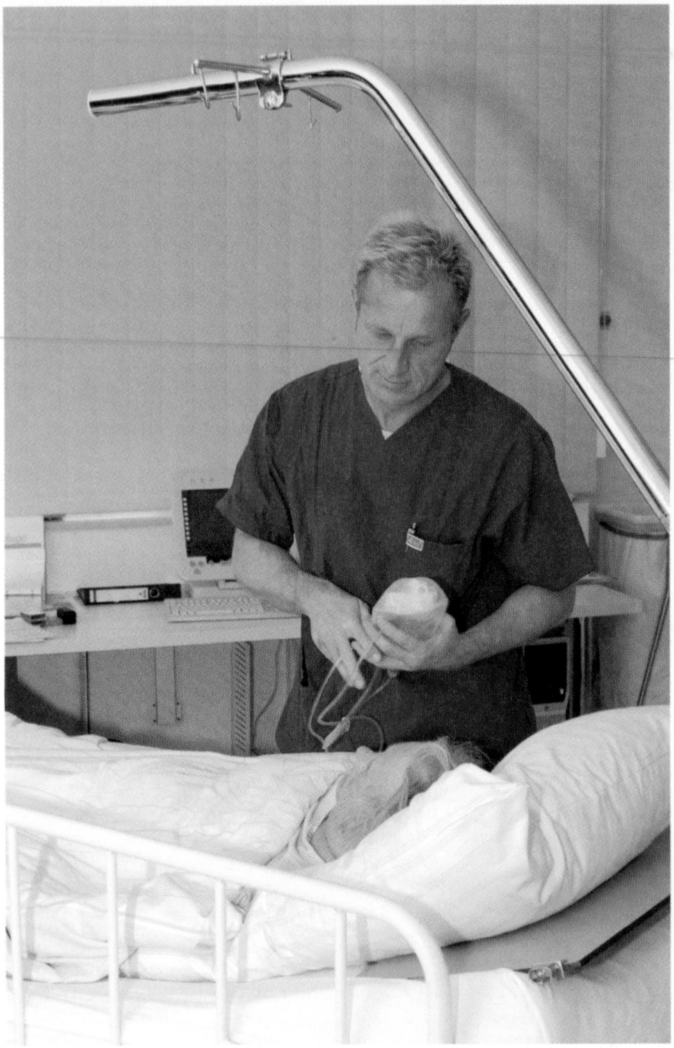

This section of the chapter is designed to help you find where each outcome is covered in the textbook and workbook.

Learning Outcomes		Text Readings, Figures, and Tables	Workbook Activities	Text and Workbook Assessments
2.1	List the levels of organization of the human body from simplest to most complex.	Overview: p. 20 Figure 2.1		*Text:* Review: 2.1 *Workbook:* Multiple Select: 10
2.2	Define the terms *matter, element, atom,* and *isotope.*	Chemical level: pp. 21–23 Figures 2.2, 2.3 Table 2.1		*Text:* Spot Check: 1 Review: 2.2
2.3	Define *molecule* and describe two methods of bonding that may form molecules.	Bonding to form molecules: pp. 23–24 Figure 2.4		*Text:* Review: 2.3 *Workbook:* Multiple Select: 3 Completion: 4 Case Study: 1
2.4	Summarize the five functions of water in the human body and give an explanation or example of each.	Water: p. 25		*Text:* Review: 2.4 *Workbook:* Multiple Select: 2
2.5	Compare solutions based on tonicity.	Solutions: p. 25 Figure 2.5		*Text:* Review: 2.5 *Workbook:* Completion: 3 Critical Thinking: 1
2.6	Determine whether a substance is an acid or a base and its relative strength if given its pH.	Acids, bases, and pH: pp. 26–27 Figures 2.6, 2.7	*Lab exercises and activities:* pH Figures 2.32–2.35	*Text:* Spot Check: 2 Review: 2.6 *Workbook:* Multiple Select: 4
2.7	Describe the four types of organic molecules in the body by giving the elements present in each, their building blocks, an example of each, the location of each example in the body, and the function of each example.	Organic molecules: pp. 27–31 Figures 2.8–2.11 Table 2.2	*Concept maps:* Chemical level Figure 2.38	*Text:* Spot Check: 3 Review: 2.7 *Workbook:* Multiple Select: 7 Matching: 1–5
2.8	Explain three factors governing the speed of chemical reactions.	Chemical reactions: pp. 31–32		*Text:* Spot Check: 4 Review: 2.8 *Workbook:* Multiple Select: 1

CHAPTER 2 MAPPING

Learning Outcomes		Text Readings, Figures, and Tables	Workbook Activities	Text and Workbook Assessments
2.9	Write the equation for cellular respiration using chemical symbols and describe it in words.	Cellular respiration: pp. 32–33		*Text:* Review: 2.9 *Workbook:* Completion: 1, 2
2.10	Explain the importance of ATP in terms of energy use in the cell.	ATP: pp. 33–34 Figure 2.12		*Text:* Spot Check: 7 Review: 2.10
2.11	Describe cell organelles and explain their functions.	Organelle level: pp. 34–38 Figures 2.13, 2.14 Table 2.3	*Coloring book:* The cell Figure 2.1 *Concept maps:* Organelle level Figure 2.39	*Text:* Spot Check: 5 Review: 2.11 *Workbook:* Matching: 6–10 Critical Thinking: 2
2.12	Compare four methods of passive transport and active transport across a cell membrane in terms of materials moved, direction of movement, and the amount of energy required.	Membrane transport: pp. 38–41 Figures 2.15–2.18	*Lab exercises and activities:* Osmosis Figure 2.36 Table 2.1 *Concept maps:* Cellular level Figure 2.40	*Text:* Spot Check: 6 Review: 2.12 *Workbook:* Multiple Select: 5, 6 Critical Thinking: 1 Case Study: 1
2.13	Describe bulk transport, including endocytosis and exocytosis.	Bulk transport: pp. 41–42 Figure 2.19	*Concept maps:* Cellular level Figure 2.40	*Text:* Review: 2.13
2.14	Describe the processes of transcription and translation in protein synthesis in terms of location and the relevant nucleic acids involved.	Protein synthesis: pp. 42–45 Figures 2.20, 2.21	*Lab exercises and activities:* Protein synthesis Figure 2.37	*Text:* Spot Check: 8 Review: 2.14 *Workbook:* Multiple Select: 8
2.15	Describe what happens to a protein after translation.	Protein synthesis: p. 45 Figure 2.22	*Lab exercises and activities:* Protein synthesis Figure 2.37	*Text:* Review: 2.15
2.16	Explain the possible consequences of mistakes in protein synthesis.	Protein synthesis: pp. 45–46	*Lab exercises and activities:* Protein synthesis Figure 2.37	*Text:* Review: 2.16
2.17	Describe the process of mitosis, including a comparison of the chromosomes in a parent cell to the chromosomes in the daughter cells.	Cell division: pp. 46–48 Figures 2.23, 2.24	*Concept maps:* Cell division Figure 2.41	*Text:* Spot Check: 9 Review: 2.17 *Workbook:* Completion: 5

Learning Outcomes		Text Readings, Figures, and Tables	Workbook Activities	Text and Workbook Assessments
2.18	Explain the possible consequences of mistakes in replication.	Cell division: p. 48	*Concept maps:* Cell division Figure 2.41	*Text:* Review: 2.18
2.19	Describe the effects of aging on cell division.	Effects of aging: pp. 48–49 Figure 2.25		*Text:* Review: 2.19
2.20	Describe the four classifications of tissues in the human body.	Tissue level: pp. 49–59 Figures 2.26–2.44	*Coloring book:* Tissues Figures 2.2–2.19 *Concept maps:* Tissue level Figure 2.42	*Text:* Review: 2.20
2.21	Describe the modes of tissue growth, change, shrinkage, and death.	Tissue growth: p. 60 Tissue change: p. 60 Tissue Shrinkage and death: p. 60	*Concept maps:* Tissue level Figure 2.42	*Text:* Review: 2.21 *Workbook:* Multiple Select: 9
2.22	Identify the human body systems and their major organs.	Organ level: pp. 60–62 Figure 2.45 Table 2.4 System level: pp. 63–65 Figure 2.46	*Coloring book:* Systems Figures 2.20–2.31 *Concept maps:* Organ and system levels Figure 2.43	*Text:* Review: 2.22

3 Introduction to Pathology

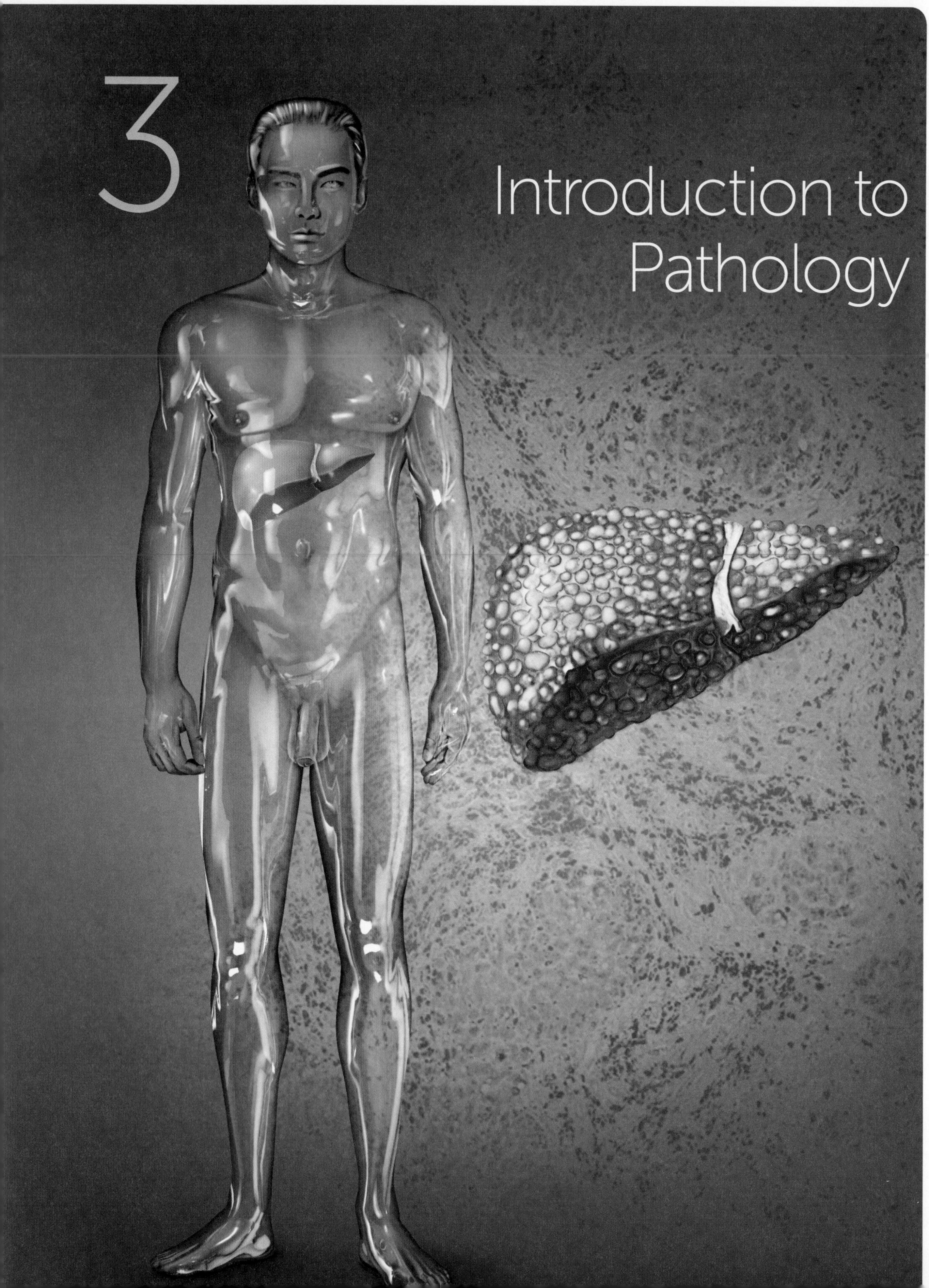

learning outcomes

This chapter of the workbook is designed to help you learn information about pathology. After completing this chapter in the text and this workbook, you should be able to:

3.1 Define *pathology.*

3.2 Define *disease* and describe the relationship between disease and homeostasis.

3.3 Define *predisposing factors* of disease and explain how specific predisposing factors affect disease.

3.4 Differentiate between signs and symptoms of disease and give an example of each.

3.5 Explain the function of pain and inflammation.

3.6 Explain six disease classifications.

3.7 Define *diagnosis* and list the steps involved in diagnosing diseases.

3.8 Define *differential diagnosis* and explain when it may be used.

3.9 Summarize four types of treatment plans.

3.10 Define *epidemiology* and explain how epidemiology affects health care.

Diagnosing Disease

Figure 3.1 shows the steps to diagnosing disease. Select one color and color the ovals in the flow chart that correspond to actions taken by the patient. In a different color, color the ovals that correspond to actions taken by a health care provider.

☐ Patient actions

☐ Health care provider actions

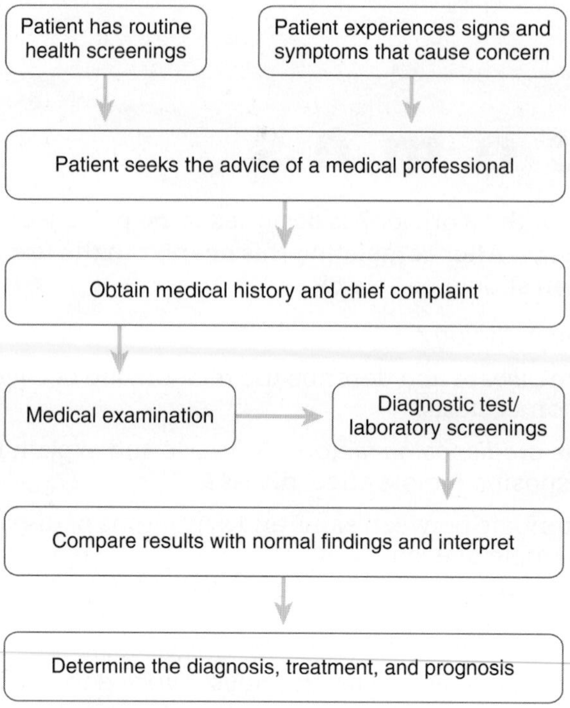

FIGURE 3.1 Steps to diagnosing disease.

Predisposing Factors of Disease

Figure 3.2 shows a pie chart with all of the predisposing factors of disease. Use one color to color all the sections of the pie that correspond to predisposing factors of disease that a person has control over. In a different color, color all of the pie sections that are predisposing factors of disease that a person cannot control.

☐ Under patient control

☐ Not under patient control

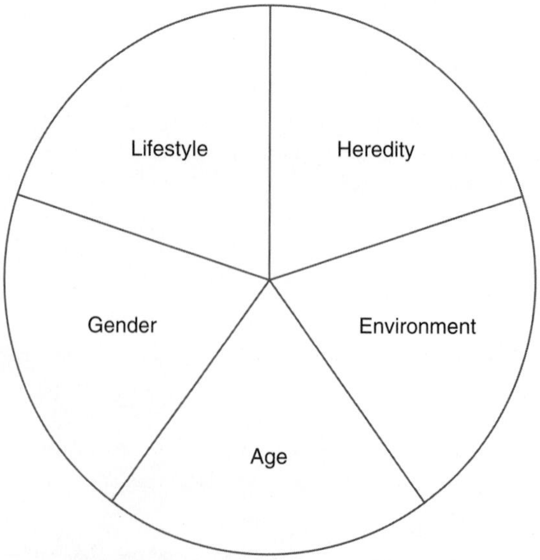

FIGURE 3.2 Predisposing factors of disease.

Family Diseases

Your family history of disease can tell you a lot about your own potential for specific diseases. This is one reason why health care providers ask you about your family's medical history when you complete the patient questionnaire. For this workbook activity, interview someone who does not mind giving you information about their family medical history. It could be a person in your own family or a friend. You can fill in the table with information about your family, if you do not mind sharing this with the class. Complete Table 3.1 below by listing the disease, which family member has it, and any risk factors the interviewee may take part in that make him or her more prone to developing the disease. After completing the table, answer the questions below.

TABLE 3.1 Family diseases.

Name of Disease	Family Member	Risk Factors for the Disease (activities that the interviewee takes part in)

1. Does the person interviewed have a better chance of developing diseases suffered by their parents or their aunts and uncles?

____

2. Explain your reasoning for your answer to question 1.

____

____

3. Is there anything they can do specifically to lessen their chance of developing a disease or condition similar to the ones listed in the table?

____

____

____

Childhood Diseases

In the textbook, you learned that age can be a predisposing factor to disease. More specifically, children are more susceptible to infectious diseases because their immune systems are still developing. Children receive a lot of vaccinations and immunizations between birth and six years of age to help defend them against harmful and potentially deadly infectious diseases. Figure 3.3 shows is a copy of the CDC's recommended immunization schedule. After reviewing the list of recommended immunizations, answer the following questions.

Recommended Immunization Schedule for Persons Aged 0 through 6 Years—United States, 2012

Vaccine / Age	Birth	1 month	2 months	4 months	6 months	9 months	12 months	15 months	18 months	19–23 months	2–3 years	4–6 years
Hepatitis B	HepB	HepB			HepB							
Rotavirus			RV	RV	RV*							
Diphtheria, Tetanus, Pertussis			DTaP	DTaP	DTaP			DTaP				DTaP
Haemophilus influenzae type b			Hib	Hib	Hib*		Hib					
Pneumococcal			PCV	PCV	PCV		PCV				PPSV	
Inactivated Poliovirus			IPV	IPV	IPV							IPV
Influenza					Influenza (Yearly)							
Measles, Mumps, Rubella							MMR					MMR
Varicella							Varicella					Varicella
Hepatitis A							Dose 1				HepA Series	
Meningococcal						MCV4						

*Six-month dose indicated only if patient does not receive a dose at 2 and 4 months.

Centers for Disease Control and Prevention. (February 10, 2012). Morbidity and mortality weekly report (MMWR). Severe Methemoglobinemia and Hemolytic Anemia from Aniline Purchased as 2C-E (4-ethyl-2,5-dimethoxyphenethylamine), a Recreational Drug, on the Internet—Oregon, 2011. Retrieved April 13, 2012 from http://www.cdc.gov/mmwr/pdf/wk/mm6105.pdf

FIGURE 3.3 Recommended immunization schedule for persons aged 0 through 6 years.

1. After considering what you read about childhood disease in Chapter 1, The Basics, and reviewing the recommended immunization schedule above, why is it recommended that children receive the influenza vaccine yearly?

____

____

____

2. What does DTaP vaccinate against?

____

3. Use the Internet to find out how prevalent polio is today compared to 50 years ago. Has the incidence of polio decreased or increased over time? What factors contribute to the number of polio cases seen today compared to how many there were in the past?

4. What common childhood illness does the Varicella vaccine help prevent? If a child can avoid getting this illness, what disease might he or she also avoid as an elderly individual?

KEY WORD CONCEPT MAPS

Key Words

The following terms are defined in the glossary of the textbook.

anaphylaxis
angiogenesis
benign
carcinogens
chromosomes
disease
epidemiology
etiologies
heredity
hypersensitivity
immunodeficiency
incidence
inflammation
malaise
malignant
mutagen
oncogenes
palliative
pathogens
pathology
prevalence
prognosis
signs
symptoms

Concept Maps

Complete the boxes in the following concept maps (**Figures 3.4 to 3.7**). Each map contains at least one key word.

Predisposing Factors of Disease

There are many predisposing factors of disease that can increase a person's risk of having a particular pathological condition.

- _______ can be considered a predisposing factor of disease.
  - The process of aging makes the _______ more susceptible to disease.
  - Children are born with partially developed _______ systems, which makes them more vulnerable to _______ diseases.
    - _______ disorders, genetic disorders, and _______ disorders can also play a role.
- _______ factors can be considered a predisposing factor of disease.
  - Genes passed down from our parents can have genetic links to diseases.
- _______ can be considered predisposing factors of disease.
  - Some diseases affect men or women only.
- _______ can be considered a predisposing factor of disease.
  - Negative habits can affect a person's overall health.
- _______ can be considered a predisposing factor of disease.
  - Chronic stress, loud noise, and poor housing conditions can affect a person's overall health.

FIGURE 3.4 Predisposing factors of disease concept map.

KEY WORD CONCEPT MAPS

Classifications of Disease

Diseases can be classified in various ways.

- ______ diseases → are caused by ______, which are disease-causing agents.
- ______ → are a class of diseases characterized by uncontrollable cell growth that forms ______. → ______, which are cancer-causing agents, and ______, which are genes that code for uncontrolled production of cellular growth factors, can cause this class of diseases.
- ______ disorders → are a class of diseases that result from the inability of the body to defend itself. → These can be caused by ______, which are allergic reactions; immunodeficiencies; or ______ disorders, which prevent the immune system from distinguishing between self and nonself.
- ______ disorders → are a class of diseases characterized by a mutation in the genetic code. → These mutations can be spontaneous, inherited, or caused by ______.
- ______ disorders → are a class of conditions of the mind.
- ______ → are a class of conditions caused by a wound or shock resulting in the body's inability to function normally.

FIGURE 3.5 Classifications of disease concept map.

Cancer

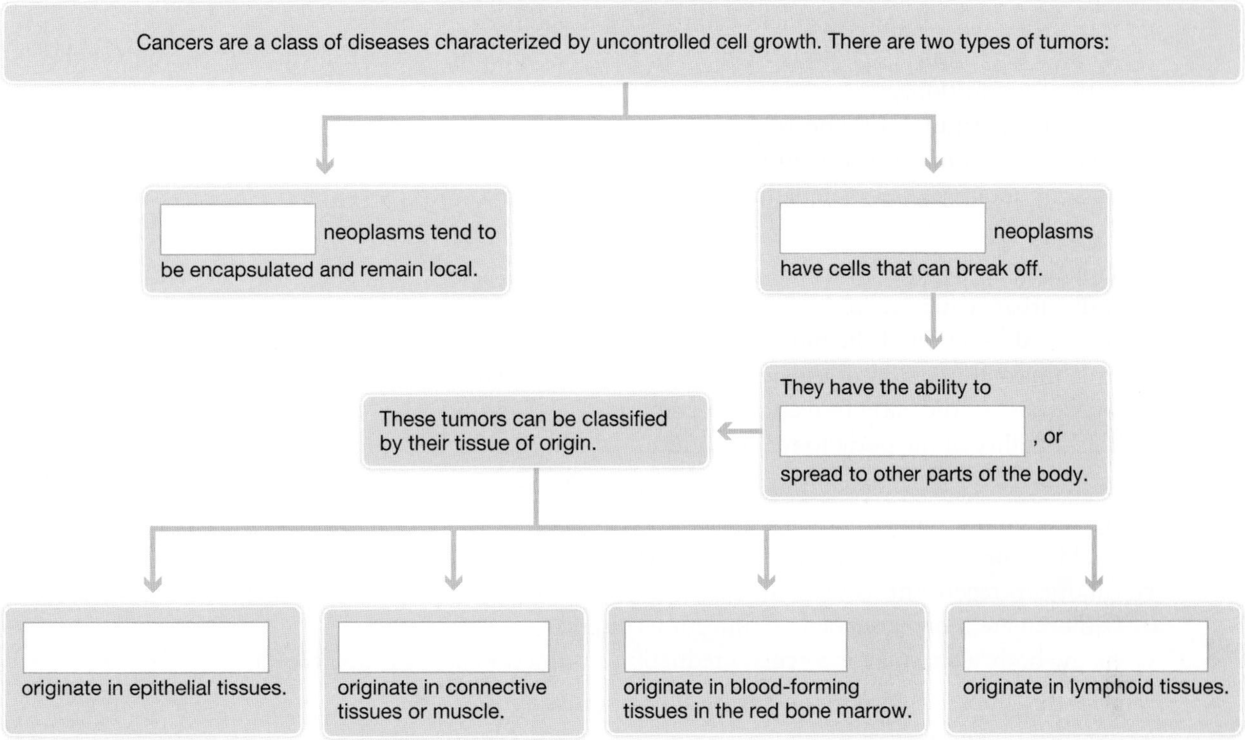

Cancers are a class of diseases characterized by uncontrolled cell growth. There are two types of tumors:
neoplasms tend to be encapsulated and remain local.
neoplasms have cells that can break off.
They have the ability to , or spread to other parts of the body.
These tumors can be classified by their tissue of origin.
originate in epithelial tissues.
originate in connective tissues or muscle.
originate in blood-forming tissues in the red bone marrow.
originate in lymphoid tissues.

FIGURE 3.6 Cancer concept map.

Immune Disorders

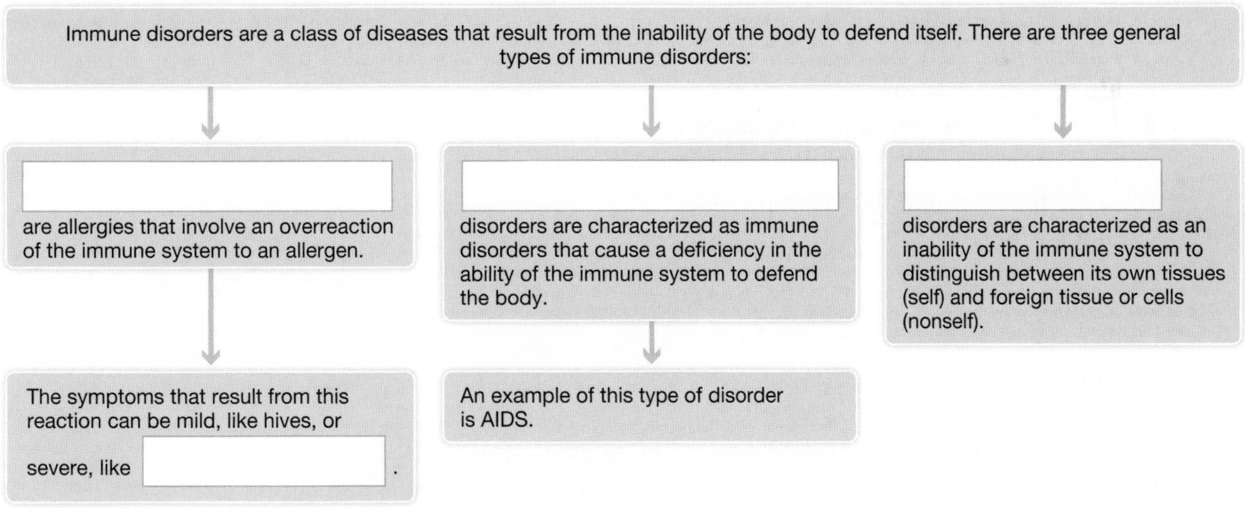

Immune disorders are a class of diseases that result from the inability of the body to defend itself. There are three general types of immune disorders:
are allergies that involve an overreaction of the immune system to an allergen.
The symptoms that result from this reaction can be mild, like hives, or severe, like .
disorders are characterized as immune disorders that cause a deficiency in the ability of the immune system to defend the body.
An example of this type of disorder is AIDS.
disorders are characterized as an inability of the immune system to distinguish between its own tissues (self) and foreign tissue or cells (nonself).

FIGURE 3.7 Immune disorders concept map.

Multiple Select: *Select the correct choices for each statement. The choices may be all correct, all incorrect, or any combination of correct and incorrect.*

1. When would a health care provider need to use a differential diagnosis?
  a. To help determine a diagnosis when there is more than one possible disease responsible for the patient's condition.
  b. To help determine a prognosis.
  c. To help determine a treatment plan.
  d. To help obtain a chief complaint.
  e. To help determine which medications to prescribe.

2. What is the definition of *disease*?
  a. A disturbance in homeostasis.
  b. A normal function of the human body.
  c. An abnormal function of the human body that results in the inability to maintain homeostasis.
  d. The ability of the body to reestablish balance.
  e. The inability of the body to maintain levels within the normal range.

3. How does age act as a predisposing factor of disease?
  a. Age does not act as a predisposing factor of disease.
  b. Children and elderly people are more susceptible to disease due to the state of their immune systems.
  c. All organs regenerate as the body gets older; therefore age does not have an effect on disease.
  d. Children are more vulnerable to infectious disease.
  e. As the body ages, there is a general reduction in function at the cellular level.

4. What is the difference in signs and symptoms of disease?
  a. Signs can be measured.
  b. Signs cannot be measured.
  c. Symptoms can be measured.
  d. Symptoms cannot be measured.
  e. Signs and symptoms can both be measured.

5. What is the function of pain?
  a. Protection of an injured body part.
  b. An alert that there is a potential problem with a particular part of the body.
  c. A defense mechanism.
  d. A signal for a person to take medication.
  e. A test of how much discomfort a person can tolerate.

6. What are the signs of inflammation?
  a. Bluish skin color
  b. Heat
  c. Swelling
  d. Pus
  e. Pain

7. When is palliative treatment used?
  a. When the patient's condition is responding to medication.
  b. When the focus of patient care is making the patient comfortable, not curing the disease.
  c. When the patient's condition can be cured.
  d. When the patient needs therapy to improve his or her condition.
  e. When the health care provider wants to teach the patient ways to avoid a particular disease.

8. What is the definition of *epidemiology*?
  a. The study of disease.
  b. The study of the prevalence of disease.
  c. The study of the incidence of disease.

d. The study of the spread of disease.
e. The study of the effect of disease on the population.

9. What classification of disease does anaphylactic shock fall under?
   a. Genetic disorder
   b. Immune disorder
   c. Hypersensitivity
   d. Trauma
   e. Cancer

10. Which of these is considered a characteristic of cancer?
   a. A benign tumor.
   b. A malignant tumor.
   c. A tumor caused by exposure to a carcinogen.
   d. A tumor caused by an oncogene.
   e. A tumor that does not metastasize.

Matching: *Match the following cancer to its tissue of origin. Choices may be used more than once.*

______	1. Cancer that originates in epithelial tissues	a. Carcinomas
______	2. Cancer that originates in muscle tissues	b. Leukemias
______	3. Cancer that originates in connective tissues	c. Lymphomas
______	4. Cancer that originates in lymphoid tissues	d. Sarcomas
______	5. Cancer that originates in blood-forming tissues	

Matching: *Match the following steps to determining a diagnosis to the purpose of that step. Choices may be used more than once.*

______	6. Obtaining a medical history	a. Allows the health care provider to establish a baseline of what is normal for that patient
______	7. Performing diagnostic tests	b. Allows the health care provider to determine whether the patient falls within the normal range
______	8. Performing a medical examination	c. Allows the health care provider to assess the patient for any abnormal findings
______	9. Performing a routine screening of the patient	d. Allows the health care provider to determine whether the patient's test results indicate the presence of disease
______	10. Comparing results with normal findings	

Completion: *Fill in the blanks to complete the following statements concerning inflammation.*

1. ______________ are released by damaged tissues to start inflammation.
2. Dilation of blood vessels during inflammation results in the symptoms of ______________, ______________, ______________, and ______________.
3. In inflammation the symptom ______________ is caused by fluid leaking out of vessels because they are dilated.
4. The extra fluid puts pressure on the nerve endings, causing ______________.
5. Dilated blood vessels bring ______________ blood flow to the area.

Critical Thinking

1. In the textbook, you learned about the Spanish flu pandemic of 1918 and you discussed various epidemics and pandemics that have affected the world. Pick one of the diseases you discussed in the chapter Spot Check 5 and answer the following: How has the epidemiological study of the disease affected the progression of the disease? What are the recommendations of health care providers to attempt to control the continuous spread of the disease?

2. In the textbook you explored how important it is for the body to maintain homeostasis to avoid disease. What are some ways that the body maintains homeostasis? In the presence of disease, what are some examples of the body not maintaining homeostasis?

Case Study

A patient comes into the clinic experiencing frequent urination and excessive thirst.

1. Describe the steps involved in determining a diagnosis for this patient.

2. What questions may be asked when obtaining the patient's chief complaint?

CHAPTER 3 REVIEW QUESTIONS

3. How can a health care provider rate the patient's level of discomfort?

____

4. If the exam and diagnostic tests do not point to one clear cause of the patient's symptoms, what kind of diagnosis would be used?

____

This section of the chapter is designed to help you find where each outcome is covered in the textbook and workbook.

Learning Outcomes		Text Readings, Figures, and Tables	Workbook Activities	Text and Workbook Assessments
3.1	Define *pathology.*	Overview: p. 72		*Text:* Review: 3.1 *Workbook:* Multiple select: 2 Critical thinking: 2
3.2	Define *disease* and describe the relationship between disease and homeostasis.	Introduction to disease: pp. 72–73 Figure 3.1		*Text:* Review: 3.2 *Workbook:* Critical thinking: 1
3.3	Define *predisposing factors* of disease and explain how specific predisposing factors affect disease.	Predisposing factors of disease: pp. 73–76 Figures 3.2–3.5	*Coloring book:* Predisposing factors of disease Figure 3.2 *Lab exercises and activities:* Family diseases Table 3.1 Childhood diseases Figure 3.3	*Text:* Review: 3.3 *Workbook:* Multiple select: 3
3.4	Differentiate between signs and symptoms of disease and give an example of each.	Signs and symptoms of disease: p. 76		*Text:* Spot check: 1 Review: 3.4 *Workbook:* Multiple select: 4
3.5	Explain the function of pain and inflammation.	Signs and symptoms of disease: pp. 77–78 Figures 3.6–3.8		*Text:* Spot check: 2 Review: 3.5 *Workbook:* Multiple select: 5, 6 Completion: 1–5 Case study: 3
3.6	Explain six disease classifications.	Six classifications of disease: pp. 79–81 Figure 3.9	*Concept maps:* Predisposing factors of disease Figure 3.4 Classifications of disease Figure 3.5 Cancer Figure 3.6 Immune disorders Figure 3.7	*Text:* Spot check: 3 Review: 3.6 *Workbook:* Multiple select: 10 Matching: 1–5

Learning Outcomes		Text Readings, Figures, and Tables	Workbook Activities	Text and Workbook Assessments
3.7	Define *diagnosis* and list the steps involved in diagnosing diseases.	Diagnosing disease: p. 82 Figures 3.10, 3.11	*Coloring book:* Diagnosing disease Figure 3.1	*Text:* Review: 3.7 *Workbook:* Matching: 6–10 Case study: 1–2
3.8	Define *differential diagnosis* and explain when it may be used.	Differential diagnosis: pp. 82–83 Figure 3.12		*Text:* Review: 3.8 *Workbook:* Multiple select: 1 Case study: 4
3.9	Summarize four types of treatment plans.	Treatment of the disease: pp. 83–84 Figures 3.13, 3.14		*Text:* Spot check: 4 Review: 3.9 *Workbook:* Multiple select: 7
3.10	Define *epidemiology* and explain how epidemiology affects health care.	Introduction to epidemiology: pp. 84–85 Figure 3.15 Table: 3.1		*Text:* Spot check: 5 Review: 3.10 *Workbook:* Multiple select: 8 Critical thinking: 1

4 The Integumentary System

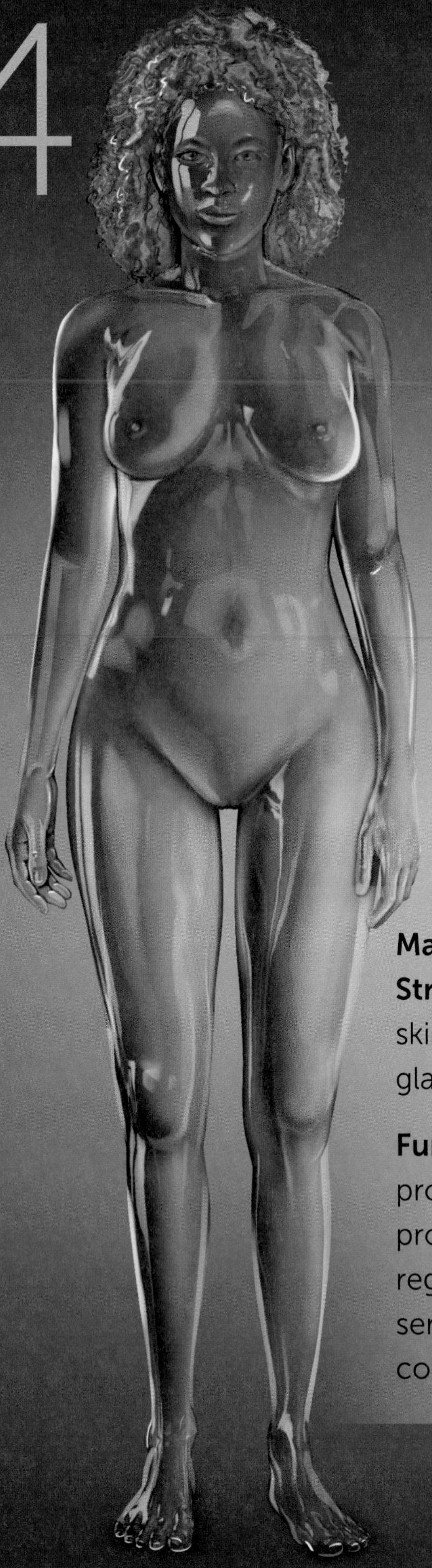

Major Organs and Structures:
skin, hair, nails, cutaneous glands

Functions:
protection, vitamin D production, temperature regulation, water retention, sensation, nonverbal communication

learning outcomes

This chapter of the workbook is designed to help you learn the anatomy and physiology of the integumentary system. As a system chapter, it includes an additional section—**Word Roots & Combining Forms.** This section lists relevant word roots and combining forms that will help you with the medical terminology for this system. You will also find a new type of chapter review question—**Word Deconstruction**—in which you will break down medical terms to their word roots, prefixes, and suffixes in addition to defining the term. After completing this chapter in the text and this workbook, you should be able to:

4.1 Use medical terminology related to the integumentary system.

4.2 Describe the histology of the epidermis, dermis, and hypodermis.

4.3 Describe the cells of the epidermis and their function.

4.4 Describe the structures of the dermis and their functions.

4.5 Compare and contrast the glands of the skin in terms of their structure, products, and functions.

4.6 Describe the histology of a hair and hair follicle.

4.7 Explain how a hair grows and is lost.

4.8 Describe the structure and function of a nail.

4.9 Explain how the layers and structures of the skin work together to carry out the functions of the system.

4.10 Explain how the skin responds to injury and repairs itself.

4.11 Compare and contrast three degrees of burns in terms of symptoms, layers of the skin affected, and method used by the body for healing.

4.12 Describe the extent of a burn using the rule of nines.

4.13 Summarize the effects of aging on the integumentary system.

4.14 Describe a diagnostic test commonly used when diagnosing integumentary system disorders.

4.15 Describe three forms of skin cancer in terms of the body area most affected, appearance, and ability to metastasize.

4.16 Describe a common integumentary disorder other than skin cancer and relate abnormal function to the pathology.

word roots & combining forms

cutane/o: skin

cyan/o: blue

derm/o: skin

dermat/o: skin

kerat/o: hard

melan/o: black

onych/o: nail

seb/o: oil

Skin

Figure 4.1 shows structures located in the two layers of the skin—the epidermis and the dermis—and in the hypodermis, which connects the skin to the rest of the body. Color the box next to each layer. Use the same color for the corresponding bar in **Figure 4.1**.

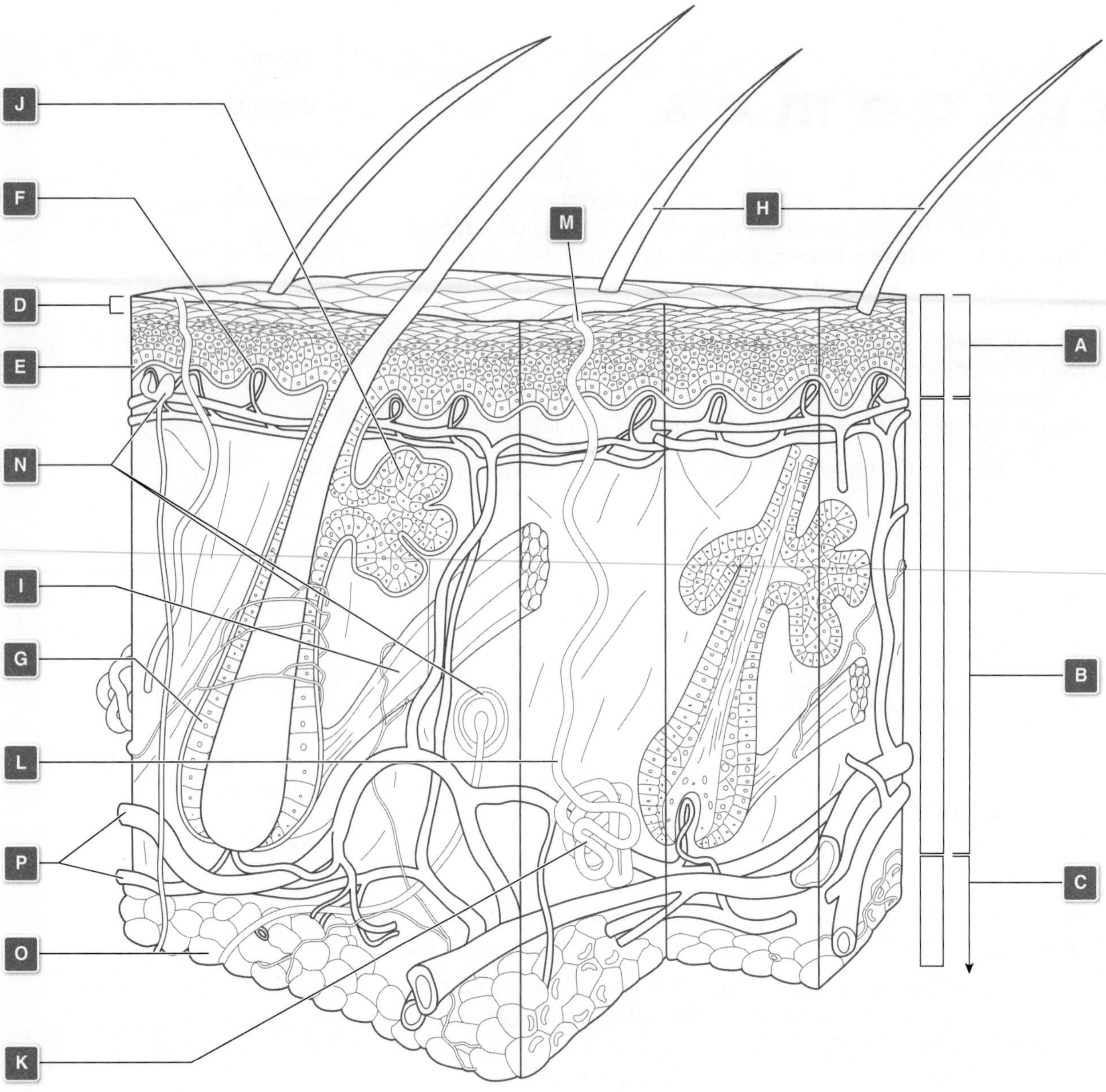

FIGURE 4.1 Skin.

☐ Epidermis(A)

☐ Dermis(B)

☐ Hypodermis(C)

COLORING BOOK

Continue to use **Figure 4.1** to color the structures listed below in each layer.

Epidermis

The epidermis is the most superficial layer of the skin. It is composed of stratified squamous epithelial tissue.

☐ Stratum corneum(D)

☐ Stratum basale(E)

Dermis

The dermis is deep to the epidermis. It is composed of loose/areolar connective tissue over dense irregular connective tissue. It contains many additional structures of the skin.

☐ Papilla(F)

☐ Hair follicle(G)

☐ Hair shaft(H)

☐ Arrector pili muscle(I)

☐ Sebaceous gland(J)

☐ Sweat gland(K)

☐ Sweat gland duct(L)

☐ Pore(M)

☐ Nerve endings(N)

Hypodermis

The hypodermis is technically not part of the skin, but it attaches the skin to the rest of the body. Composed mostly of adipose connective tissue, the hypodermis contains numerous blood vessels that extend into the dermis to supply its structures with the necessary nutrients to carry out their functions and carry the wastes away.

☐ Adipocyte(O)

☐ Blood vessels(P)

Hair and Hair Follicle

Hair follicles are located in thin skin. They are important in the production of hair and in the healing of skin in cases of injury. Color the box next to each term. Use the same color for the corresponding structure in **Figure 4.2**.

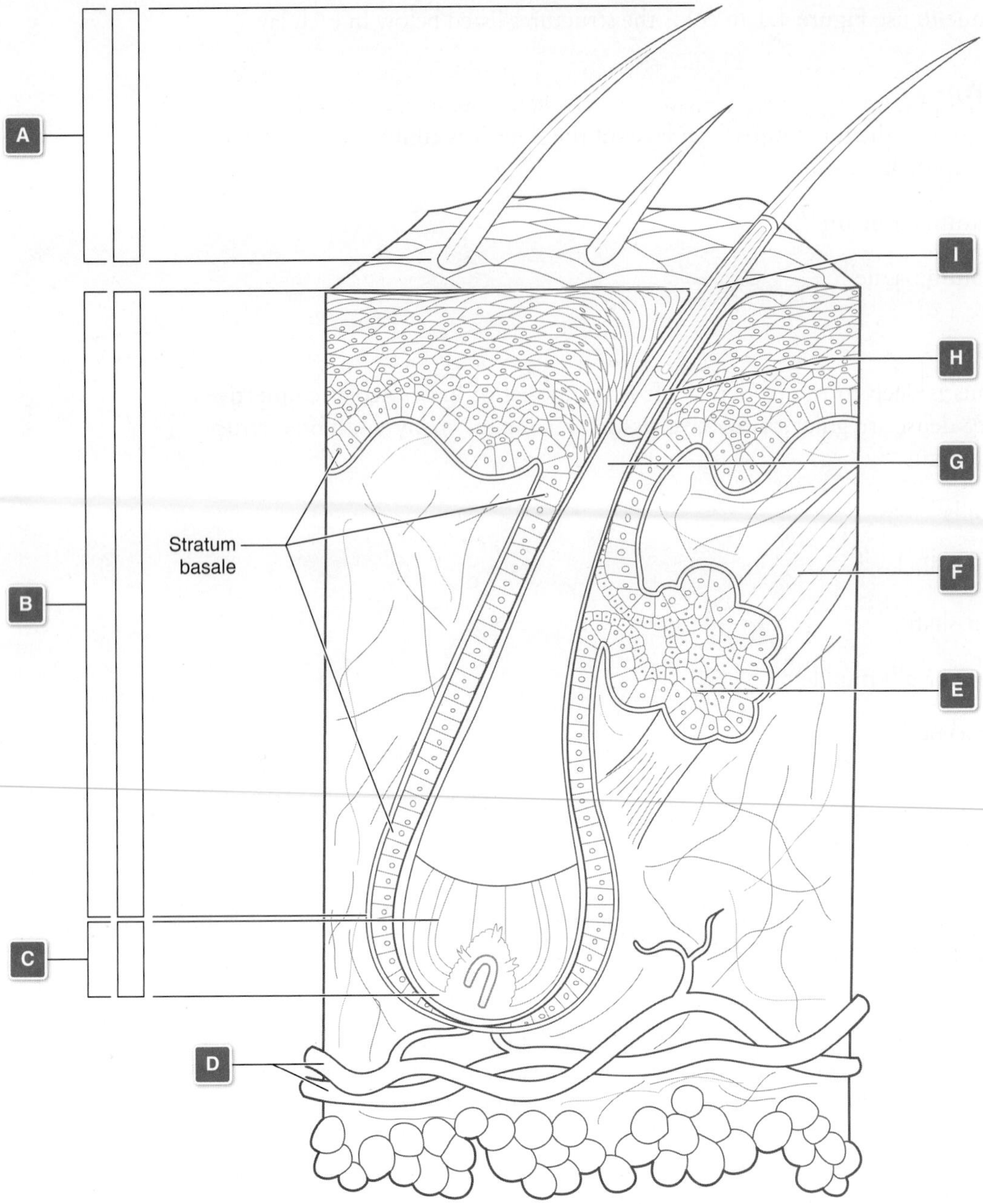

FIGURE 4.2 Hair and hair follicle. The hair follicle is formed by an extension of the stratum basale of the epidermis.

- ☐ Hair shaft(A)
- ☐ Hair root(B)
- ☐ Hair bulb(C)
- ☐ Blood vessels(D)
- ☐ Sebaceous gland(E)
- ☐ Arrector pili muscle(F)
- ☐ Cuticle(G)
- ☐ Cortex(H)
- ☐ Medulla(I)

COLORING BOOK

Nails

Nails protect the ends of the fingers and toes and are used for grasping, manipulating objects, and scratching. They are composed of hard keratin. Color the box next to each term. Use the same color for the corresponding structure in **Figure 4.3**.

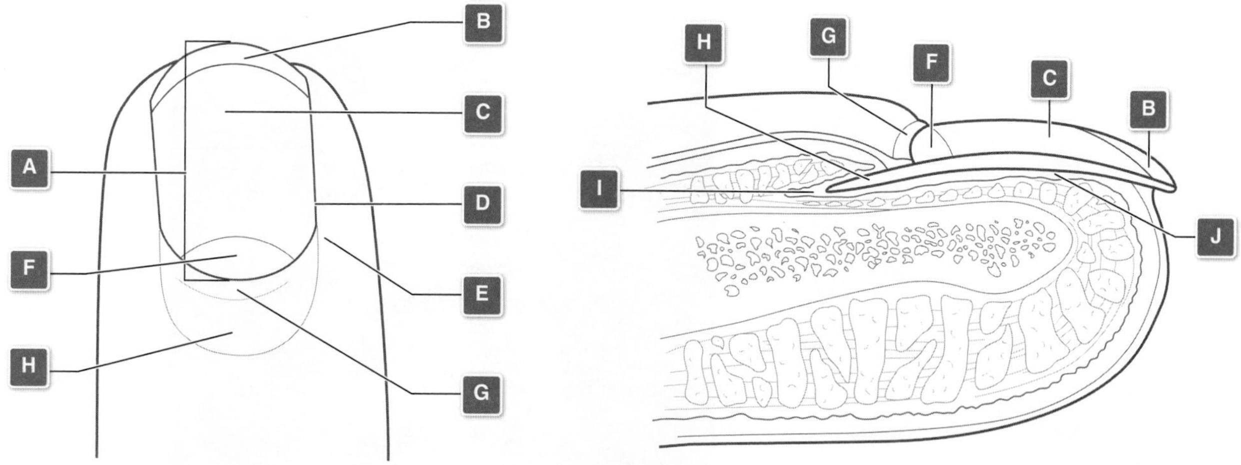

FIGURE 4.3 Anatomy of a nail.

- ☐ Nail plate(A)
- ☐ Free edge(B)
- ☐ Nail body(C)
- ☐ Nail groove(D)
- ☐ Nail fold(E)
- ☐ Lunule(F)
- ☐ Eponychium(G)
- ☐ Nail root(H)
- ☐ Nail matrix(I)
- ☐ Nail bed(J)

COLORING BOOK

Skin Observation 1

One of the best resources you have for studying human anatomy and physiology is your own body and the bodies of those around you. The integumentary system is the most visible system of the human body. Look at Figure 4.4, which shows the anterior forearm and carpal region of a 57-year-old female.

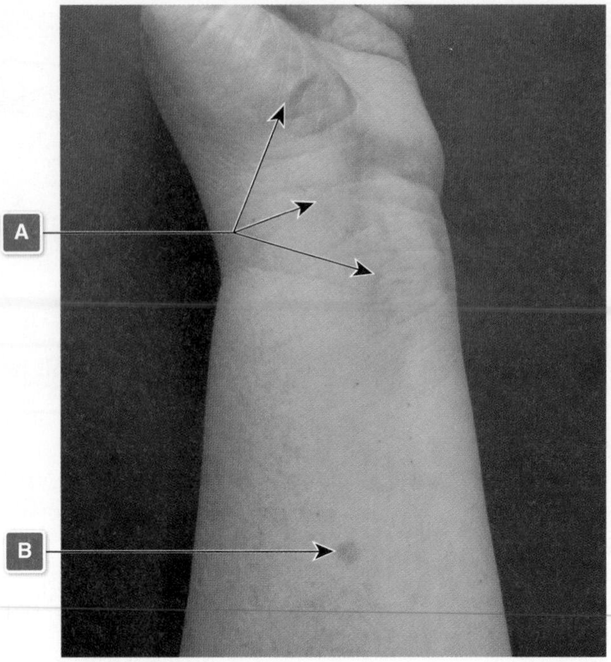

FIGURE 4.4 Skin observation.

1. Identify the faint blue structures labeled "A" on the photograph.

   ____

2. Explain the reasoning for your answer for question 1.

   ____

   ____

3. Why are the structures labeled "A" blue?

   ____

   ____

4. Identify the brown mark labeled "B" on the photograph. This mark was not visible a year ago.

   ____

5. Why do you think this mark appeared in the last year?

   ____

   ____

6. Based on what is causing this brown mark, does this brown mark require medical attention? Justify your answer in terms of anatomy and physiology.

____

____

____

warning

You may have identified the blue structures as veins, but you should be careful identifying structures deep to the skin on the basis of color. The structures labeled "A" only appear blue. Collagen in the dermis causes dark things deep to it to appear blue. Blood vessels that dilate in the superficial dermis appear red because you are not looking through much collagen to see them. The collagen is deep to those superficial vessels. The structures marked "A" are deep in the skin. Does this change your answer to question 1?

The adipose tissue of the hypodermis cancels out this effect. You cannot see objects deep to the hypodermis. Blood vessels superficial in the hypodermis will appear blue. Blood vessels deep to the hypodermis are not visible depending upon the amount of hypodermis. You will learn more about this in the cardiovascular system chapter on blood.

7. A bruise is caused by bleeding from broken blood vessels. Explain the coloration of a bruise. Include the layer of skin in your explanation.

____

____

Skin Observation 2

You will need to find an adult whose age is at least 20 years different from yours. For example, if you are 20, you will need to find someone at least 40 years old or older. If you are 50, you may find someone 30 years old or younger or someone at least 70 years old.

Pinch the skin on the posterior surface of your hand (see **Figure 4.5**), and observe what happens when you let go.

8. Record your age. ______________________

9. What did you observe when you let go of the skin?

____

____

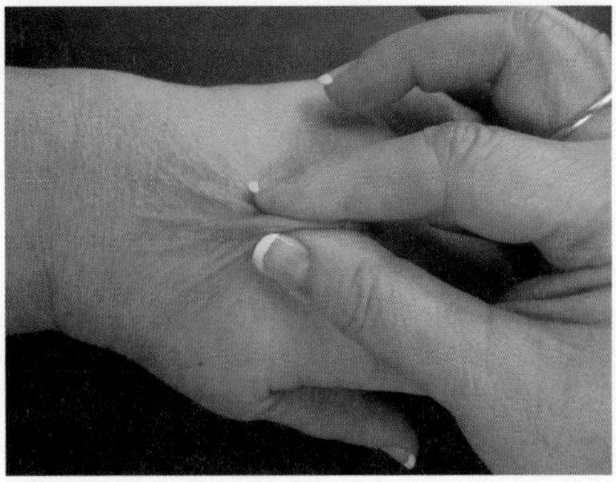

FIGURE 4.5 Pinch.

Pinch the skin of the person who differs in age from you by at least 20 years.

10. Record the person's age.

____

11. What did you observe when you let go of the skin?

____

____

12. How does this observation compare with the observation of your skin?

____

13. In terms of anatomy and physiology, what is responsible for the difference in your observations?

____

____

____

Skin Observation 3

Nerve endings for fine touch are located in the stratum basale, while warmth, cold, pressure, and pain are detected by nerve endings in the dermis. Nerve endings are also associated with the hair follicle to detect movement of the hair. Nerve endings, however, are not distributed evenly in all regions of the skin.

You will again need to find a partner for this exercise. You will also need two pointed objects, such as two toothpicks or two pencil leads.

The point of this exercise is to test two different areas of the skin to see how sensitive they are to fine touch. The areas to be tested are the *anterior fingertip* and the *posterior neck or shoulder.* The person being tested should close his or her eyes. The person conducting the test should gently touch the area of the skin with either one pointed object or two pointed objects at the same time. See **Figure 4.6.** The person being tested should say whether he or she felt one or two touches.

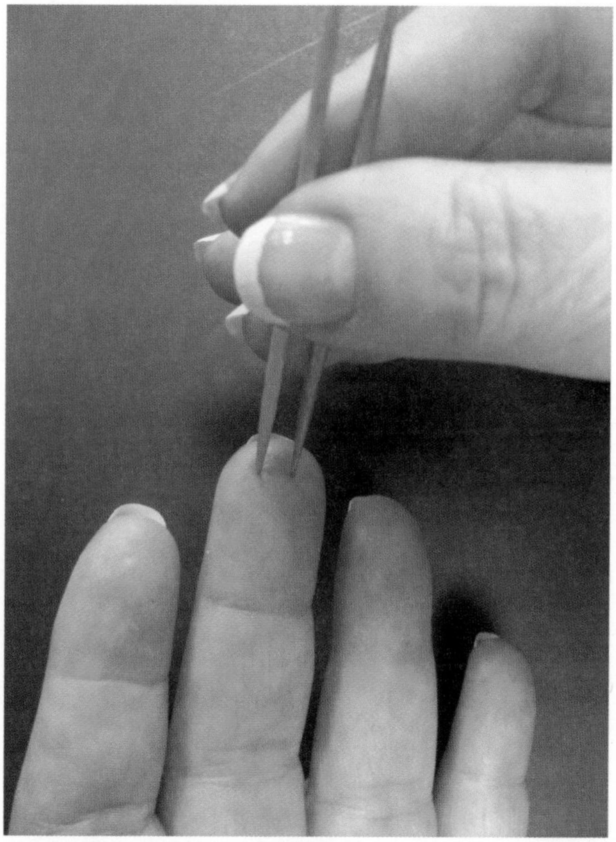

FIGURE 4.6 Two-point discrimination.

> **warning**
>
> *Gently!* There is no need to pierce the skin. You are trying to map fine touch.

Vary the use of one or two touches so that the person being tested does not know which to expect. You are looking for how close together you can get the two pointed objects and still have the person being tested report it as two touches instead of one touch.

14. Which nerve endings are being tested?

____

15. In which layer of skin are they located?

____

16. In which area of skin are you able to have the two points the closest and still have the person being tested report that they are two touches?

____

____

17. What conclusion can you draw from this exercise concerning the distribution of this type of nerve ending for these two regions?

____

____

Key Words

The following terms are defined in the glossary of the textbook.

acne
contact inhibition
cornification
cutaneous
dermatitis
epidermis
exocrine glands
fibrosis
first-degree burn
keratin
mediators of inflammation
melanocytes
papillae
pathogens
regeneration
scales
sebum
stratum basale
subcutaneous layer
sweat glands
thin skin
urticaria
vesicles
wound contracture

Concept Maps

Complete the boxes in the following concept maps (**Figures** 4.7 to 4.10). Each map contains at least one key word.

Layers of the Skin

The integumentary system is composed of skin, hair, nails, and cutaneous glands. The skin has two layers.

The epidermis is the most superficial layer. It is composed of stratified squamous epithelial tissue. This layer of skin is divided into strata.

The dermis is the deeper layer of skin. It is composed of loose/areolar connective tissue over dense irregular connective tissue.

The hypodermis is not a layer of skin but lies deep to the dermis. It serves to attach skin to the rest of the body. It is composed mostly of adipose tissue.

The deepest stratum is the basale.

This layer contains:

- cuboidal, which grow and divide.
- melanocytes, which produce pigment.
- Nerve endings, which detect fine touch.

The stratum lucidum is found only in thick skin (examples are lips, palms, and plantar surfaces of the feet).

The stratum corneum is the most superficial layer. It is composed of dead cells that are filled with keratin.

This layer of skin contains:

- Papillae, which are bumps at the superficial edge of the dermis that contain blood vessels.
- Fibroblasts, which are cells that produce collagen and elastic fibers.
- Nerve cells that serve as receptors.
- Cutaneous glands such as sebaceous and sweat glands.
- Sebum, which contain hair.

FIGURE 4.7 Layers of the skin concept map.

Cutaneous Glands

Cutaneous glands of the skin are located throughout the dermis. They are exocrine glands, meaning they use ducts to deliver their products to the proper place. There are two types of cutaneous glands.

The sebaceous glands produce sebum, an oily, lipid-rich substance that moisturizes the skin and hair.

There are various types of sweat glands located throughout the dermis. They differ in location, function, and product.

merocrine sweat glands are the most numerous. They are located all over the body. These glands produce a(n) watery sweat, which functions in the regulation of body temperature by keeping the body cool through evaporation.

apocrine sweat glands are found in the axillary region, the pubic region, and the beard. These glands are connected to hair follicles. The lipid-rich sweat produced by these glands has a scent that is designed to influence the behavior of others. Bacteria feeding on this type of sweat produce body odor.

ceruminous glands are located in the ear. They produce cerumen, also called earwax. This substance keeps the eardrum flexible, waterproofs the ear canal, and protects the ear from insects, bacteria, and debris.

mamary glands are located in the breast. These glands produce breast milk, which contains fat and protein to nourish an infant.

FIGURE 4.8 Cutaneous glands concept map.

KEY WORD CONCEPT MAPS

Burns

Burns can be categorized into three degrees based on the layers of skin involved.

First-degree burns are the most common. They involve the most superficial layer of skin, the epidermis. Symptoms are redness, pain, and swelling.

Second-degree burns involve the epidermis and the dermis. They are also referred to as partial-thickness burns. Symptoms are redness, pain, swelling, and blisters.

3rd-degree, or full-thickness, burns are the most serious. They involve the epidermis, the dermis, and the hypodermis. Symptoms are charring and no pain due to the destruction of nerve endings in the dermis.

FIGURE 4.9 Burns concept map.

Skin Cancer

Skin cancer is the most common form of cancer. There are three types of skin cancer.

Basal cell carcinoma is the most common form of skin cancer. It starts in the cells of the stratum basale. This type of cancer rarely metastasizes.

squamous cell carcinoma is a type of skin cancer that starts with the abnormal growth of keratinocytes in the stratum spinosum of the epidermis. This type of cancer can metastasize.

malignant melanoma is the rarest form of skin cancer. It is also considered to be the most deadly. It is the abnormal growth of melanocytes usually in a preexisting mole. This type of cancer can occur anywhere on the body and metastasizes easily.

FIGURE 4.10 Skin cancer concept map.

Word Deconstruction: *In the textbook, you built words to fit a definition using combining forms, prefixes, and suffixes. Here you are to break down the term into its parts (prefixes, roots, and suffixes) and give a definition. Prefixes and suffixes can be found inside the back cover of the textbook.*

FOR EXAMPLE Dermatitis: dermat/itis—inflammation of the skin

1. Onychoma: ______________________________
2. Cyanotic: ______________________________
3. Seborrhea: ______________________________
4. Keratosis: ______________________________
5. Subcutaneous: ______________________________

Multiple Select: *Select the correct choices for each statement. The choices may be all correct, all incorrect, or any combination of correct and incorrect.*

1. Which of the following statements correctly match(es) the infection with the pathogen causing it?
   a. Jock itch is caused by bacteria.
   b. Tinea infections are caused by a virus.
   c. Cellulitis is most often caused by a virus.
   d. Ringworm is caused by a fungus.
   e. Warts are caused by a virus.
2. Which of the following statements correctly match(es) the layer of the skin to the tissue composing it?
   a. The stratum corneum is composed of simple cuboidal epithelial tissue.
   b. The dermis is composed of dense regular connective tissue over dense irregular connective tissue.
   c. The dermis is composed of loose/areolar connective tissue over dense regular connective tissue.
   d. The epidermis is composed of stratified squamous epithelial tissue.
   e. The hypodermis is composed of adipose connective tissue.
3. Which of the following statements correctly state(s) the location or function of the cells in the epidermis and dermis?
   a. Keratinocytes are located in the stratum basale in the epidermis and the hair follicle of the dermis.
   b. Melanocytes are located in the stratum basale in the epidermis and the hair follicle of the dermis.
   c. Dendritic cells are located in the stratum basale in the epidermis and the hair follicle of the dermis.
   d. Fibroblasts are located in the stratum basale in the epidermis and the hair follicle of the dermis.
   e. Dead squamous cells filled with keratin are located in the stratum basale in the epidermis and the hair follicle of the dermis.
4. Which of the following statements is (are) true about hair?
   a. Vellus hair starts to develop during the last trimester of fetal development.
   b. Terminal hair is visible on the hair of a newborn's eyebrows and eyelashes.
   c. Hair grows at the rate of approximately 1 inch per month.
   d. The growth stage for terminal hair varies by location.
   e. Lanugo hair can be found in the axillary region of an adult.

5. Which of the following statements is (are) true concerning the anatomy of a nail?
   a. The lunule is visible at the distal end of all nails.
   b. The nail plate is made of soft keratin.
   c. The nail plate is made of lipids.
   d. The cuticle of a nail has interlocking plates.
   e. The eponychium is composed of stratum basale cells.
6. Which of the following statements describe(s) a function of the integumentary system?
   a. The skin prevents loss of body fluids.
   b. Sweat and sebaceous glands are important in temperature regulation.
   c. The skin contains a precursor molecule for vitamins A and C.
   d. The skin contains keratin that protects underlying structures from UV light.
   e. The immune system is alerted to pathogens entering through the skin by dendritic cells.
7. Which of the following statements is (are) true about aging?
   a. Sagging and wrinkling result because skin becomes more elastic with age.
   b. The number of melanocytes decreases, but the production of some melanocytes increases with age.
   c. Fibroblasts produce less collagen fibers with age.
   d. Sebaceous glands produce less sebum with age, causing the hair and skin to be drier.
   e. The hypodermis thins with age.
8. Which of the following statements is (are) true concerning skin cancer?
   a. People with dark skin are more likely to develop malignant melanoma.
   b. People with light skin are most likely to develop skin cancer.
   c. Melanin protects the underlying tissues from UV light.
   d. Sunburns as a child have nothing to do with developing skin cancer later in life.
   e. All moles need to be removed.
9. Which of the following statements is (are) true concerning the skin's response to injuries?
   a. Normal function returns when the wound is healed by fibrosis.
   b. Regeneration results in scar tissue.
   c. The type of healing depends upon how far the edges are apart.
   d. It is important to bring the wound edges together so that the stratum corneum can regenerate new epidermis.
   e. Stratum basale cells reach contact inhibition if the wound heals by regeneration.
10. Which of the following statements is (are) true about burns?
   a. Second-degree burns produce blisters.
   b. The most common first-degree burn is a sunburn.
   c. First-degree burns damage just the epidermis.
   d. Third-degree burns are the most serious.
   e. The serious concerns in third-degree burns are infection and fluid loss.

Matching: *Match the following cutaneous glands to the description of what they produce.*

_______	1. Lipid-rich sweat that is used for a scent	a. Mammary glands
_______	2. Waxy substance	b. Ceruminous glands
_______	3. Watery sweat used for cooling	c. Apocrine glands
_______	4. Modified sweat with proteins, lipids, and carbohydrates	d. Sebaceous glands
_______	5. Lipid-rich substance that moisturizes hair and skin	e. Merocrine glands

Matching: *Match the following structures to their descriptions.*

_______ **6.** Smooth muscle attached to a hair follicle

_______ **7.** Opening of a duct for a sweat gland at the skin surface

_______ **8.** Part of a hair out of the skin.

_______ **9.** Soft keratin layer inside the hair

_______ **10.** Interlocking plates on a hair surface

a. Arrector pili
b. Cuticle
c. Cortex
d. Medulla
e. Root
f. Shaft
g. Pore

Completion: *Fill in the blanks to complete the following statements concerning disorders of the integumentary system.*

1. _______ _______ can be used to diagnose fungal infections and parasitic infections.

2. Examining the skin using a Wood's Light helps locate _______ changes in the skin.

3. A(n) _______ is a diagnostic test usually performed when skin cancer is suspected.

4. The extent of a burn can be determined by the _______ _______ _______.

5. When determining the extent of a burn, the body is divided into _______ areas, each of which represents approximately 9 percent of the body's surface area.

Critical Thinking

1. Consider two second-degree burns of the same severity. One is on the posterior surface of the hand. The other is on the palmar surface of the hand. Given that both wounds are treated the same, which burn should heal faster? Justify your answer in terms of anatomy and physiology.

____

____

____

2. You are doing some filing and receive a paper cut that does not bleed but does hurt. Explain how the cut will heal on the basis of the layer of skin that must be involved.

____

____

____

____

Case Study

A 65-year-old woman visits her doctor with a complaint of an itchy rash on the side of her face, near her eye. The rash seems to travel in a line extending back past her hair line onto her scalp. The rash is red with small raised fluid-filled vesicles. Her symptoms began yesterday and they do seem to be getting worse, with the addition of throbbing pain at the site of the vesicles. When asked about her medical history, she cannot remember coming in contact with any allergen in the past 24 hours, nor has she had any recent changes in diet or medication. She did have chicken pox when she was seven years old.

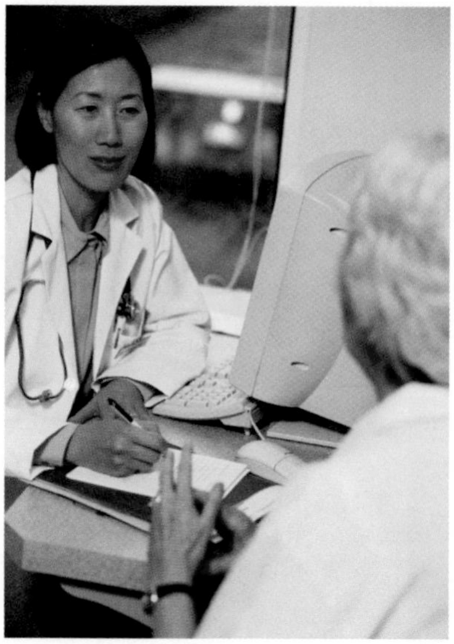

1. What is the likely diagnosis of this woman and what is the organism that has caused this condition?

2. Can she be cured of her condition? Explain your answer.

3. What is the recommended method of treatment?

4. Why does the rash have what appears to be a linear pattern starting on her face and continuing onto her scalp?

5. Why is her medical history important to her diagnosis?

CHAPTER 4 MAPPING

This section of the chapter is designed to help you find where each outcome is covered in the textbook and workbook.

Learning Outcomes		Text Readings, Figures, and Tables	Workbook Activities	Text and Workbook Assessments
4.1	Use medical terminology related to the integumentary system.	Word roots & combining forms: p. 89	Word roots & combining forms	*Text:* Review: 4.1 *Workbook:* Word deconstruction: 1–5
4.2	Describe the histology of the epidermis, dermis, and hypodermis.	Anatomy of the skin, hair, and nails: pp. 90–92 Figures 4.2–4.4	*Coloring book:* Skin Figure 4.1 *Concept maps:* Layers of the skin Figure 4.7	*Text:* Spot check: 1 Review: 4.2 *Workbook:* Multiple select: 2
4.3	Describe the cells of the epidermis and their function.	Cells of the epidermis: pp. 92–93 Figures 4.4–4.6	*Concept maps:* Layers of the skin Figure 4.7	*Text:* Review: 4.3 *Workbook:* Multiple select: 3 Critical thinking: 2
4.4	Describe the structures of the dermis and their functions.	Dermis: pp. 94–95 Figure 4.7	*Coloring book:* Skin Figure 4.1 *Lab exercises and activities:* Skin observations 1–3 Figures 4.4–4.6 *Concept maps:* Layers of the skin Figure 4.7	*Text:* Spot check: 2 Review: 4.4 *Workbook:* Multiple select: 3
4.5	Compare and contrast the glands of the skin in terms of their structure, products, and functions.	Cutaneous glands: pp. 95–96 Table 4.1 Figures 4.7–4.8	*Concept maps:* Cutaneous glands Figure 4.8	*Text:* Review: 4.5 *Workbook:* Matching: 1–5
4.6	Describe the histology of a hair and hair follicle.	Hair follicles: pp. 97–98 Figures 4.9–4.10	*Coloring book:* Hair and hair follicle Figure 4.2	*Text:* Review: 4.6 *Workbook:* Matching: 6–10
4.7	Explain how a hair grows and is lost.	Hair: p. 99		*Text:* Spot check: 3 Review: 4.7 *Workbook:* Multiple select: 4
4.8	Describe the structure and function of a nail.	Nails: pp. 99–100 Figure 4.11	*Coloring book:* Nails Figure 4.3	*Text:* Review: 4.8 *Workbook:* Multiple select: 5

Learning Outcomes		Text Readings, Figures, and Tables	Workbook Activities	Text and Workbook Assessments
4.9	Explain how the layers and structures of the skin work together to carry out the functions of the system.	Physiology of the integumentary system: pp. 100–102 Figure 4.12	*Lab exercises and activities:* Skin observation 3 Figure 4.6	*Text:* Spot check: 4 Review: 4.9 *Workbook:* Multiple select: 6
4.10	Explain how the skin responds to injury and repairs itself.	Regeneration vs. fibrosis: pp. 103–104 Figure 4.13		*Text:* Spot check: 5 Review: 4.10 *Workbook:* Multiple select: 9 Critical thinking: 1
4.11	Compare and contrast three degrees of burns in terms of symptoms, layers of the skin affected, and method used by the body for healing.	Burns: pp. 104–105 Figure 4.14	*Concept maps:* Burns Figure 4.9	*Text:* Spot check: 6 Review: 4.11 *Workbook:* Multiple select: 10 Critical thinking: 1
4.12	Describe the extent of a burn using the rule of nines.	Burns: p. 106 Figure 4.15		*Text:* Spot check: 6 Review: 4.12
4.13	Summarize the effects of aging on the integumentary system.	Effects of aging: pp. 106–107		*Text:* Spot check: 7 Review: 4.13 *Workbook:* Multiple select: 7
4.14	Describe a diagnostic test commonly used when diagnosing integumentary system disorders.	Diagnostic test for integumentary system disorders: p. 107 Table 4.2		*Text:* Review: 4.14 *Workbook:* Completion 1–3
4.15	Describe three forms of skin cancer in terms of the body area most affected, appearance, and ability to metastasize.	Skin cancer: pp. 107–109 Figure 4.16	*Lab exercises and activities:* Skin observation 1 Figure 4.4 *Concept maps:* Skin cancer Figure 4.10	*Text:* Review: 4.15 *Workbook:* Multiple select: 8
4.16	Describe a common integumentary disorder other than skin cancer and relate abnormal function to the pathology.	Skin infections: pp. 109–115 Table 4.3 Figures 4.17–4.25		*Text:* Review: 4.16 *Workbook:* Multiple select: 1 Case study: 1–5

5 The Skeletal System

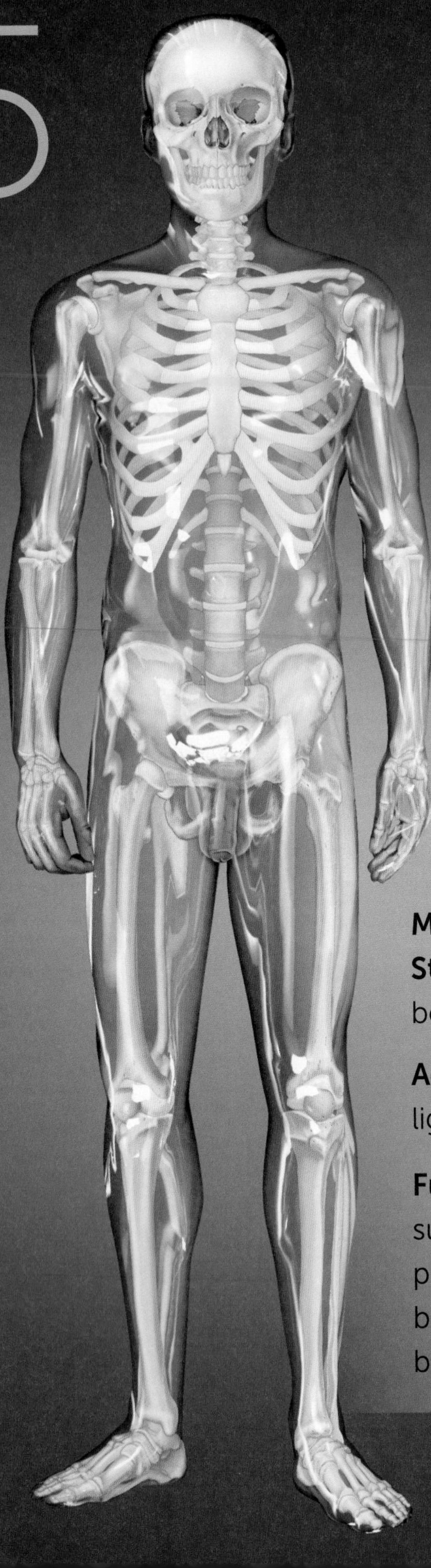

Major Organs and Structures:
bones

Accessory Structures:
ligaments, cartilages

Functions:
support, movement, protection, acid-base balance, electrolyte balance, blood formation

learning outcomes

This chapter of the workbook is designed to help you learn the anatomy and physiology of the skeletal system. After completing this chapter in the text and this workbook, you should be able to:

- **5.1** Use medical terminology related to the skeletal system.
- **5.2** Distinguish between the axial skeleton and the appendicular skeleton.
- **5.3** Describe five types of bones classified by shape.
- **5.4** Identify bones, markings, and structures of the axial skeleton and appendicular skeleton.
- **5.5** Describe the cells, fibers, and matrix of bone tissue.
- **5.6** Compare and contrast the histology of compact and cancellous bone.
- **5.7** Compare and contrast the histology of hyaline, elastic, and fibrocartilage connective tissues.
- **5.8** Describe the anatomy of a long bone.
- **5.9** Distinguish between two types of bone marrow in terms of location and function.
- **5.10** Describe three major structural classes of joints and the types of joints in each class.
- **5.11** Differentiate between rheumatoid arthritis and osteoarthritis.
- **5.12** Explain how minerals are deposited in bone.
- **5.13** Compare and contrast endochondral and intramembranous ossification.
- **5.14** Compare and contrast endochondral and appositional bone growth.
- **5.15** Explain how bone is remodeled by reabsorption.
- **5.16** Explain the nutritional requirements of the skeletal system.
- **5.17** Describe the negative-feedback mechanisms affecting bone deposition and reabsorption by identifying the relevant glands, hormones, target tissues, and hormone functions.
- **5.18** Summarize the six functions of the skeletal system and give an example or explanation of each.
- **5.19** Summarize the effects of aging on the skeletal system.
- **5.20** Classify fractures using descriptive terms.
- **5.21** Explain how a fracture heals.
- **5.22** Describe a diagnostic test commonly used when diagnosing skeletal system disorders.
- **5.23** Describe skeletal system disorders and relate abnormal function to the pathology.

word roots & combining forms

ankyl/o: bent, crooked

arthr/o: joint

burs/o: sac

carp/o: wrist

chondr/o: cartilage

condyl/o: condyle

cost/o: rib

crani/o: head, skull

femor/o: femur, bone of the thigh

fibul/o: fibula, lateral bone of the lower leg

humer/o: humerus, bone of the upper arm

ili/o: ilium, bone of the hip

ischi/o: ischium, bone of the hip

lumb/o: lower back

maxill/o: maxilla, upper jaw

myel/o: bone marrow, spinal cord

orth/o: straight

oste/o: bone

patell/o: patella, kneecap

phalang/o: phalanges, bones of the fingers and toes

pub/o: pubis, bone of the hip

stern/o: sternum, breastbone

synov/i: synovial fluid, joint, or membrane

tars/o: tarsals, foot

tibi/o: tibia, medial bone of the lower leg

Axial versus Appendicular Skeleton

A human skeleton can be divided into two categories—the axial skeleton and the appendicular skeleton. The bones in each skeleton are listed below and shown in Figure 5.1.

Axial Skeleton

The axial skeleton is composed of bones located in the head, neck, and trunk.

- Cranial bones: frontal, occipital, temporal, parietal, ethmoid, and sphenoid.
- Facial bones: nasal, lacrimal, zygomatic, inferior nasal concha, maxilla, palatine, mandible, vomer.
- Spinal column: 7 cervical vertebrae, 12 thoracic vertebrae, 5 lumbar vertebrae, sacrum, coccyx.
- Sternum.
- Ribs: Of the 12 pairs of ribs, 7 pairs are true ribs with individual costal cartilages. Five pairs are false ribs. Of the 5 pairs of false ribs, 2 pairs are floating because they do not have or share a costal cartilage.
- Hyoid bone.

Appendicular Skeleton

The appendicular skeleton is composed of the bones of the limbs and the bones of the girdles that connect the limbs to the axial skeleton.

- Pectoral girdle: clavicle and scapula.
- Bones of the upper limb:
  - Humerus
  - Radius
  - Ulna
  - Carpal bones
  - Metacarpals
  - Phalanges
- Pelvic girdle: ilium, ischium, pubis.
- Bones of the lower limb:
  - Femur
  - Tibia
  - Fibula
  - Patella
  - Tarsal bones
  - Metatarsals
  - Phalanges

Color the box next to each skeleton term (axial, appendicular). Use the same color for all of the corresponding bones located in each skeleton. See Figure 5.1.

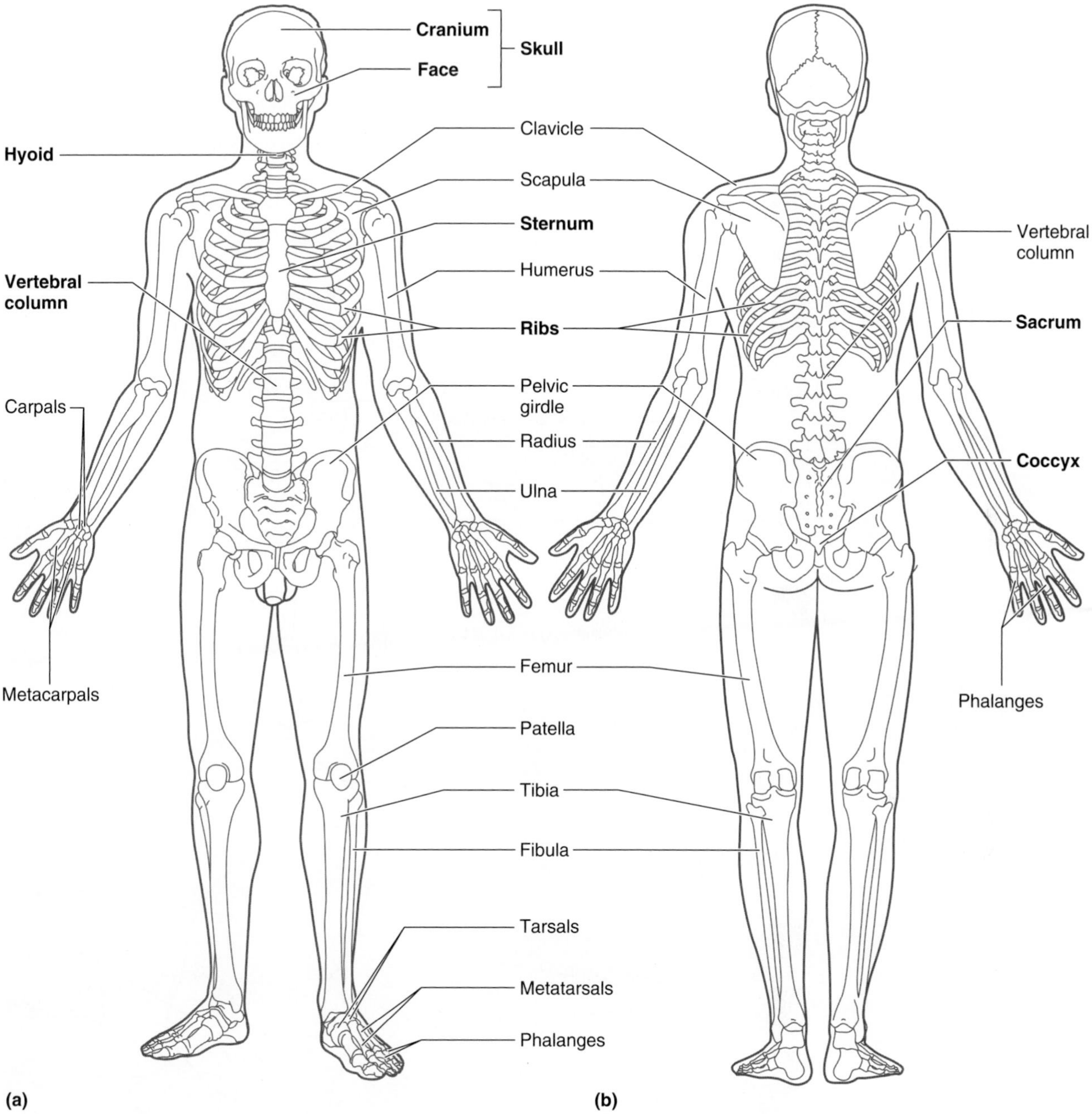

FIGURE 5.1 Axial and appendicular skeletons: (a) anterior, (b) posterior.

☐ Axial skeleton

☐ Appendicular skeleton

Skull

Figure 5.2 shows four views of the skull's bones and special features. Color the box next to each term. Use the same color for the corresponding structure on each of the different views of the skull. See **Figure 5.2**.

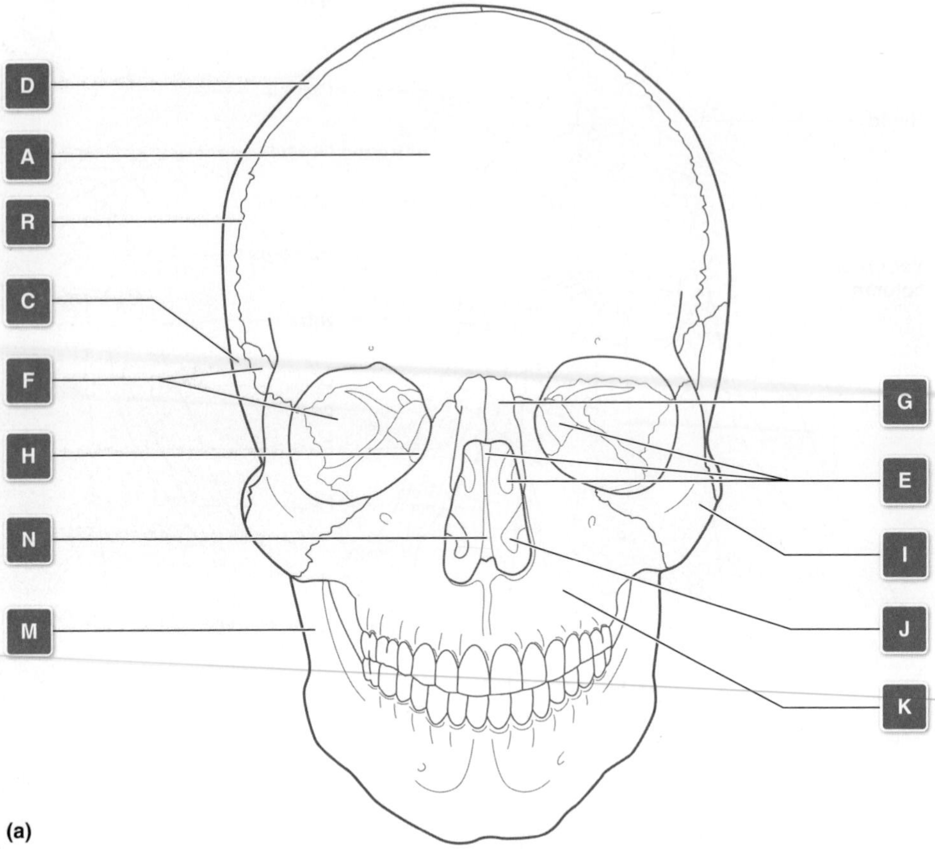

(a)

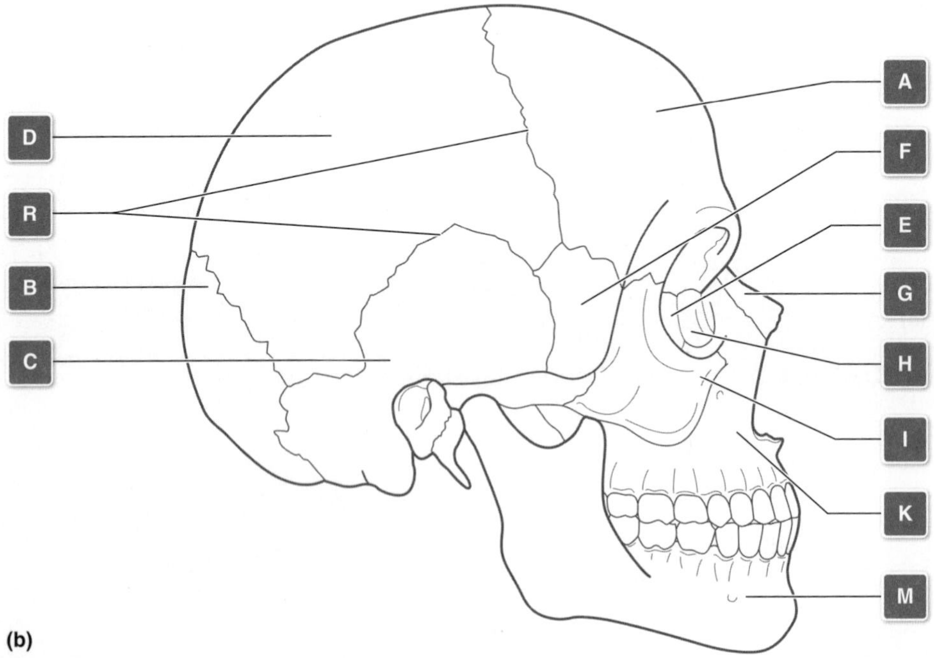

(b)

FIGURE 5.2 The skull: (a) anterior view, (b) lateral view, (c) inferior view, (d) superior view of the cranial floor.

COLORING BOOK

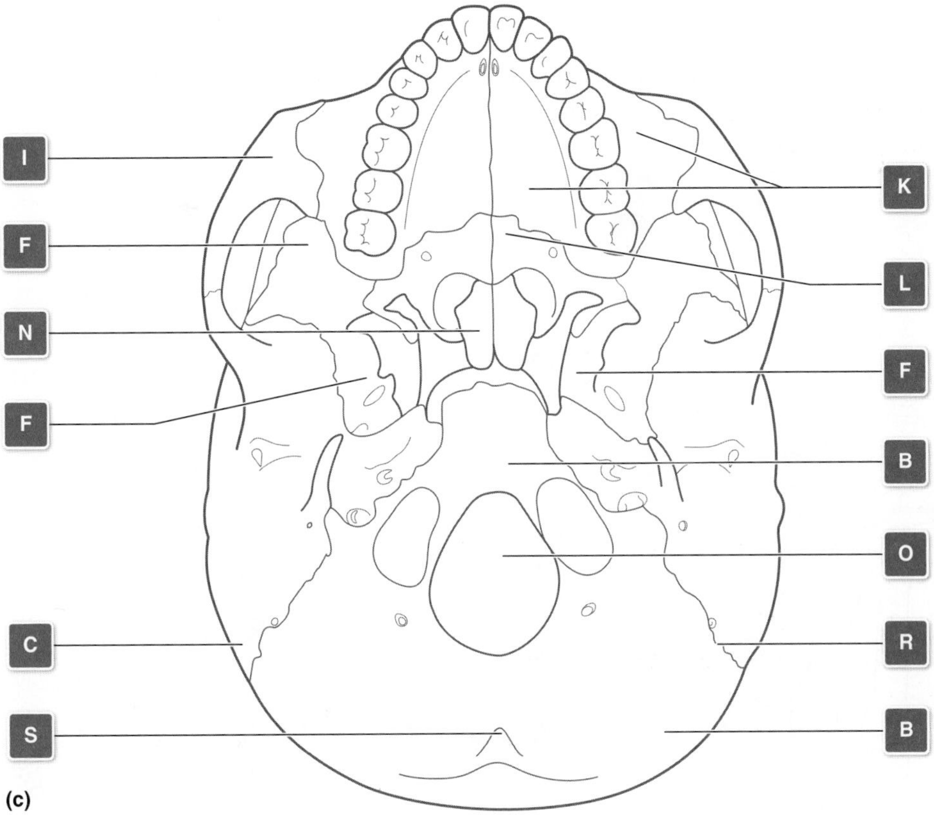

(c)

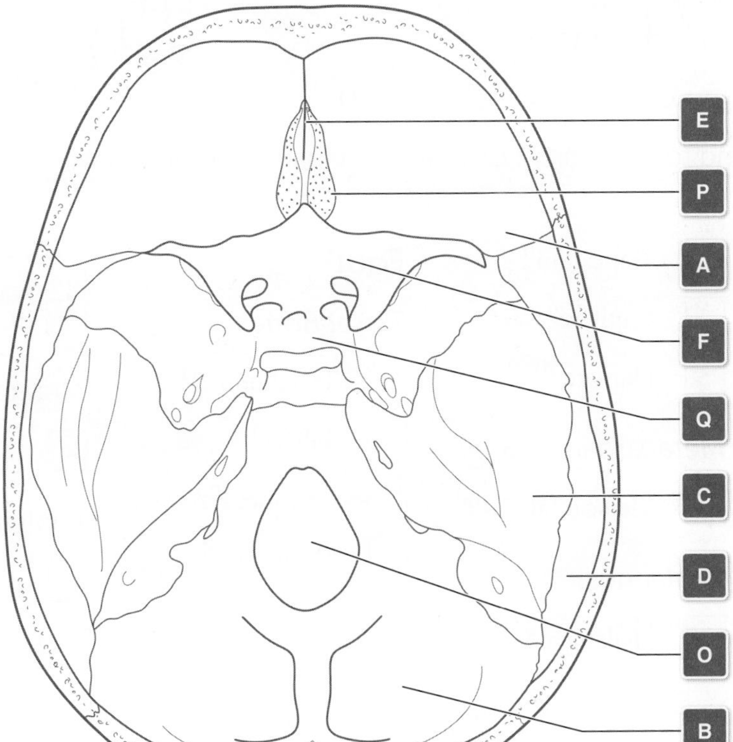

(d)

Bones

- ☐ Frontal bone(A)
- ☐ Occipital bone(B)
- ☐ Temporal bone(C)
- ☐ Parietal bone(D)
- ☐ Ethmoid bone(E)
- ☐ Sphenoid bone(F)
- ☐ Nasal bone(G)
- ☐ Lacrimal bone(H)
- ☐ Zygomatic bone(I)
- ☐ Inferior nasal concha(J)
- ☐ Maxilla(K)
- ☐ Palatine(L)
- ☐ Mandible(M)
- ☐ Vomer(N)

Special Features of the Skull

- ☐ Foramen magnum(O)
- ☐ Cribriform plate(P)
- ☐ Sella turcica(Q)
- ☐ Suture(R)
- ☐ External occipital protuberance(S)

FIGURE 5.2 concluded

Hand and Foot

Figure 5.3 shows the bones of the hand (with the wrist) and the bones of the foot. Color the box next to each term. Use the same color for the corresponding bone in Figure 5.3.

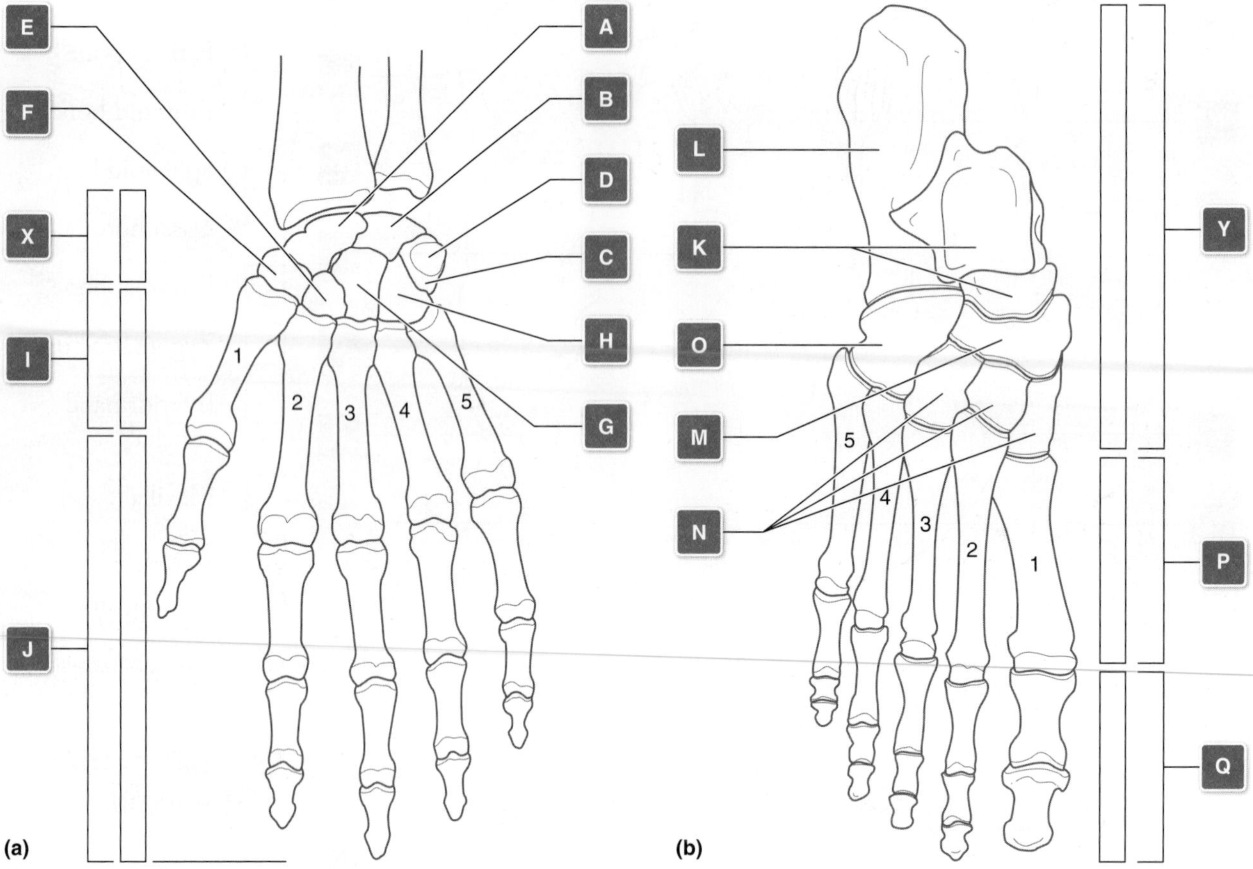

FIGURE 5.3 The hand and foot: (a) anterior view of the hand, (b) superior view of the foot.

Hand and Wrist

Carpal Bones

- ☐ Carpal bones(X)
- ☐ Scaphoid(A)
- ☐ Lunate(B)
- ☐ Triquetrum(C)
- ☐ Pisiform(D)
- ☐ Trapezoid(E)
- ☐ Trapezium(F)
- ☐ Capitate(G)
- ☐ Hamate(H)

Metacarpals 1–5

- ☐ Metacarpals 1–5(I)

Phalanges

- ☐ Phalanges(J)

Foot

Tarsal Bones

- ☐ Tarsal bones(Y)
- ☐ Talus(K)
- ☐ Calcaneus(L)
- ☐ Navicular(M)
- ☐ Cuneiforms(N)
- ☐ Cuboid(O)

Metatarsals 1–5

- ☐ Metatarsals 1–5(P)

Phalanges

- ☐ Phalanges(Q)

COLORING BOOK

Skeleton Assembly

In this activity you will construct a skeleton. First, label the bones in **Figure 5.4**, and then cut out the bones from **Figure 5.4** to assemble a skeleton.

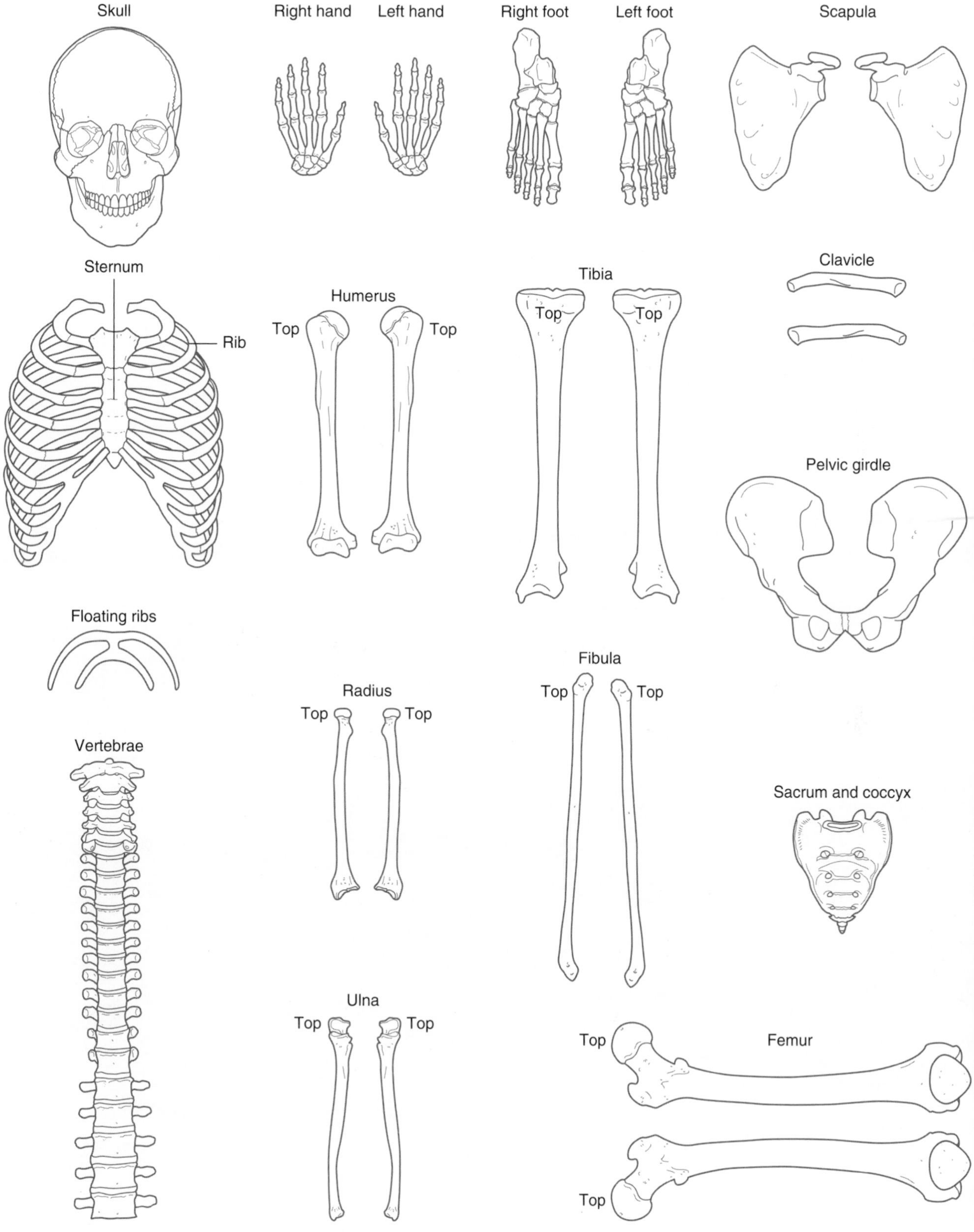

FIGURE 5.4 Bones of the skeleton.

LABORATORY EXERCISES AND ACTIVITIES

Right or Left?

Figures 5.5 to **5.9** show bones as they might be placed on a lab table. You are to identify the following bones and their structures to determine whether the bone belongs on the right or left side of the body. See **Figures 5.5** to **5.9**.

Use the image in **Figure 5.5** for the following questions:

1. Identify the bone in **Figure 5.5**. ____________________
2. Identify the structure labeled "A." ____________________
3. Identify the structure marked "B." ____________________
4. Does this bone belong to the right or left side of the body? ____________
5. What knowledge of these structures helped you determine right or left? ____________________

____

____

____

FIGURE 5.5

Use the images in **Figure 5.6** for the following questions:

6. Identify the bone in **Figure 5.6**. ____________________
7. Identify the structure labeled "C." ____________________
8. Identify the structure labeled "D." ____________________
9. Identify the structure marked "E." ____________________
10. What bone articulates with this one at structure D? ____________
11. What bone articulates with this one at structure E? ____________
12. Does this bone belong to the right or left side of the body? ____________

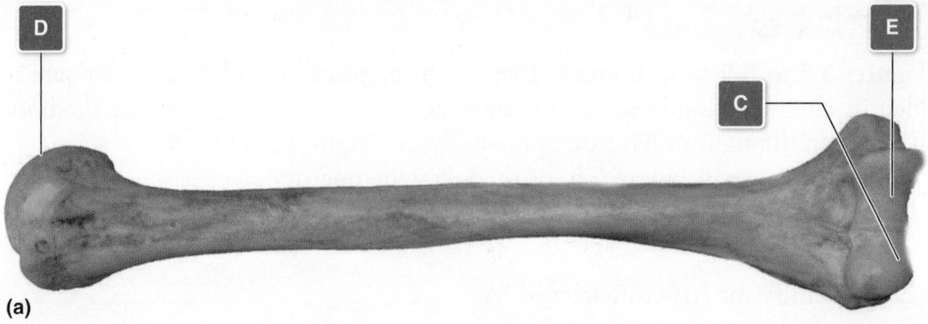

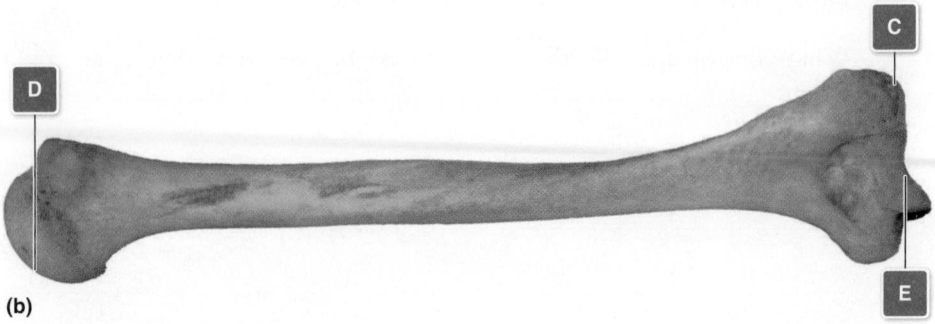

FIGURE 5.6 Views (a) and (b) are of the same bone.

13. What knowledge of these structures helped you determine right or left? ______

______

______

______

Use the images in **Figure 5.7** for the following questions:

14. Identify bone G in **Figure 5.7.** ______

15. Identify bone H. ______

16. Identify bone I. ______

17. What is the term for these three bones together? ______

18. What joint will be formed with another bone at J? ______

19. Identify the structure labeled "K." ______

20. What bone articulates with structure K? ______

21. Do these bones belong to the right or left side of the body? ______

22. What knowledge of these structures helped you determine right or left? ____________________

____

____

____

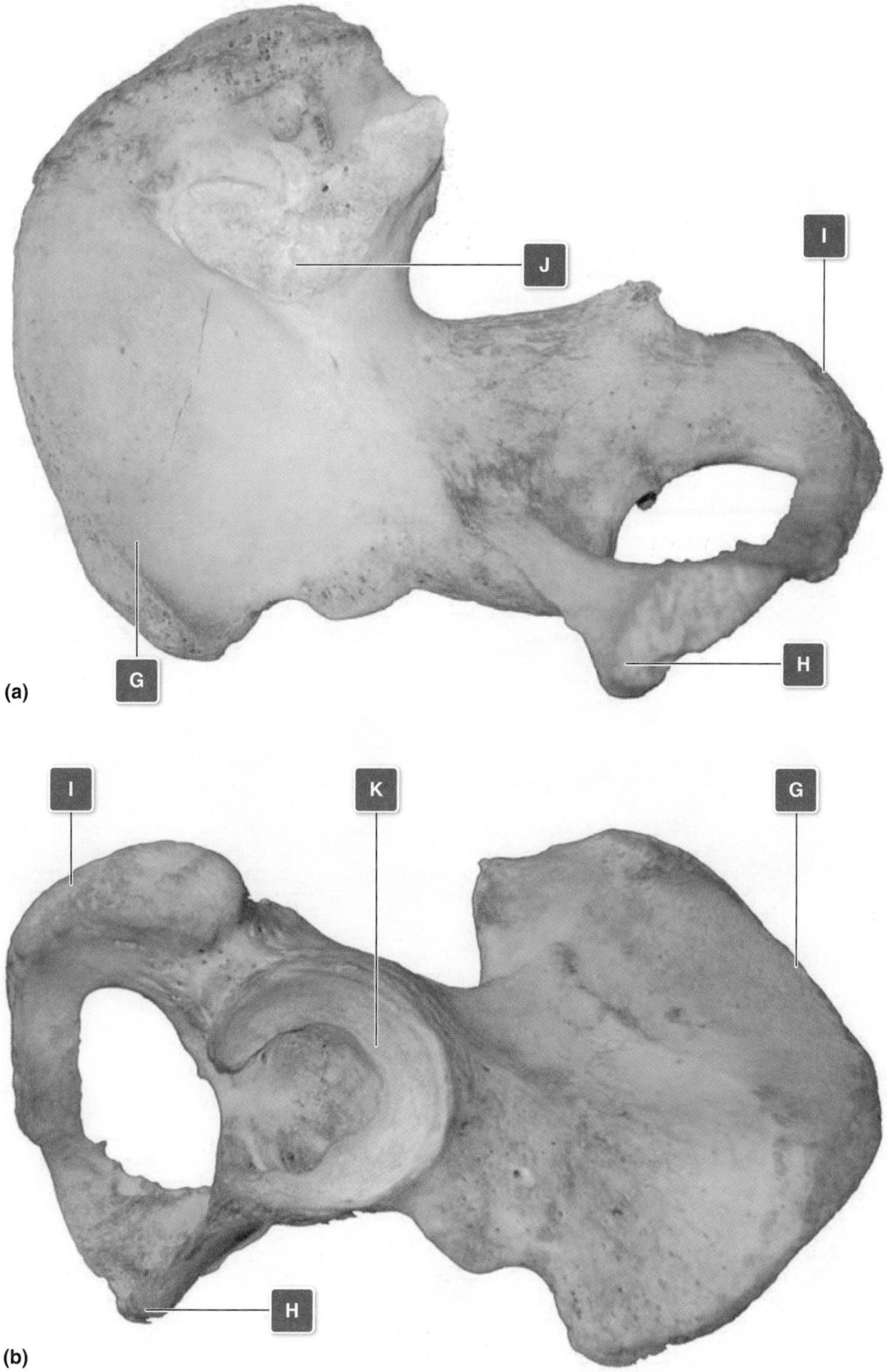

FIGURE 5.7 Views (a) and (b) are of the same bones.

Use the images in **Figure 5.8** for the following questions:

23. Identify bone A in **Figure 5.8.** ______________________________

24. Identify structure B. ______________________________

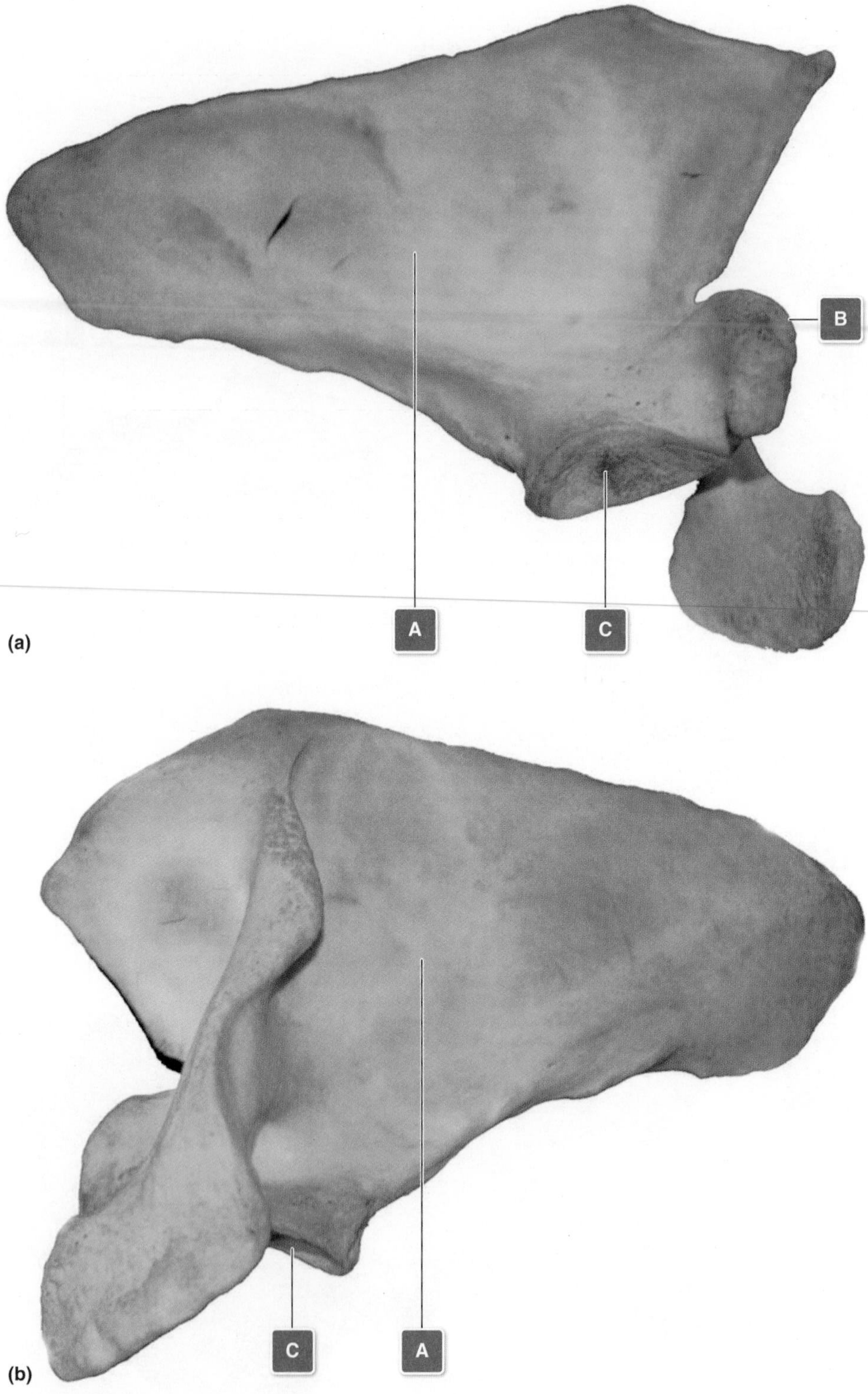

FIGURE 5.8 Views (a) and (b) are of the same bone.

25. Identify structure C. ______

26. What bone articulates with structure C? ______

27. Does this bone belong to the right or left side of the body? ______

28. What knowledge of these structures helped you determine right or left? ______

______

______

______

Use the images in **Figure 5.9** for the following questions:

29. Identify bone L in **Figure 5.9**. ______

30. Identify bone M. ______

31. Identify bone N. ______

32. Identify the joint. ______

33. What class of joint is this? ______

34. What type of joint is this in that class? ______

35. Identify structure O. ______

36. Identify ligament P. ______

37. Identify ligament Q. ______

38. Identify ligament R. ______

39. Identify structure S (rough spot). ______

40. What bone is missing from this joint? ______

41. Does this specific joint belong to the right or left side of the body? ______

42. What knowledge of these structures helped you determine right or left? ______

______

______

______

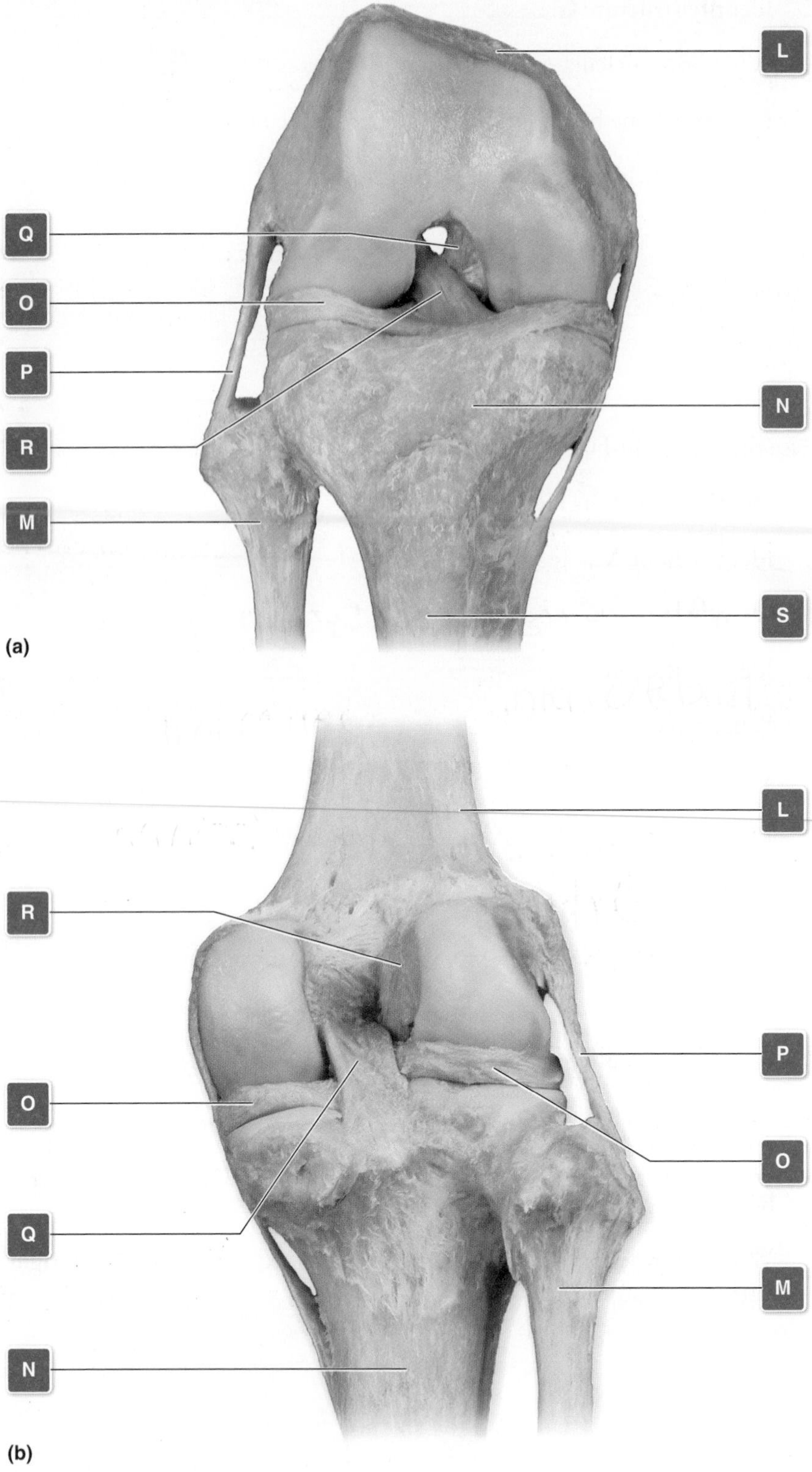

FIGURE 5.9 Views (a) and (b) are of the same joint. One bone of this joint is missing in the dissection.

Key Words

The following terms are defined in the glossary of the textbook.

absorption
appendicular skeleton
appositional bone growth
axial skeleton
cancellous bone
chondrocyte
cleft palate
comminuted
compact bone
deposition
diaphysis
endochondral ossification
epiphyseal plate
fontanelle
foramen magnum
hydroxyapatite
hyperuricemia
mastoiditis
meniscus
metatarsophalangeal joint
osteon (Haversian system)
reabsorption
synovial membrane
trabeculae

Concept Maps

Complete the boxes in the following concept maps (**Figures** 5.10 to 5.16). Each map contains at least one key word.

Composition of the Skeletal System

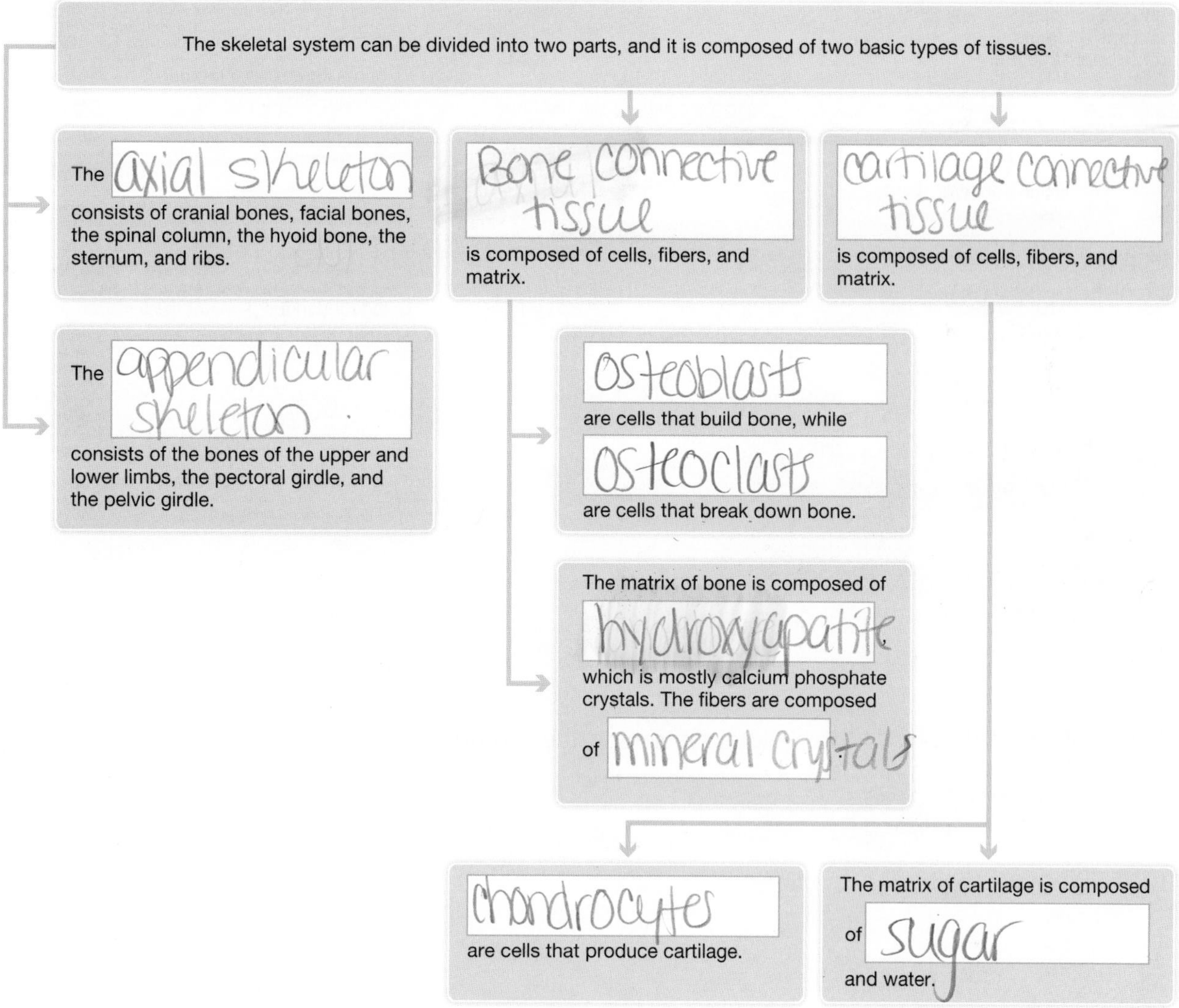

FIGURE 5.10 Composition of the skeletal system concept map.

Histology of the Skeletal System

The skeletal system is composed of two types of bone and three types of cartilage.

Compact bones is found in the shafts of long bones and on the surfaces of flat and irregular bones.

cancellous bones is found in the end of long bones and the middle of flat and irregular bones.

hyaline is found in the nose, costal cartilages, and the ribs covering the ends of long bones.

Compact bone is arranged in a series of osteons, also called Haversian systems, that resemble targets. They are composed of several structures:

- haversian canal, which contains blood vessels and a nerve.
- lacunae, which are spaces that house osteocytes.
- lamallae, which are layers of matrix.
- canaliculi, which allow osteocytes access to the central canal.

space between appears spongy because it is arranged in slivers and plates called trabeculae, with spaces between the bone.

Matrix is found in these spaces.

Fibrocartilage connective tissue is found in the intervertebral disks, the intervertebral of the knee, and the pubic symphysis.

elastic is found in the pinna of the ear and the epiglottis.

FIGURE 5.11 Histology of the skeletal system concept map.

Bone Development

Bone development occurs in a fetus through a process called ossification. There are two types of bone ossification.

↓

Intramembranous ossification forms the flat bones of the developing fetus.

Osteoblasts deposit bone in the membranes present at the site of future flat bones.

↓

Endochondral ossification forms the long bones of the developing fetus.

osteoblasts deposit bone in cartilage models.

↓

The primary site of this type of ossification is the diaphysis, and the secondary sites are the epiphysis.

↓

The epiphyseal plate is the cartilage between the primary and secondary sites of ossification. This area continues to produce osteoblasts as bones continue to grow longer.

FIGURE 5.12 Bone development concept map.

Bone Growth and Remodeling

Bone growth occurs through two different processes.

endochondral growth occurs in long bones to increase their length.

This process continues until puberty, when estrogen and testosterone speed the deposition of bone, so bone is produced faster than cartilage in the epiphyseal line. This causes the epiphyseal plates to close and bone lengthening to cease.

Appositional bone growth occurs in all bones to make them more massive.

This process continues as long as there is a sufficient supply of trabeculae for osteoblasts and a significant amount of stress applied to the bone.

Bone remodeling is a constant process that occurs daily.

If there is more calcium than necessary in the blood, appositional bone growth will take place in the bone. On the other hand, if there is less blood calcium than necessary, calcium will be taken from bone.

FIGURE 5.13 Bone growth and remodeling concept map.

Hormonal Regulation of Bone

Hormones regulate bone deposition and reabsorption.

estrogen and testosterone speed up the osteoblasts of bone by hormones at puberty to cause the epiphyseal plates to close.

The thyroid gland secretes calcitonin when blood calcium levels are too high. This causes osteoblasts to deposit the excess calcium into bone.

The parathyroid gland produces parathyroid hormone when blood calcium levels are too low. This hormone tells osteoblasts to reabsorb bone. The hormone also signals kidneys to reabsorb calcium during urine formation and signals the small intestine to absorb calcium during digestion. Active vitamin D, also called ______, is necessary for ______ to take place.

FIGURE 5.14 Hormonal regulation of bone concept map.

Joints

Joints or articulation are the areas where two bones meet. They are classified on the basis of their anatomy. They may or may not allow movement.

Fibrous joints have fibrous tissue between bones. Types of joints in this class include:

- gomphoses
  Example: a tooth in a socket
- syndesmoses
  Example: joint between the radius and ulna
- sutures
  Example: joint between the flat bones of the skull

Cartilaginous joints have cartilage between bones. Types of joints in this class include:

- symphyses
  Example: joint between pubic bones
- synchondroses
  Example: epiphyseal plate

Synovial joints are lined with Synovial membranes and contain synovial fluid in the joint space. Types of joints in this class include:

- Hinge
  Example: joints between phalanges
- ball & socket
  Example: hip joint
- gliding
  Example: wrist joints that allow an up-and-down wave motion
- Ellipsoid
  Example: wrist joints that allow a side-to-side wave motion
- Saddle
  Example: joint between metacarpal I and wrist that allows the thumb to touch the palm
- Pivot
  Example: joint between the atlas and axis that allows the head to turn to the side

FIGURE 5.15 Joints concept map.

Fractures and How They Heal

A fracture is a break in bone that results from injury or trauma, as in a fall, or it can result from a disease process that weakened the bone. There are several ways to classify fractures.

→ A(n) closed fracture does not cause a break in the skin.

→ A(n) open fracture breaks through the skin.

→ A(n) complete fracture occurs when the bone breaks into two or more pieces.

→ A(n) Displaced fracture causes the bone to no longer be in proper alignment.

→ A(n) hairline fracture causes a crack in the bone.

→ A(n) greenstich fracture occurs when the bone has broken through one side but not completely through the other side.

→ A(n) depressed fracture occurs when the bone has been dented.

→ A(n) Spiral fracture occurs when the bone is twisted to the point of fracture.

→ A(n) transverse fracture occurs when the bone is broken perpendicular to its length.

→ A(n) oblique fracture occurs when the break in the bone is at an angle.

→ A(n) comminuted fracture occurs when the bone is broken into three or more pieces.

→ A(n) compression fracture occurs when cancellous bone has been compressed. This can occur in vertebrae.

There is a process to the healing of a fracture.

The pieces of bone must first be put in proper alignment. A(n) closed reduction sets the ends of the bones in proper alignment without surgery. A(n) open reduction is a surgical procedure that uses screws and plates to keep the broken pieces of bone in alignment.

1. When the bone breaks, it bleeds.
2. The blood then clots.
3. Stem cells from the periosteum form a soft callus of fibrocartilage.
4. Osteoblasts continue forming bone in the break while osteoclasts remodel the bone to reestablish the marrow cavity.

FIGURE 5.16 Fractures and how they heal concept map.

KEY WORD CONCEPT MAPS

Word Deconstruction: *In the textbook, you built words to fit a definition using combining forms, prefixes, and suffixes. Here you are to break down the term into its parts (prefixes, roots, and suffixes) and give a definition. Prefixes and suffixes can be found inside the back cover of the textbook.*

FOR EXAMPLE Dermatitis: dermat/itis—inflammation of the skin

1. Osteoarthropathy: ______________________
2. Arthroscopic: ______________________
3. Orthopedic: ______________________
4. Osteodystrophy: ______________________
5. Chondrodynia: ______________________

CHAPTER 5 REVIEW QUESTIONS

Multiple Select: *Select the correct choices for each statement. The choices may be all correct, all incorrect, or any combination of correct and incorrect.*

1. Which of the following statements is (are) true concerning thoracic vertebrae?
   - **a.** A thoracic vertebra protects the neck.
   - **b.** A thoracic vertebra has extra foramen.
   - **c.** The thoracic vertebrae lack intervertebral disks.
   - **d.** A thoracic vertebra has facets for the attachment of ribs.
   - **e.** A thoracic vertebra has a spine and transverse processes for the attachment of ligaments and tendons.
2. Which of the following structures belong(s) to the appendicular skeleton?
   - **a.** Intervertebral disks
   - **b.** Long bones
   - **c.** Sutures
   - **d.** Clavicle
   - **e.** Ossa coxae
3. Which of the following statements is (are) true about cartilage?
   - **a.** Hyaline cartilage can be found in articular cartilage.
   - **b.** Fibrocartilage can be found in the pubic symphysis.
   - **c.** Fibrocartilage connective tissue has fibers running in all directions.
   - **d.** Elastic cartilage can be found in a syndesmosis.
   - **e.** Costal cartilage is composed of elastic cartilage connective tissue.
4. Which of the following statements is (are) true about bone marrow?
   - **a.** Red bone marrow produces red blood cells but not white blood cells.
   - **b.** Red bone marrow can be found in flat bones.
   - **c.** Red and yellow bone marrow can be found in a long bone such as the tibia.
   - **d.** Yellow marrow is found in cancellous bone.
   - **e.** Yellow marrow is fatty tissue.
5. Which of the following is (are) the result of aging on the skeletal system?
   - **a.** The development of rheumatoid arthritis.
   - **b.** Increased reabsorption.
   - **c.** Osteoclasts that are more efficient.
   - **d.** Increase in bone deposition.
   - **e.** Decrease in bone mass and increase in bone density.

6. Which of the following statements is (are) true about fractures?
   a. A callus is usually visible on an x-ray after the fracture has healed.
   b. A closed reduction attaches the broken ends of the bones together with screws and plates.
   c. Osteoblasts deposit bone in the fibrocartilage of the break.
   d. Cells of the periosteum are important to heal the break.
   e. Osteoclasts from the endosteum reestablish an epiphysis.

7. Which of the following statements is (are) true about the knee?
   a. The knee is a cartilaginous joint because it has a meniscus.
   b. The knee has bursae, which are fluid-filled extensions of the synovial membrane.
   c. The knee contains a sesamoid bone.
   d. The knee is stabilized laterally by cruciate ligaments.
   e. The knee is an example of a hinge joint.

8. Which of the following statements is (are) true concerning joint classification?
   a. Fibrous joints can be found in the skull.
   b. A cartilaginous joint can be found in the pelvis.
   c. Gomphoses are found in the appendicular skeleton.
   d. Synovial joints provide a wide range of motion.
   e. Synchondroses decrease with age.

9. Which of the following statements is (are) true concerning feedback mechanisms in the skeletal system?
   a. A positive-feedback mechanism occurs when the blood calcium level is too high and osteoclasts deposit bone to lower the blood calcium level.
   b. A positive-feedback mechanism is seen during deposition when crystals form on a seed crystal.
   c. A negative-feedback mechanism occurs when the blood calcium level is too low and osteoblasts reabsorb bone to raise the blood calcium level.
   d. Negative-feedback mechanisms are used to maintain an electrolyte balance of calcium in the blood.
   e. Hormones from the thyroid gland and parathyroid gland use positive-feedback mechanisms to regulate bone deposition and reabsorption.

10. Which of the following statements is (are) possible concerning a fracture of the tibia?
   a. An epiphyseal fracture is possible at the proximal end of the tibia of a child.
   b. A displaced transverse fracture of the shaft of the tibia would break the endosteum and compact bone.
   c. An open fracture of the tibia would cause a break in the periosteum.
   d. It is possible to have a depressed fracture of the diaphysis of the tibia.
   e. A comminuted fracture of the tibia is likely to need an open reduction.

Matching: *Match the bone to its class as defined by shape.*

______	**1.** Patella	**a.** Long bone
______	**2.** Lumbar vertebra	**b.** Short bone
______	**3.** Calcaneus	**c.** Sesamoid bone
______	**4.** Floating rib	**d.** Flat bone
______	**5.** Ulna	**e.** Irregular bone

Matching: *Match the following hormones or chemicals to their target tissues in the skeletal system. Some questions may have more than one answer. Some choices may be used more than once.*

_______ **6.** Calcitriol

_______ **7.** Calcitonin

_______ **8.** Estrogen

_______ **9.** Testosterone

_______ **10.** Parathyroid hormone

a. Kidneys

b. Osteoclasts

c. Osteoblasts

d. Small intestine

Completion: *Fill in the blanks to complete the following statements.*

1. The number of bones may vary between two individuals because one individual may have additional _______________ bones growing in tendons at a joint.

2. The _______________, the _______________, and the _______________ form the acetabulum.

3. Chondrocytes are housed in _______________.

4. Hydroxyapatite is mainly _______________ crystals.

5. Osteoclasts produce _______________ to reabsorb bone.

Critical Thinking

1. Describe the best environment, lifestyle, and nutrition that would lead to a healthy skeletal system in old age. Justify your choices.

2. Predict which skeletal system disorder would result from the following situations: (a) a lactose-intolerant pregnant woman with no prenatal care, (b) an individual who has played professional sports for 15 years, (c) an open fracture of the tibia, and (d) a woman (age 27) with premature onset of menopause.

3. Achondroplastic dwarfism results when all of the epiphyseal plates close prematurely. Pituitary dwarfism results when there is insufficient pituitary growth hormone to make all of the bones grow normally. Given this information, how would the proportions of an achondroplastic dwarf differ from those of a pituitary dwarf? (**Hint:** What bones have epiphyseal plates?)

Case Study

David was taken to the hospital after having an accident on his bike. One of the long bones in his right leg appears to be fractured, although it is difficult to tell the extent of David's injury until he has been seen by his health care provider.

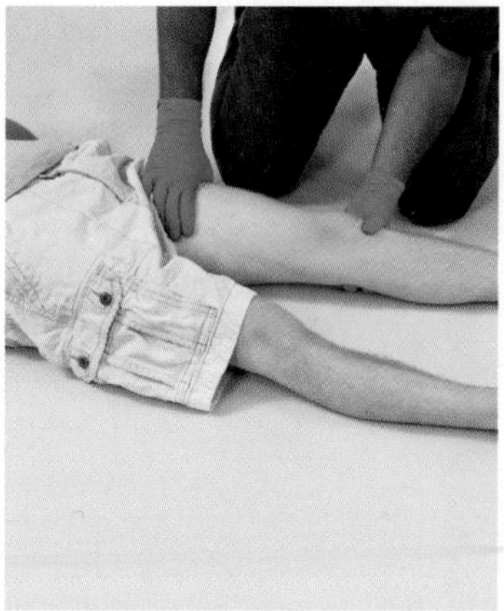

1. What diagnostic test will mostly likely be done to David's leg? What will this test show? Explain how this test works to help health care providers determine a diagnosis.

2. If David's leg is fractured, what are potential ways the fracture can be treated? Explain the different methods.

3. Explain the process of how long bones grow. What type of fracture can occur in a child that may affect the growth of a long bone? If David were a young boy with this type of fracture, what could potentially be affected if the fracture is not treated properly?

CHAPTER 5 MAPPING

This section of the chapter is designed to help you find where each outcome is covered in the textbook and workbook.

Learning Outcomes		Text Readings, Figures, and Tables	Workbook Activities	Text and Workbook Assessments
5.1	Use medical terminology related to the skeletal system.	Word roots & combining forms: p. 121 Overview: pp. 122–123	Word roots & combining forms	*Text:* Review: 5.1 *Workbook:* Word Deconstruction: 1–5
5.2	Distinguish between the axial skeleton and the appendicular skeleton.	Anatomy of the skeletal system: p. 123 Figure 5.2	*Coloring book:* Axial versus appendicular skeleton Figure 5.1 *Concept maps:* Composition of the skeletal system Figure 5.10	*Text:* Review: 5.2 *Workbook:* Multiple Select: 2
5.3	Describe five types of bones classified by shape.	Classification of bones: pp. 124–125 Figures 5.3–5.6		*Text:* Review: 5.3 *Workbook:* Matching: 1–5 Completion: 1
5.4	Identify bones, markings, and structures of the axial skeleton and appendicular skeleton.	Axial skeleton: pp. 125–136 Figures 5.7–5.24 Appendicular skeleton: pp. 137–147 Figures 5.25–5.37	*Coloring book:* Axial versus appendicular; skeleton; Skull; Hand and foot Figures 5.1–5.3 *Lab exercises and activities:* Skeleton assembly; Right or left? Figures 5.4–5.9 *Concept maps:* Composition of the skeletal system Figure 5.10	*Text:* Spot Check: 1–4 Review: 5.4 *Workbook:* Multiple Select: 1, 2 Completion: 2
5.5	Describe the cells, fibers, and matrix of bone tissue.	Bone connective tissue: p. 147	*Concept maps:* Composition of the skeletal system; Histology of the skeletal system Figures 5.10–5.11	*Text:* Spot Check: 5, 6 Review: 5.5 *Workbook:* Completion: 4
5.6	Compare and contrast the histology of compact and cancellous bone.	Bone connective tissue: pp. 147–149 Figure 5.38	*Concept maps:* Histology of the skeletal system Figure 5.11	*Text:* Review: 5.6

Learning Outcomes		Text Readings, Figures, and Tables	Workbook Activities	Text and Workbook Assessments
5.7	Compare and contrast the histology of hyaline, elastic, and fibrocartilage connective tissues.	Cartilage connective tissue: pp. 149–150 Figure 5.39	*Concept maps:* Histology of the skeletal system Figure 5.11	*Text:* Spot Check: 5, 6 Review: 5.7 *Workbook:* Multiple Select: 3 Completion: 3
5.8	Describe the anatomy of a long bone.	Anatomy of a long bone: pp. 150–152 Figure 5.40	*Concept maps:* Histology of the skeletal system Figure 5.11	*Text:* Spot Check: 7 Review: 5.8 *Workbook:* Multiple Select: 4
5.9	Distinguish between two types of bone marrow in terms of location and function.	Red bone marrow: p. 152		*Text:* Review: 5.9 *Workbook:* Multiple Select: 4
5.10	Describe three major structural classes of joints and the types of joints in each class.	Joints: pp. 152–158 Figures 5.41–5.42 Table 5.1	*Lab exercises and activities:* Right or left? Figure 5.9 *Concept maps:* Joints Figure 5.15	*Text:* Spot Check: 8 Review: 5.10 *Workbook:* Multiple Select: 7, 8
5.11	Differentiate between rheumatoid arthritis and osteoarthritis.	Joints: pp. 158–160 Figures 5.43–5.44		*Text:* Review: 5.11
5.12	Explain how minerals are deposited in bone.	Mineral deposition: p. 160		*Text:* Review: 5.12 *Workbook:* Multiple Select: 9
5.13	Compare and contrast endochondral and intramembranous ossification.	Bone development: pp. 161–163 Figures 5.46–5.47 Table 5.2	*Concept maps:* Bone development Figure 5.12	*Text:* Spot Check: 9 Review: 5.13
5.14	Compare and contrast endochondral and appositional bone growth.	Bone growth: pp. 163–164 Figures 5.48–5.50	*Concept maps:* Bone growth and remodeling Figure 5.13	*Text:* Spot Check: 10 Review: 5.14 *Workbook:* Critical Thinking: 3 Case Study: 3

Learning Outcomes		Text Readings, Figures, and Tables	Workbook Activities	Text and Workbook Assessments
5.15	Explain how bone is remodeled by reabsorption.	Bone remodeling: pp. 164–165 Figure 5.51		*Text:* Review: 5.15 *Workbook:* Completion: 5
5.16	Explain the nutritional requirements of the skeletal system.	Nutritional requirements of the skeletal system: pp. 165 Figure 5.52		*Text:* Spot Check: 10 Review: 5.16 *Workbook:* Critical Thinking: 1
5.17	Describe the negative-feedback mechanisms affecting bone deposition and reabsorption by identifying the relevant glands, hormones, target tissues, and hormone functions.	Hormonal regulation of bone deposition and reabsorption: pp. 165–167 Figures 5.53–5.54 Table 5.3	*Concept maps:* Hormonal regulation of bone Figure 5.14	*Text:* Review: 5.17 *Workbook:* Multiple Select: 9 Matching: 6–10
5.18	Summarize the six functions of the skeletal system and give an example or explanation of each.	Functions of the skeletal system: pp. 167–168		*Text:* Review: 5.18
5.19	Summarize the effects of aging on the skeletal system.	Effects of aging on the skeletal system: pp. 168–169		*Text:* Review: 5.19 *Workbook:* Multiple Select: 5 Critical Thinking: 1
5.20	Classify fractures using descriptive terms.	Fractures: pp. 169–170 Figure 5.55	*Concept maps:* Fractures and how they heal Figure 5.16	*Text:* Spot Check: 11 Review: 5.20 *Workbook:* Multiple Select: 10 Case Study: 2

Learning Outcomes		Text Readings, Figures, and Tables	Workbook Activities	Text and Workbook Assessments
5.21	Explain how a fracture heals.	Fracture healing: pp. 170–171 Figures 5.56–5.57	*Concept maps:* Fractures and how they heal Figure 5.16	*Text:* Review: 5.21 *Workbook:* Multiple Select: 6
5.22	Describe a diagnostic test commonly used when diagnosing skeletal system disorders.	Diagnostic tests for skeletal system disorders: p. 172 Table 5.4		*Text:* Spot Check: 12 Review: 5.22 *Workbook:* Case Study: 1
5.23	Describe skeletal disorders and relate abnormal function to the pathology.	Skeletal system disorders: pp. 172–176 Figures 5.58–5.61 Table 5.5		*Text:* Spot Check: 13 Review: 5.23 *Workbook:* Critical Thinking: 2, 3

6 The Muscular System

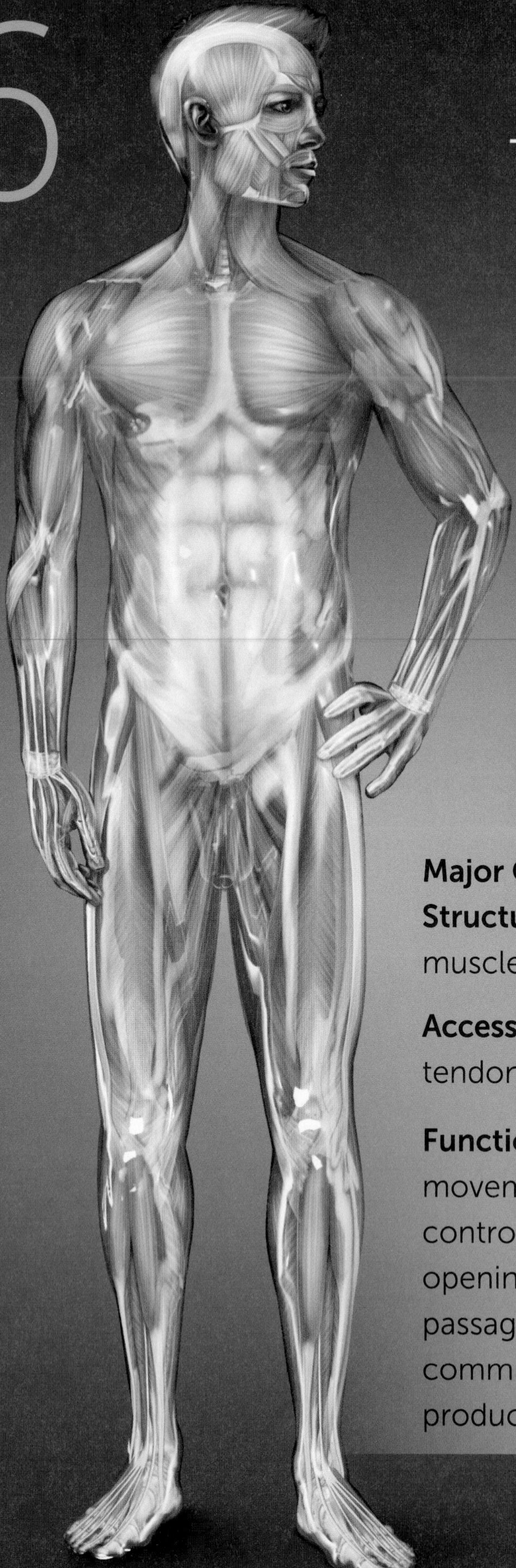

Major Organs and Structures:
muscles

Accessory Structures:
tendons

Functions:
movement, stability, control of body openings and passages, communication, heat production

learning outcomes

This chapter of the workbook is designed to help you learn the anatomy and physiology of the muscular system. After completing this chapter in the text and this workbook, you should be able to:

6.1 Use medical terminology related to the muscular system.

6.2 Define terms concerning muscle attachments and the ways muscles work in groups to aid, oppose, or modify each other's actions.

6.3 Demonstrate actions caused by muscles.

6.4 Identify muscles, giving the origin, insertion, and action.

6.5 Describe the structural components of a muscle, including the connective tissues.

6.6 Describe the structural components of a skeletal muscle fiber, including the major proteins.

6.7 Explain the five physiological characteristics of all muscle tissue.

6.8 Explain how a nerve stimulates a muscle cell at a neuromuscular junction.

6.9 Describe a muscle contraction at the molecular level.

6.10 Compare and contrast a muscle twitch and tetany with regard to the steps of a muscle contraction at the molecular level.

6.11 Define *motor unit* and explain the effect of recruitment.

6.12 Compare and contrast isotonic and isometric contractions.

6.13 Describe an example of a lever system in the human body, giving the resistance, effort, and fulcrum.

6.14 Compare aerobic and anaerobic respiration in terms of amount of ATP produced, speed, and duration.

6.15 Explain the basis of muscle fatigue and soreness.

6.16 Compare and contrast skeletal, cardiac, and smooth muscle tissue in terms of appearance, structure, type of nerve stimulation, type of respiration, and location.

6.17 Explain the nutritional requirements of the muscular system.

6.18 Summarize the five functions of the muscular system and give an example or explanation of each.

6.19 Summarize the effects of aging on the muscular system.

6.20 Describe a common diagnostic test used to diagnose disorders of the muscular system.

6.21 Describe muscle disorders and relate abnormal function to pathology.

word **roots** & combining **forms**

muscul/o: muscle

my/o: muscle

sarc/o: flesh

sthen/o: strength

Skeletal Muscles

Figures 6.1 to 6.7 show the skeletal muscles of the body by region. Color the box next to each term. Use the same color for the corresponding structures in the figures.

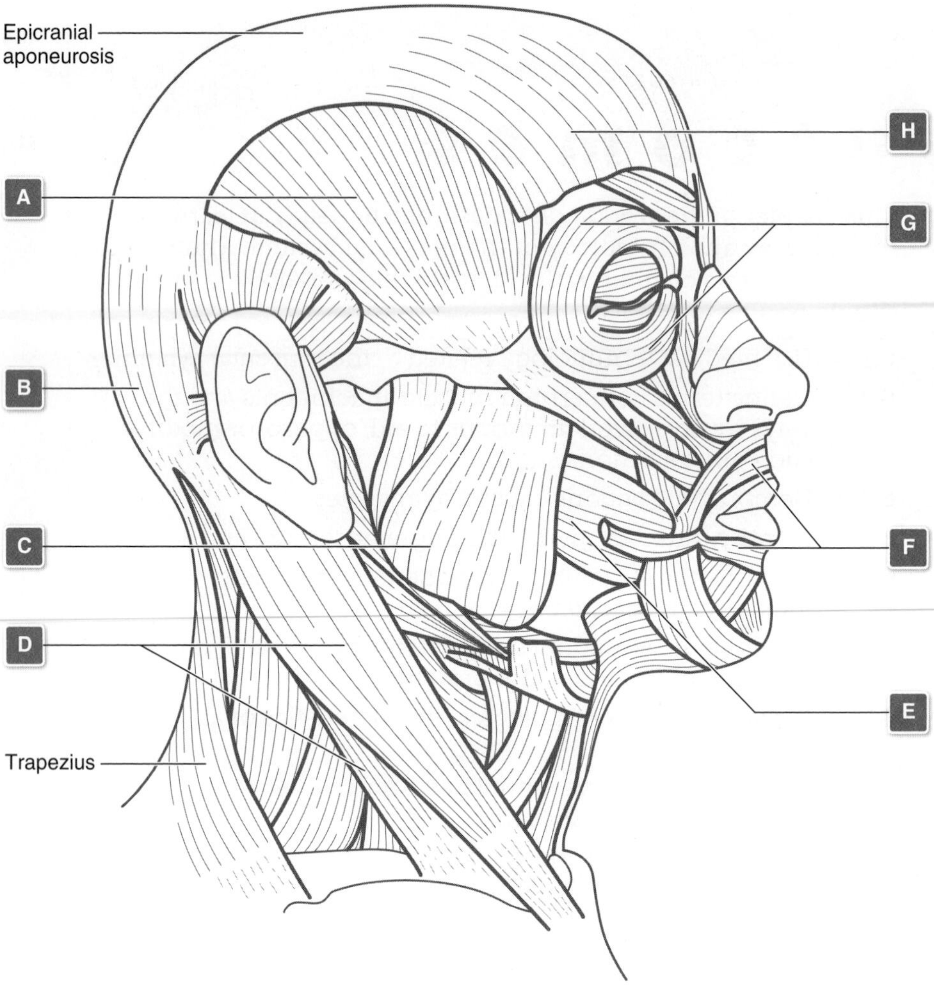

Epicranial aponeurosis
A
B
C
D
Trapezius
H
G
F
E

FIGURE 6.1 Muscles of the head and neck.

Muscles of the Head and Neck

- ☐ Temporalis(A)
- ☐ Occipitalis(B)
- ☐ Masseter(C)
- ☐ Sternocleidomastoid(D)
- ☐ Buccinator(E)
- ☐ Orbicularis oris(F)
- ☐ Orbicularis oculi(G)
- ☐ Frontalis(H)

COLORING BOOK

Muscles of the Thorax and Abdomen

- ☐ Pectoralis major(A)
- ☐ Pectoralis minor(B)
- ☐ Serratus anterior(C)
- ☐ Rectus abdominus(D)
- ☐ External abdominal obliques(E)
- ☐ Internal abdominal obliques(F)
- ☐ Transverse abdominal(G)
- ☐ Internal intercostals(H)
- ☐ External intercostals(I)
- ☐ Diaphragm(J)

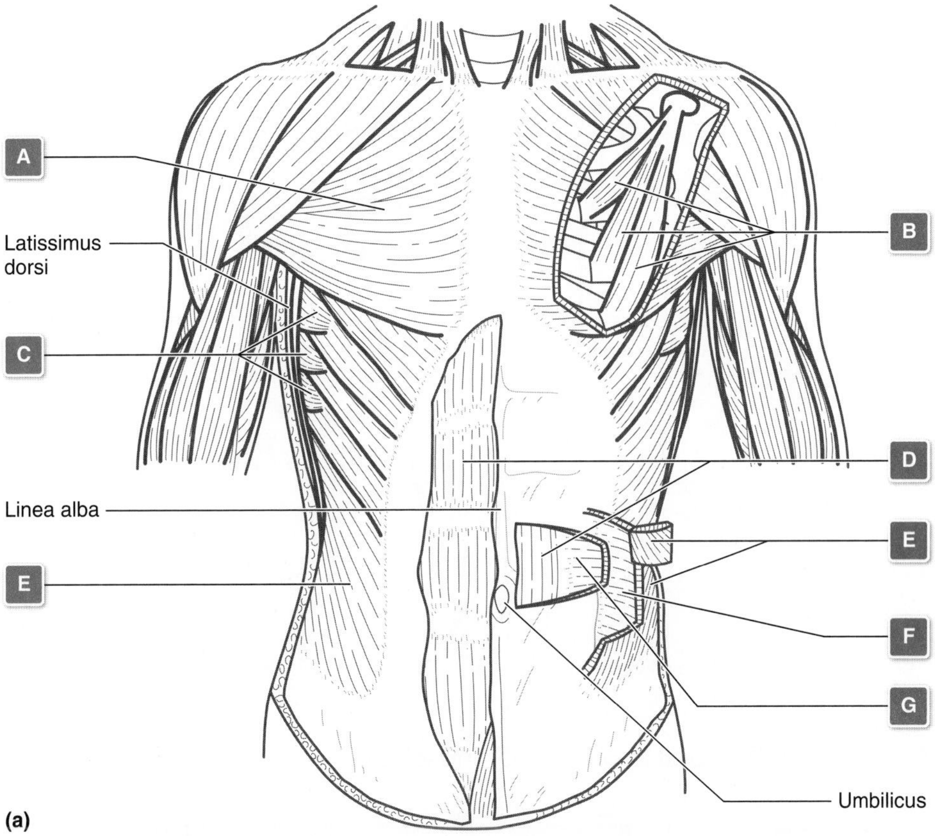

A
Latissimus dorsi
C
Linea alba
E
B
D
E
F
G
Umbilicus
(a)

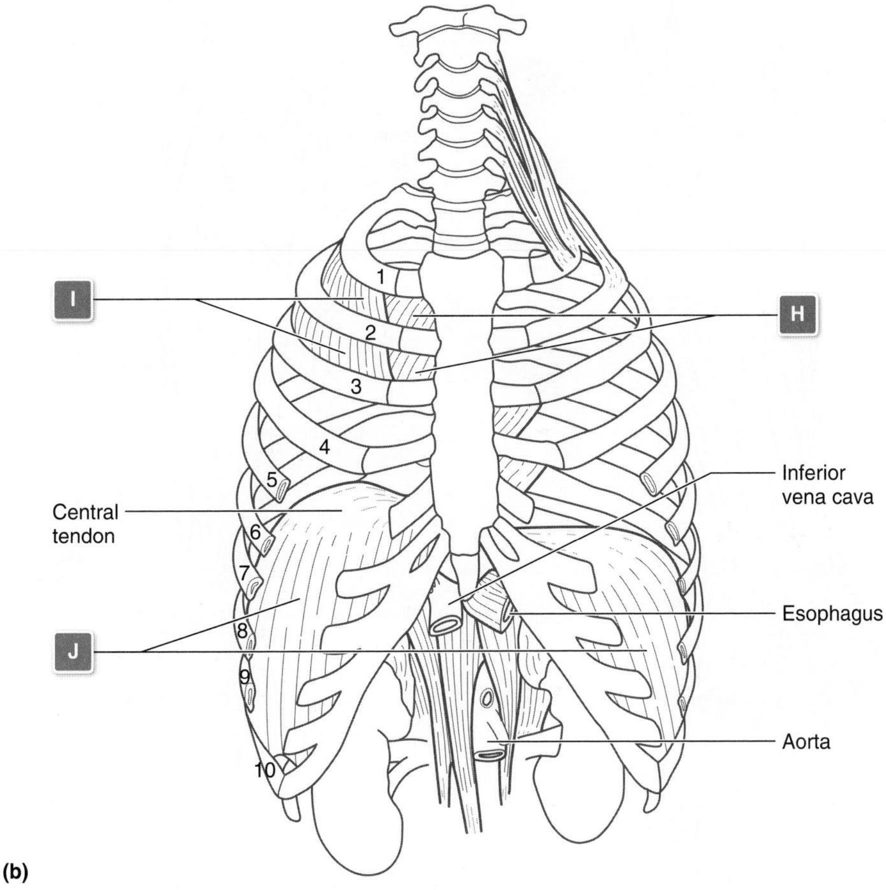

I
H
1
2
3
4
5
6
7
8
9
10
Central tendon
J
Inferior vena cava
Esophagus
Aorta
(b)

FIGURE 6.2 Muscles of the thorax and abdomen: (a) anterior view of superficial and deep muscles, (b) anterior view of deep muscles.

COLORING BOOK

COLORING BOOK

Muscles of the Arm

- [ ] Deltoid (T)
- [ ] Pectoralis major (U)
- [ ] Triceps brachii (V)
- [ ] Biceps brachii (W)
- [ ] Brachialis (X)
- [ ] Brachioradialis (Y)

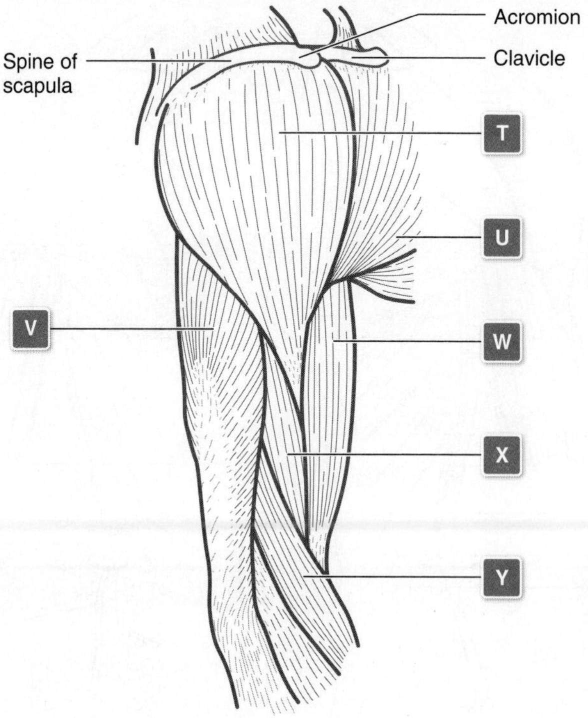

FIGURE 6.3 Pectoral and brachial muscles.

Muscles of the Forearm

- [ ] Brachialis (A)
- [ ] Brachioradialis (B)
- [ ] Flexor carpi radialis (C)
- [ ] Palmaris longus (D)
- [ ] Flexor carpi ulnaris (E)
- [ ] Extensor carpi radialis (F)
- [ ] Extensor carpi ulnaris (G)
- [ ] Extensor digitorum (H)

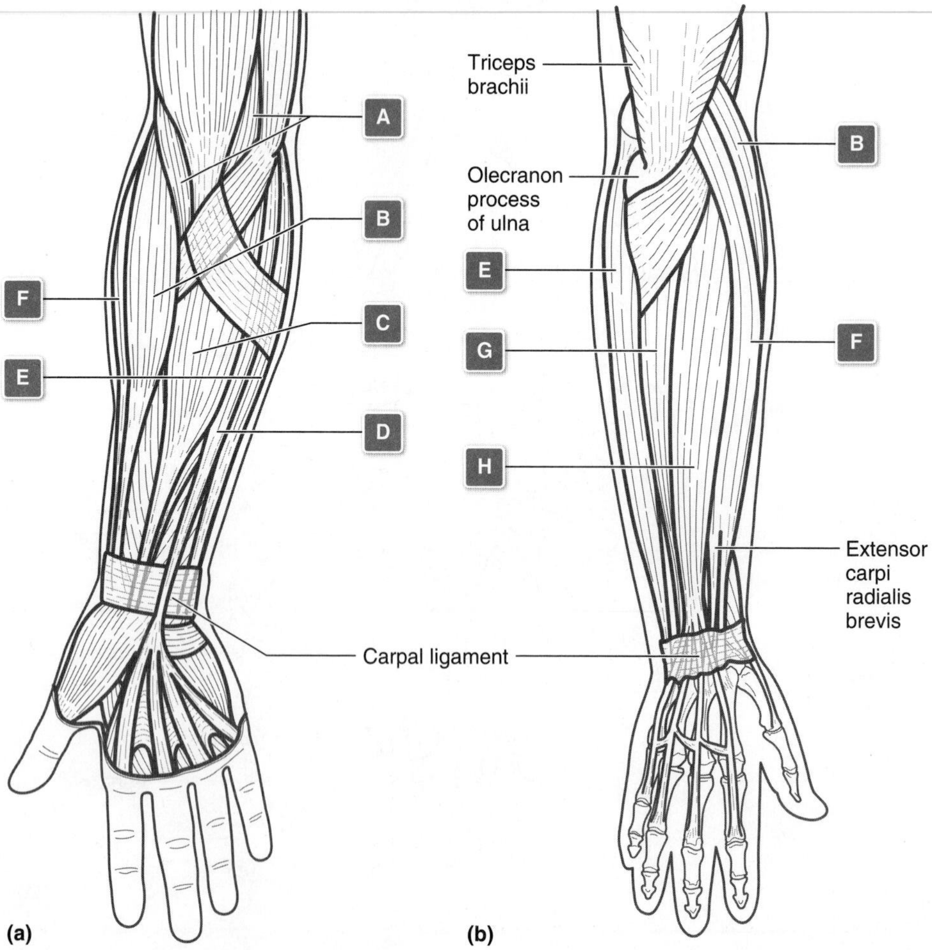

FIGURE 6.4 Muscles of the forearm: (a) anterior, (b) posterior.

Muscles of the Back

- [ ] Sternocleido-mastoid (L)
- [ ] Trapezius (M)
- [ ] Deltoid (N)
- [ ] Erector spinae (O)
- [ ] Latissimus dorsi (P)
- [ ] Gluteus medius (Q)
- [ ] Gluteus maximus (R)

COLORING BOOK

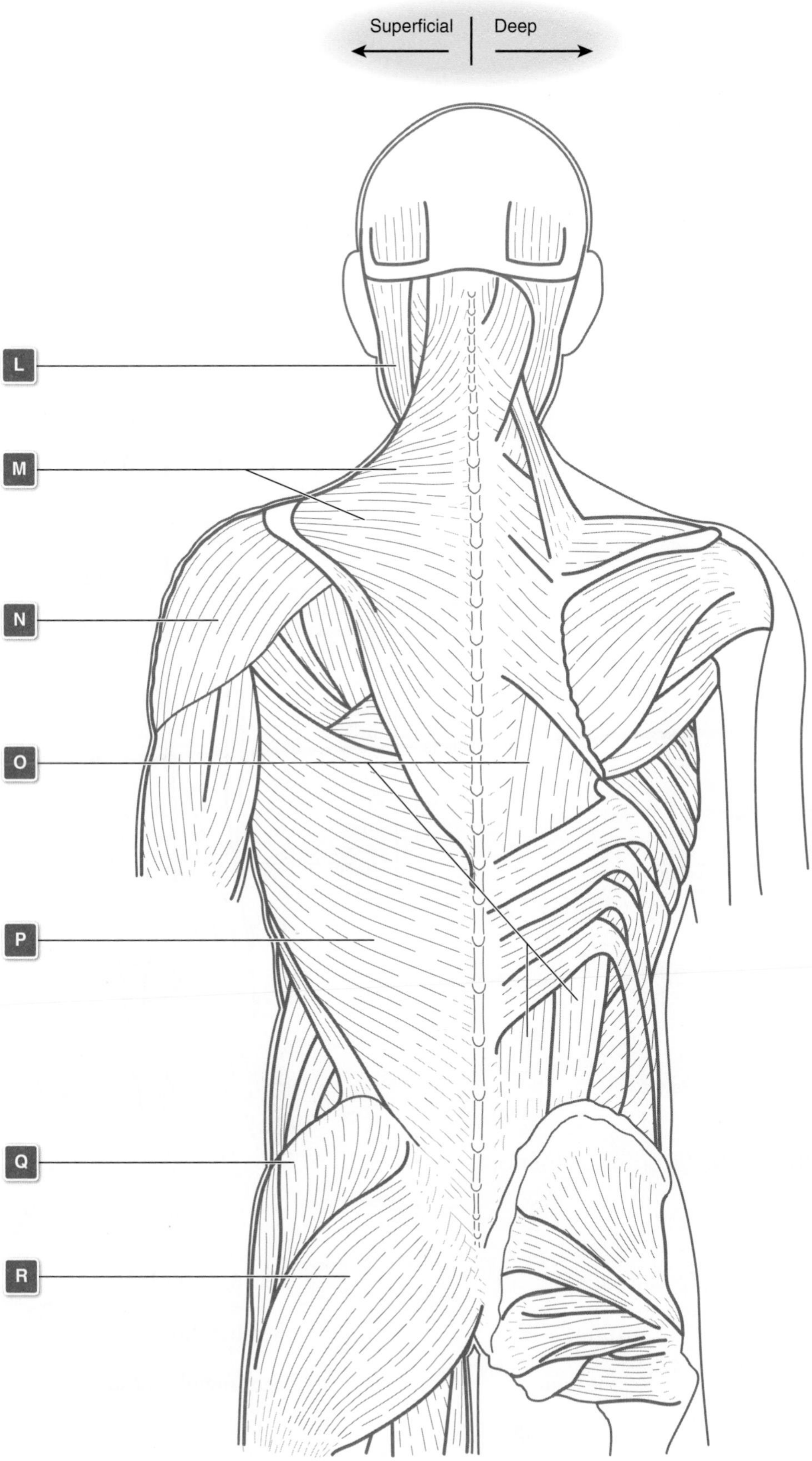

Superficial
Deep
L
M
N
O
P
Q
R

FIGURE 6.5 Back and gluteal muscles.

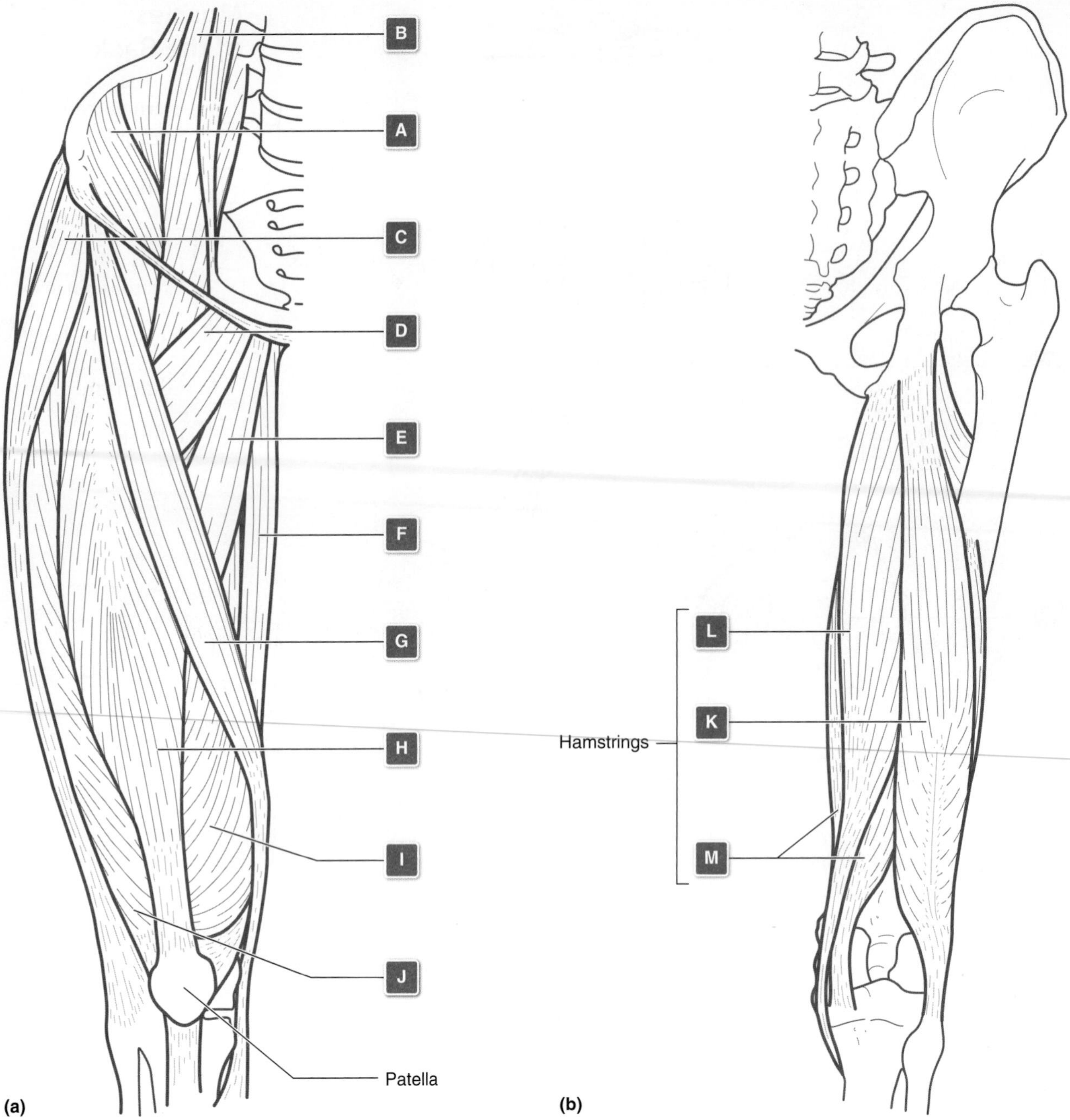
B
A
C
D
E
F
G
H
I
J
Patella
(a)
L
K
Hamstrings
M
(b)

FIGURE 6.6 Muscles of the thigh: (a) anterior, (b) posterior.

Muscles of the Thigh

- ☐ Iliacus(A)
- ☐ Psoas major(B)
- ☐ Tensor fasciae latae(C)
- ☐ Pectineus(D)
- ☐ Adductor longus(E)
- ☐ Gracilis(F)
- ☐ Sartorius(G)
- ☐ Rectus femoris(H)
- ☐ Vastus medialis(I)
- ☐ Vastus lateralis (J)
- ☐ Biceps femoris(K)
- ☐ Semitendinosus(L)
- ☐ Semimembranosus(M)

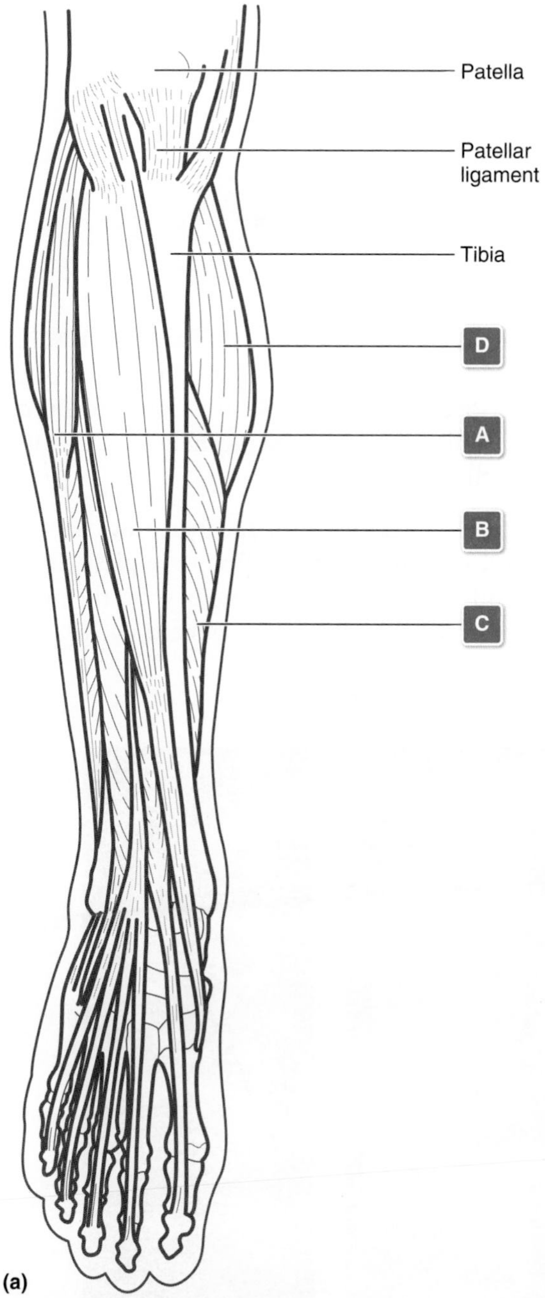
Patella
Patellar ligament
Tibia
D
A
B
C
(a)

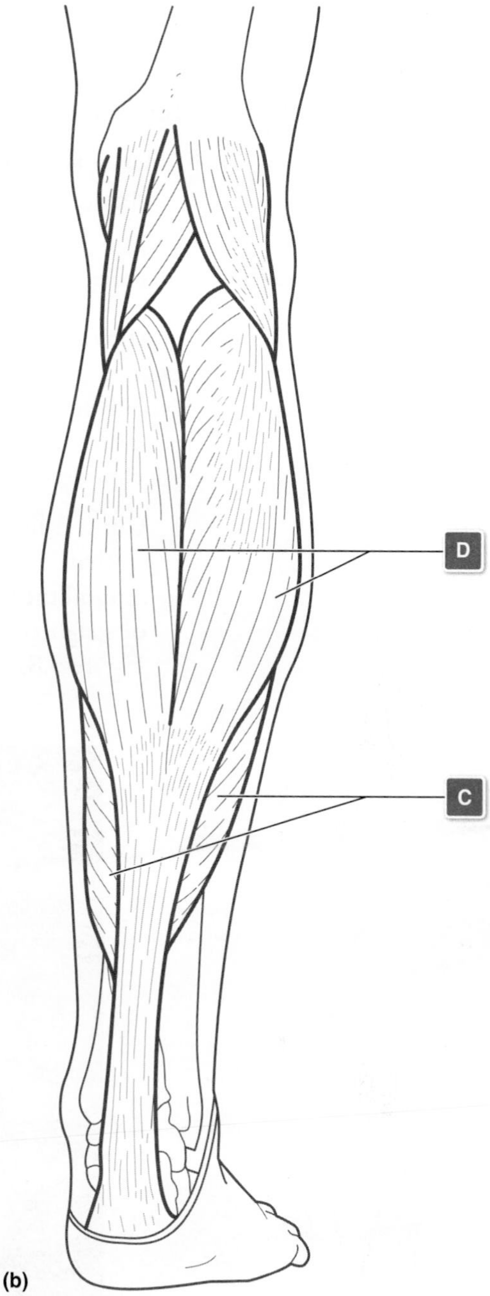
D
C
(b)

FIGURE 6.7 Muscles of the leg: (a) anterior, (b) posterior.

COLORING BOOK

Muscles of the Leg

- ☐ Fibularis(A)
- ☐ Tibialis anterior(B)
- ☐ Soleus(C)
- ☐ Gastrocnemius(D)

Muscle Model

In this activity you will build a model of a muscle with all of its connective tissues. You will need:

- Uncooked spaghetti noodles
- Plastic wrap
- Wax paper
- Aluminum foil

(If you do not have any of the above, you may substitute similar materials. For example, notebook paper can be substituted for plastic wrap as long as the edges extend beyond the spaghetti noodles. See step 1 of the instructions. You must have three different wraps.) You may want to find a partner to help you assemble your model. Follow the directions carefully:

1. Cut a 2-inch strip of plastic wrap. Place six pieces of spaghetti on the wrap so that the ends of the wrap extend past the spaghetti. Wrap the plastic wrap around the spaghetti, leaving the ends of the wrap extending beyond the pieces of spaghetti. Do this step 12 times so that you form 12 spaghetti packets. See Figure 6.8.

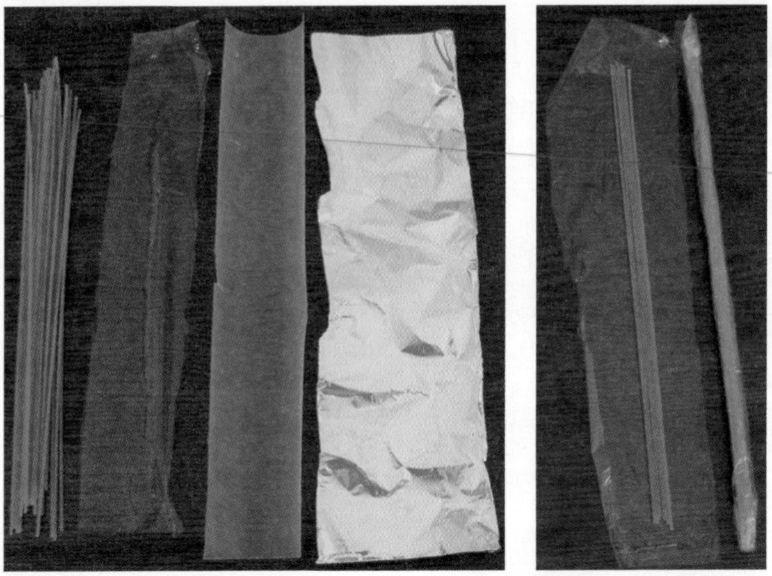

(a) (b)

FIGURE 6.8 Muscle model: (a) spaghetti, plastic wrap, wax paper, aluminum foil, (b) step 1.

2. Cut a 3-inch strip of wax paper. Place three of your spaghetti packets on the wax paper with the ends of the wax paper matching up to the ends of the plastic wrap. Wrap the three spaghetti packets in the wax paper. Do this step four times so that all of the spaghetti packets are wrapped.
3. Cut a 4-inch piece of aluminum foil. Place the four wax-paper packets on the aluminum foil with the ends of the wax paper matching the ends of the aluminum foil. Wrap the wax-paper packets in the aluminum foil. Pinch the ends of the aluminum foil, wax paper, and plastic wrap together.

Your model of a muscle is now complete. Given that a single piece of spaghetti represents a myofibril in your model, answer the following questions:

1. What does a spaghetti packet represent? ___
2. What connective tissue does the plastic wrap represent?
___
3. What does a wax paper packet represent? ___
4. What connective tissue does the wax paper represent?
___
5. What does the aluminum foil packet represent? ___
6. What connective tissue does the aluminum foil represent?
___
7. What do the pinched ends of aluminum foil, wax paper, and plastic wrap together represent?
___
8. What would you need to do to represent fascia in your model? ___
___
___

Muscle Twitch

A muscle twitch is the contraction of a muscle cell in response to one nerve impulse. Refer to the graph in **Figure 6.9** to answer the following questions concerning a muscle twitch:

1. What amount of stimulus is necessary for this twitch to occur?
___
2. What phase of a muscle twitch is indicated by "G" on the graph?
___
3. Which steps of a muscle contraction at the molecular level happen during phase G?
___
___
___
___
___
___

LABORATORY EXERCISES AND ACTIVITIES

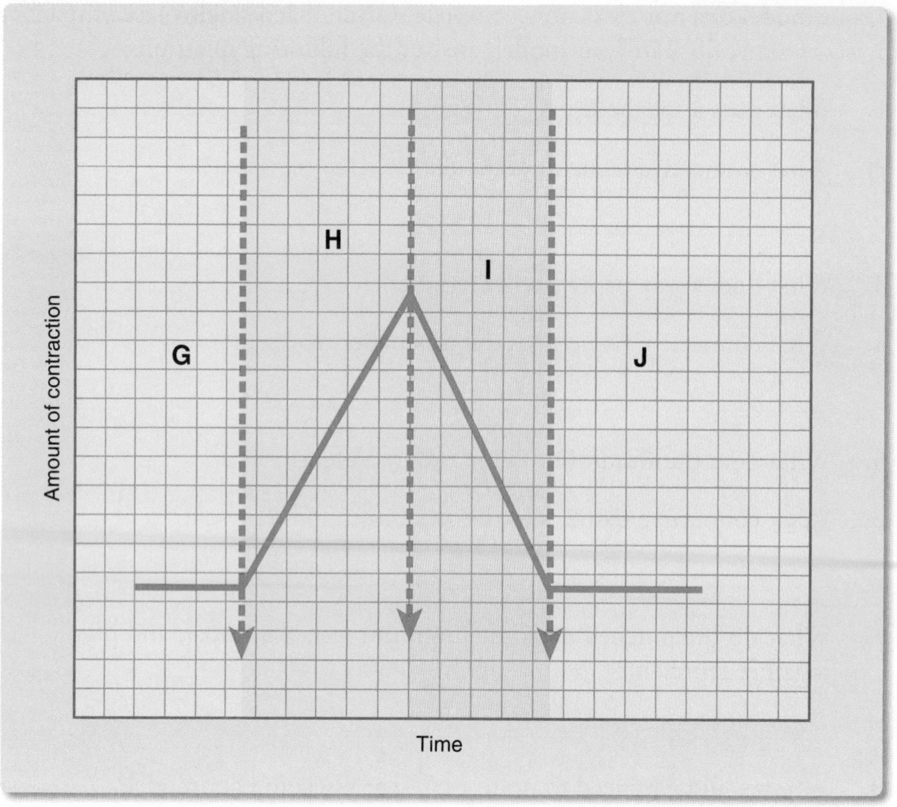

FIGURE 6.9 Muscle twitch graph.

4. What phase of a muscle twitch is indicated by "H" on the graph?

____

5. Which steps of a muscle contraction at the molecular level happen during phase H?

____

____

____

____

____

____

6. What phase of a muscle twitch is indicated by "I" on the graph?

____

7. Which steps of a muscle contraction at the molecular level happen during phase I?

8. What phase of a muscle twitch is indicated by "J" on the graph?

9. Which steps of a muscle contraction at the molecular level happen during phase J?

10. Which phases require the addition of ATP?

11. What determines whether a series of nerve impulses results in a series of twitches or tetany?

Levers

The muscular system uses muscles to move bones in lever systems. All lever systems have three parts: a fulcrum (F) that acts as a pivot point on the lever; an effort (E), where force is applied to the lever; and a resistance (R), which is the weight the lever is to lift. In the human body, bones act as levers. The insertion of a muscle is the effort applied, and the fulcrum is a joint. The order of R, E, and F determines the class of lever: First-class levers have the F in the middle, second-class levers have the R in the middle, and third-class levers have the E in the middle.

Examine **Figure 6.10**. Answer the following questions about the lever system used when the masseter muscle elevates the jaw.

1. Which letter in **Figure 6.10** represents the fulcrum? ________________
2. Which letter in **Figure 6.10** represents the effort? ________________
3. Which letter in **Figure 6.10** represents the resistance? ________________
4. What class of lever system is used when the masseter elevates the jaw?

____

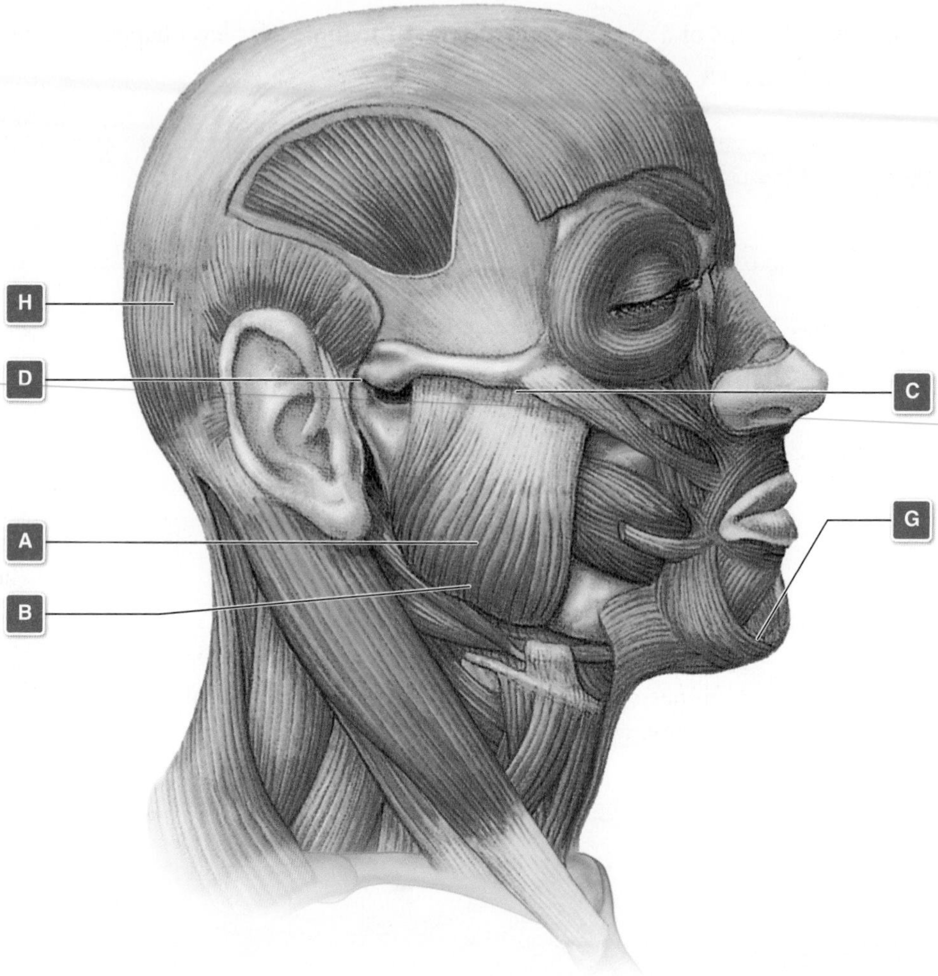

FIGURE 6.10 Lever system for the masseter muscle.

LABORATORY EXERCISES AND ACTIVITIES

Key Words

The following terms are defined in the glossary of the textbook.

- acetylcholine (ACh)
- aerobic respiration
- anaerobic respiration
- antagonist
- atrophy
- extension
- fascicle
- fatigue
- flexion
- insertion
- isometric
- isotonic
- lever
- motor unit
- muscle twitch
- myalgia
- origin
- recruitment
- sarcomere
- sliding filament theory
- sprain
- strain
- synergists
- tetany

Concept Maps

Use key words and other bold words from the chapter to complete the following concept maps (**Figures 6.11** to **6.18**).

Anatomy of a Muscle

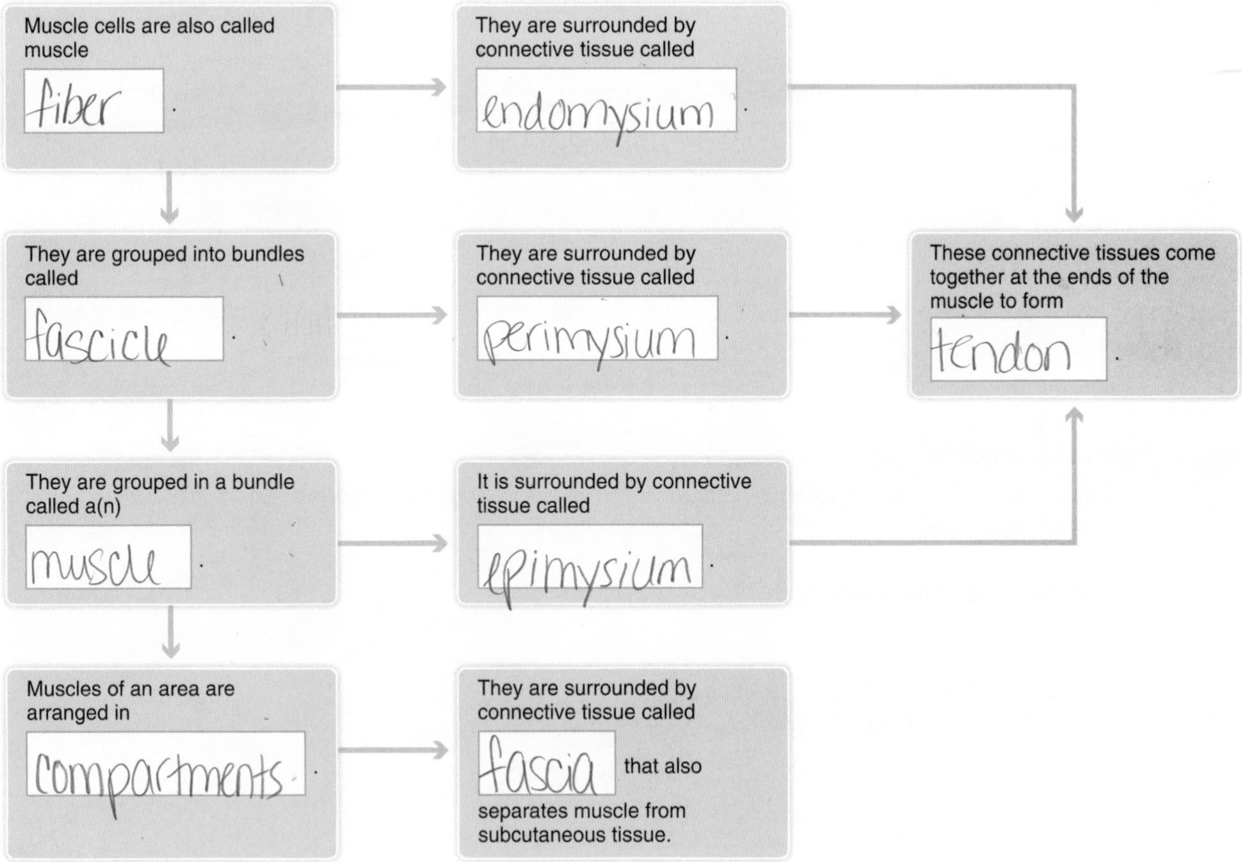
Muscle cells are also called muscle
fiber
They are surrounded by connective tissue called
endomysium
They are grouped into bundles called
fascicle
They are surrounded by connective tissue called
perimysium
These connective tissues come together at the ends of the muscle to form
tendon
They are grouped in a bundle called a(n)
muscle
It is surrounded by connective tissue called
epimysium
Muscles of an area are arranged in
compartments
They are surrounded by connective tissue called
fascia
that also separates muscle from subcutaneous tissue.

FIGURE 6.11 Anatomy of a muscle concept map.

Muscle Cell Anatomy

A muscle cell is a bundle of Z lines containing thick and thin myofilaments.

Thick myofilaments are composed of the protein myosins, which has a pincherlike end for grabbing, called a cross-bridge.

Thin myofilaments are composed of three proteins:

- Actin, which appears as a double, twisted chain.
- tropomyosin, which appears as a thread running through the chain.
- troponin, which is a small protein along the chain that is the binding site for calcium.

The myofilaments are arranged in sarcomeres.

Thick myofilaments are not attached to Z lines.

Thin myofilaments are attached to Z lines at the ends of the sarcomere.

darkness visible on a muscle cell are caused by the overlapping of thick and thin myofilaments.

FIGURE 6.12 Muscle cell anatomy concept map.

Characteristics of Muscle Tissue

Muscle cells display different characteristics.

The ability of a muscle cell to contract when stimulated by a nerve is excitability.

The traveling of that nerve stimulus along the length of the muscle cell is conductivity.

extensibility is the ability of a muscle to be stretched. Muscles do not do their work by getting longer.

Once stretched or shortened, the muscle cell has the ability to return to its original shape. This is elasticity.

contractility is the ability of a muscle to shorten with force. Muscles work by pulling, not pushing.

FIGURE 6.13 Characteristics of muscle tissue concept map.

Neuromuscular Junction

A nerve meets a muscle cell at the neuromuscular junction.

An electrical signal travels to the knob at the end of the nerve cell. It triggers the release of a chemical signaling protein called a(n) neurotransmitter.

Acetylcholine is the specific signaling protein released at a neuromuscular junction for skeletal muscle.

The signaling protein fits into receptors on the muscle cell.

The minimum amount of signaling protein needed to cause a contraction is the threshold.

This is a(n) all or nothing effect. Having more signaling protein will not cause a bigger contraction.

FIGURE 6.14 Neuromuscular junction concept map.

KEY WORD CONCEPT MAPS

Muscle Contraction

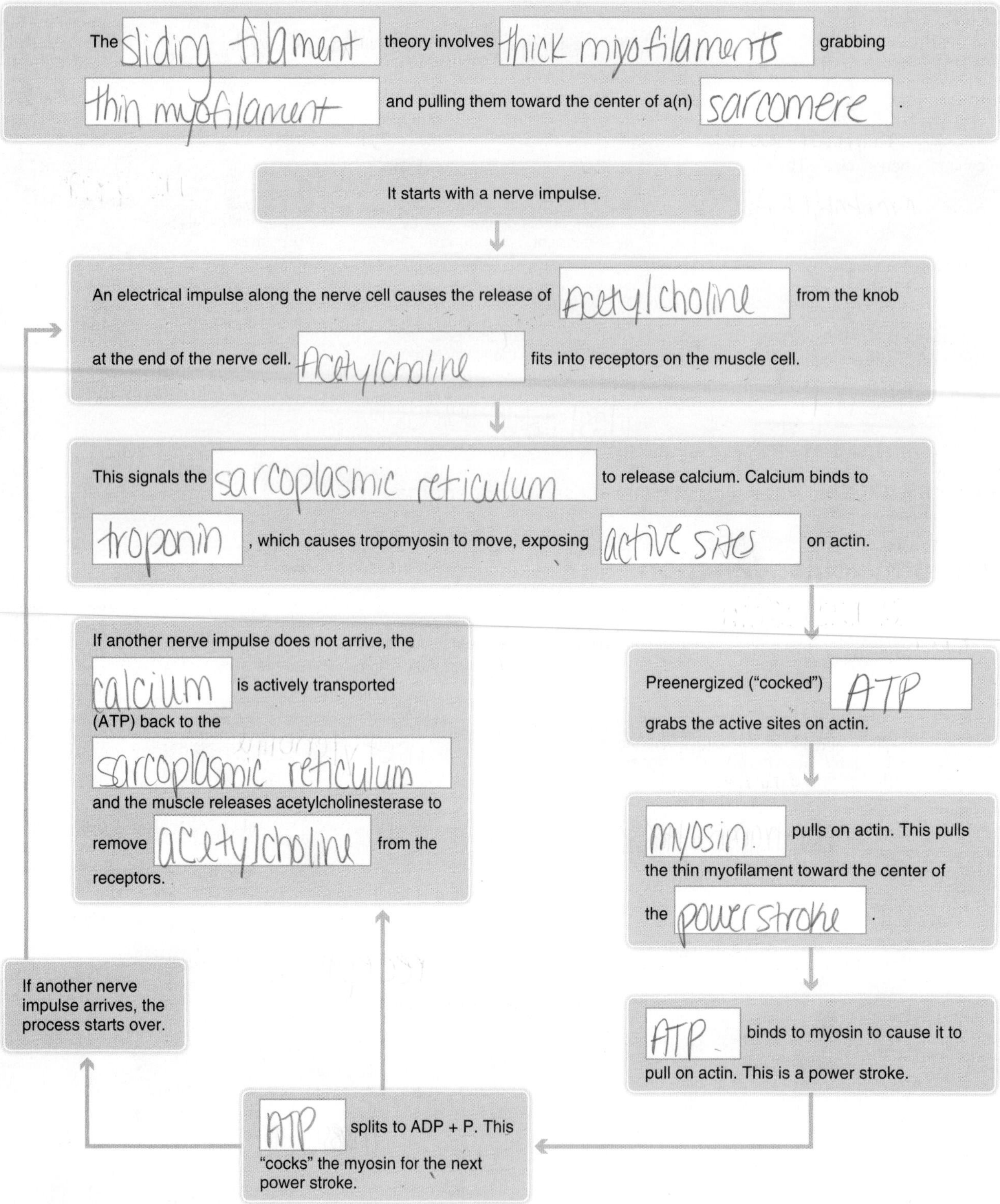
The sliding filament theory involves thick myofilaments grabbing
thin myofilament and pulling them toward the center of a(n) sarcomere .
It starts with a nerve impulse.
An electrical impulse along the nerve cell causes the release of Acetylcholine from the knob
at the end of the nerve cell. Acetylcholine fits into receptors on the muscle cell.
This signals the sarcoplasmic reticulum to release calcium. Calcium binds to
troponin , which causes tropomyosin to move, exposing active sites on actin.
If another nerve impulse does not arrive, the
calcium is actively transported
(ATP) back to the
sarcoplasmic reticulum
and the muscle releases acetylcholinesterase to
remove acetylcholine from the
receptors.
Preenergized ("cocked") ATP
grabs the active sites on actin.
myosin pulls on actin. This pulls
the thin myofilament toward the center of
the powerstroke .
If another nerve impulse arrives, the process starts over.
ATP binds to myosin to cause it to
pull on actin. This is a power stroke.
ATP splits to ADP + P. This
"cocks" the myosin for the next
power stroke.

FIGURE 6.15 Muscle contraction concept map.

Muscle Metabolism

mitochondria in muscle cells can produce energy in two ways:

aerobic respiration

anaerobic respiration

Both processes use glucose.

Both processes start with glycolysis.

Aerobic respiration:

Requires oxygen.

Produces carbon dioxide as a waste product and water.

This many-step process produces enough energy to form 36 ATPs from each glucose molecule.

________ fibers specialize in aerobic respiration.

Anaerobic respiration:

Requires no oxygen.

Produces lactic acid as a waste product.

This short process produces enough energy to form 2 ATPs.

________ fibers specialize in anaerobic respiration.

The buildup of lactic acid results in fatigue, which is the inability of a muscle to fully respond to a nerve stimulus.

Oxygen must be used to remove lactic acid.

Oxygen debt is the amount of oxygen necessary to remove the lactic acid.

FIGURE 6.16 Muscle metabolism concept map.

Fatigue

Fatigue is the inability of a muscle cell to respond to a nerve stimulus.

Causes of fatigue include:

- Buildup of lactic acid.
- Insufficient glucose to start the process.
- Lack of acetylcholine from the nerve cell.

fast-twitch fibers fatigue quickly due to the buildup of lactic acid.

lactic acid is a waste product that must be removed by adding oxygen.

The amount of oxygen needed to remove the lactic acid is called the oxygen debt.

slow-twitch fibers are specially adapted to not fatigue.

These fibers use aerobic respiration.

These fibers contain myeglobin, a protein that stores oxygen.

These fibers have large mitochondria to process the energy.

These fibers contain glycogen, which is a storage molecule for glucose.

These fibers tend to have a better blood supply to supply oxygen and glucose to the muscle.

The ratio of the amount of these fibers in any given muscle is genetically determined.

FIGURE 6.17 Fatigue concept map.

Types of Muscle Tissue

There are three types of muscle tissue.

Skeletal muscle tissue is attached to bone. It has long, striated cells with many nuclei per cell pushed off to the side.

Cardiac muscle tissue is located in the heart. It has branched, striated cells with a single nucleus and specialized junctions between cells called intercalated disks.

Smooth muscle tissue is found in hollow internal organs and blood vessel walls. It has spindle-shaped cells with no striations and a single nucleus per cell.

FIGURE 6.18 Types of muscle tissue concept map.

Word Deconstruction: *In the textbook, you built words to fit a definition using combining forms, prefixes, and suffixes. Here you are to break down the term into its parts (prefixes, roots, and suffixes) and give a definition. Prefixes and suffixes can be found inside the back cover of the textbook.*

FOR EXAMPLE Dermatitis: dermat/itis—inflammation of the skin

1. Myasthenia: ______________________________
2. Cardiomyopathy: ______________________________
3. Fibromyalgia: ______________________________
4. Sarcocarcinoma: ______________________________
5. Muscular dystrophy: ______________________________

Multiple Select: *Select the correct choices for each statement. The choices may be all correct, all incorrect, or any combination of correct and incorrect.*

1. What happens during a muscle contraction?
   - **a.** Myofibrils slide past each other during a contraction.
   - **b.** Tension increases in the muscle in an isotonic contraction.
   - **c.** The origin may be determined by another muscle.
   - **d.** The placement of the insertion determines the mechanical advantage.
   - **e.** There is rapid recruitment in a boxer's punch.
2. Which of the following statements is (are) true concerning the rectus femoris?
   - **a.** It is the prime mover for knee flexion.
   - **b.** It is a synergist to the vastus intermedius for hip flexion.
   - **c.** It is a synergist to the vastus lateralis for knee extension.
   - **d.** It is an antagonist to the semimembranosus.
   - **e.** It is part of the hamstrings.
3. Which of the following statements is (are) true concerning the platysma?
   - **a.** It is intrinsic to the head.
   - **b.** It is extrinsic to the thorax.
   - **c.** It gives off heat.
   - **d.** It is used for communication.
   - **e.** It is a synergist to the temporalis muscle.
4. Which of the following statements is (are) true about skeletal muscle tissue?
   - **a.** Skeletal muscle appears striated due to the presence of Z lines.
   - **b.** Skeletal muscle can either push or pull at the insertion.
   - **c.** Skeletal muscle cells have a single nucleus.
   - **d.** Skeletal muscle cells are spindle-shaped.
   - **e.** Skeletal muscle cells branch.
5. Which step(s) of a muscle contraction would *not* be completed during tetany?
   - **a.** Active transport of calcium.
   - **b.** Release of calcium.
   - **c.** Release of acetylcholinesterase.
   - **d.** Release of acetylcholine.
   - **e.** Binding of calcium to troponin to expose active sites on actin.
6. Which of the following possibly cause(s) fatigue?
   - **a.** Lack of glucose.
   - **b.** Lack of ATP.
   - **c.** Lack of acetylcholine.
   - **d.** Buildup of glycogen.
   - **e.** Buildup of lactic acid.

7. Which of the following is (are) the result of aging on the muscular system?
   a. Lean muscle mass decreases.
   b. Fat is deposited in muscle.
   c. Gait shortens.
   d. Movement slows.
   e. Fatigue happens more rapidly.

8. Which of the following statements is (are) accurate concerning muscular system disorders?
   a. Muscular dystrophy is hereditary.
   b. Myasthenia gravis is an autoimmune disease.
   c. A cramp can be caused by dehydration.
   d. A hiatal hernia involves a loop of intestine pushing through the abdominal wall.
   e. Compartment syndrome is a problem with the endomysium of a muscle.

9. Which of the following statements accurately describe(s) the connective tissues of muscle?
   a. Endomysium surrounds a fascicle.
   b. Perimysium surrounds a muscle.
   c. Fascia surrounds muscle of an area.
   d. A tendon is composed of epimysium, endomysium, and perimysium but not fascia.
   e. Epimysium surrounds myofibrils.

10. What is true about making a fist?
   a. The frequency of nerve impulses matters if the fist is to be held.
   b. The frequency of nerve impulses determines how tight the fist is made.
   c. The number of motor units determines how tight the fist is made.
   d. The number of nerve cells used determines the number of motor units used.
   e. Making a fist requires flexor muscles that originate on the medial epicondyle of the humerus.

Matching: *Match the step of a muscle contraction at the molecular level to the phase in which it would occur in a twitch. Some answers may be used more than once.*

_______	**1.** Acetylcholinesterase is released.	**a.** Latent phase
_______	**2.** Myosin grabs an active site on actin.	**b.** Contraction phase
_______	**3.** Myosin pulls (power stroke).	**c.** Relaxation phase
_______	**4.** Calcium is released.	**d.** Refractory phase
_______	**5.** Calcium is actively transported.	

Matching: *Match the description to the type of muscle tissue. Some answers may be used more than once. Some questions have more than one answer.*

_______	**6.** Striated	**a.** Cardiac muscle tissue
_______	**7.** Voluntary	**b.** Skeletal muscle tissue
_______	**8.** Autorhythmic	**c.** Smooth muscle tissue
_______	**9.** Multiple nuclei per cell	
_______	**10.** Branching	

Completion: *Fill in the blanks to complete the following statements about the physiological characteristics of muscle.*

1. ______________________________ is the ability to be stretched.

2. ______________________________ is the ability to return to shape if stretched.

3. ______________________ is the ability to shorten with force.

4. ______________________ is the ability to pass on a stimulus to the rest of the cell.

5. ______________________ is the ability to be stimulated by a nerve.

Critical Thinking

1. Some insecticides work by interfering with acetylcholinesterase. What would be the consequences to muscles if the insecticide was inhaled or absorbed through the skin? What would you expect to see? Explain in terms of muscle contractions at the molecular level.

2. If you have ever eaten chicken, you know the breasts and wings are considered white meat and the legs and thighs are considered dark meat. There are few blood vessels in the breasts, but many blood vessels in the thighs and legs. If you have seen live chickens, you have observed that they can fly for only short distances but can walk around all day. How are fast- and slow-twitch fibers distributed in a chicken as compared to in a human? Explain. What type of respiration is done by each type of muscle fiber? How does the type of respiration influence their behavior?

3. Describe an exercise activity that works muscles of the (a) anterior arm, (b) posterior arm, (c) abdominal muscles, and (d) thigh muscles.

Case Study

Asha is on the track team. She decided to increase the intensity of her training to ensure she won the upcoming race. She took a different route on her practice run that included more hills and valleys than what she is used to. Asha found herself in excruciating pain and unable to finish her run. Her pain was located in the front of her lower leg.

1. What muscular disorder was Asha most likely suffering from? Explain why you came to this conclusion.

2. What muscle and other structures in the lower leg are affected by the disorder listed in your first answer?

3. Could Asha be experiencing any other muscular disorders; if so, which ones? Explain the difference in the additional disorders you listed.

4. What impact will the muscular disorder most likely have on her ability to participate in the race in two weeks?

CHAPTER 6 REVIEW QUESTIONS

This section of the chapter is designed to help you find where each outcome is covered in the textbook and workbook.

Learning Outcomes		Text Readings, Figures, and Tables	Workbook Activities	Text and Workbook Assessments
6.1	Use medical terminology related to the muscular system.	Word roots & combining forms, pronunciation key: pp. 183–184 Overview: pp. 184–185 Anatomy of the muscular system: p. 185	Word roots & combining forms	*Text:* Review: 6.1 *Workbook:* Word deconstruction: 1–5
6.2	Define terms concerning muscle attachments and the ways muscles work in groups to aid, oppose, or modify each other's actions.	Anatomical terms: p. 185 Figure 6.10		*Text:* Review: 6.2 *Workbook:* Multiple Select: 1, 3
6.3	Demonstrate actions caused by muscles.	Muscle actions: pp. 186–189 Figures 6.2–6.9		*Text:* Review: 6.3 *Workbook:* Multiple Select: 10 Critical Thinking: 3
6.4	Identify muscles, giving the origin, insertion, and action.	Muscles by region: pp. 189–203 Figures 6.10–6.20 Tables 6.1–6.7	*Coloring book:* Skeletal muscles Figures 6.1–6.7	*Text:* Spot Check: 1, 2 Review: 6.4 *Workbook:* Multiple Select: 2, 3
6.5	Describe the structural components of a muscle, including the connective tissues.	Anatomy of a skeletal muscle: pp. 204–205 Figure 6.21	*Lab exercises and activities:* Muscle model Figure 6.8 *Concept maps:* Anatomy of a muscle Figure 6.11	*Text:* Review: 6.5 *Workbook:* Multiple Select: 9
6.6	Describe the structural components of a skeletal muscle fiber, including the major proteins.	Anatomy of a skeletal muscle cell: pp. 205–207 Figures 6.22–6.24	*Concept maps:* Muscle cell anatomy Figure 6.12	*Text:* Review: 6.6 *Workbook:* Multiple Select: 4
6.7	Explain the five physiological characteristics of all muscle tissue.	Physiological characteristics of muscle tissue: p. 207	*Concept maps:* Characteristics of muscle tissue Figure 6.13	*Text:* Review: 6.7 *Workbook:* Completion: 1–5 Multiple Select: 4
6.8	Explain how a nerve stimulates a muscle cell at a neuromuscular junction.	Neuromuscular junction: pp. 207–208 Figure 6.25	*Concept maps:* Neuromuscular junction Figure 6.14	*Text:* Review: 6.8 *Workbook:* Critical Thinking: 1 Multiple Select: 10

CHAPTER 6 MAPPING

Learning Outcomes		Text Readings, Figures, and Tables	Workbook Activities	Text and Workbook Assessments
6.9	Describe a muscle contraction at the molecular level.	Muscular contraction at a molecular level: pp. 208–210 Figure 6.26	*Lab exercises and activities:* Muscle twitch Figure 6.9 *Concept maps:* Muscle contraction Figure 6.15	*Text:* Spot Check: 3 Review: 6.9 *Workbook:* Critical Thinking: 1 Multiple Select: 1 Matching: 1–5
6.10	Compare and contrast a muscle twitch and tetany with regard to the steps of a muscle contraction at the molecular level.	Types of muscle contractions: pp. 210–212 Figures 6.27–6.28		*Text:* Review: 6.10 *Workbook:* Multiple Select: 5
6.11	Define *motor unit* and explain the effect of recruitment.	Motor units and recruitment: p. 212		*Text:* Spot Check: 4 Review: 6.11 *Workbook:* Multiple Select: 1, 10
6.12	Compare and contrast isotonic and isometric contractions.	Isotonic and isometric contractions: p. 212		*Text:* Spot Check: 5 Review: 6.12 *Workbook:* Multiple Select: 1
6.13	Describe an example of a lever system in the human body, giving the resistance, effort, and fulcrum.	Levers: pp. 213–214 Figure 6.29	*Lab exercises and activities:* Levers Figure 6.10	*Text:* Review: 6.13 *Workbook:* Multiple Select: 1
6.14	Compare aerobic and anaerobic respiration in terms of amount of ATP produced, speed, and duration.	Muscle metabolism: pp. 214–216 Figure 6.30 Table 6.8	*Concept maps:* Muscle metabolism Figure 6.16	*Text:* Review: 6.14 *Workbook:* Critical Thinking: 2
6.15	Explain the basis of muscle fatigue and soreness.	Fatigue: p. 216	*Concept maps:* Fatigue Figure 6.17	*Text:* Spot Check: 6 Review: 6.15 *Workbook:* Multiple Select: 6
6.16	Compare and contrast skeletal, cardiac, and smooth muscle tissue in terms of appearance, structure, type of nerve stimulation, type of respiration, and location.	Comparison of muscle tissues: pp. 216–217 Table 6.9	*Concept maps:* Types of muscle tissue Figure 6.18	*Text:* Review: 6.16 *Workbook:* Matching: 6–10

Learning Outcomes		Text Readings, Figures, and Tables	Workbook Activities	Text and Workbook Assessments
6.17	Explain the nutritional requirements of the muscular system.	Nutritional requirements of muscle tissue: pp. 217–218 Table 6.10		*Text:* Spot Check: 7 Review: 6.17
6.18	Summarize the five functions of the muscular system and give an example or explanation of each.	Functions of the muscular system: p. 219 Figure 6.31		*Text:* Review: 6.18 *Workbook:* Multiple Select: 3
6.19	Summarize the effects of aging on the muscular system.	Effects of aging on the muscular system: pp. 219–220		*Text:* Spot Check: 8 Review: 6.19 *Workbook:* Multiple Select: 7
6.20	Describe a common diagnostic test used to diagnose disorders of the muscular system.	Diagnostic test for muscular system disorders: p. 220 Table 6.11		*Text:* Spot Check: 9 Review: 6.20
6.21	Describe muscle disorders and relate abnormal function to pathology.	Muscular system disorders: pp. 220–224 Figures 6.32–6.34 Table 6.12		*Text:* Spot Check: 10, 11 Review: 6.21 *Workbook:* Multiple Select: 8 Case Study: 1–4

7 The Nervous System

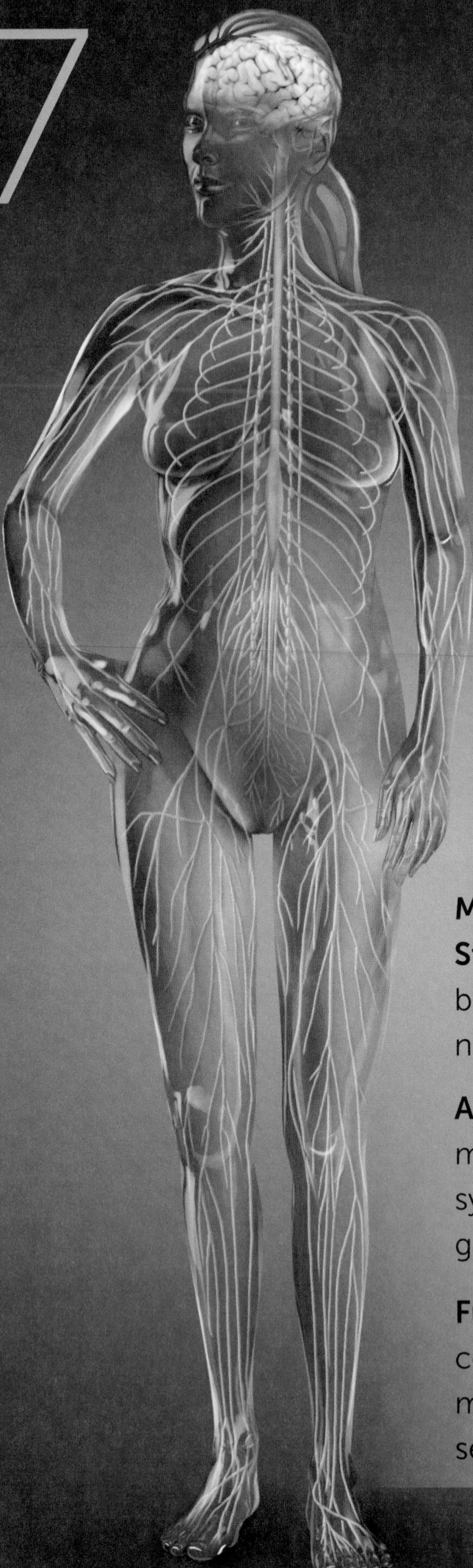

Major Organs and Structures:
brain, spinal cord, nerves

Accessory Structures:
meninges, sympathetic chain of ganglia

Functions:
communication, motor control, sensation

learning outcomes

This chapter of the workbook is designed to help you learn the anatomy and physiology of the nervous system. After completing this chapter in the text and this workbook, you should be able to:

7.1 Use medical terminology related to the nervous system.

7.2 Describe the organization of the nervous system in regard to structure and function.

7.3 Describe the anatomy of a neuron.

7.4 Differentiate multipolar, bipolar, and unipolar neurons in terms of anatomy, location, and direction of nerve impulses.

7.5 Describe neuroglial cells and state their function.

7.6 Describe the meninges covering the brain and spinal cord.

7.7 Explain the importance of cerebrospinal fluid, including its production, circulation, and function.

7.8 Describe the major landmarks and subdivisions of the brain and state their functions.

7.9 Describe the spinal cord.

7.10 Describe the anatomy of a nerve and its connective tissues.

7.11 List the cranial nerves in order, stating their function and whether they are sensory, motor, or both.

7.12 Describe the attachment of nerves to the spinal cord.

7.13 Compare the parasympathetic and sympathetic divisions of the autonomic nervous system in terms of anatomy and function.

7.14 Describe a resting membrane potential.

7.15 Compare and contrast a local potential and an action potential.

7.16 Describe a specific reflex and list the components of its reflex arc.

7.17 Explain the difference between short-term and long-term memory.

7.18 Differentiate between Broca's area and Wernicke's area in regard to their location and function in speech.

7.19 Explain the function of the nervous system by writing a pathway for a sensory message sent to the brain to be processed for a motor response.

7.20 Explain the nutritional requirements of the nervous system.

7.21 Explain the effects of aging on the nervous system.

7.22 Describe common diagnostic tests used to diagnose disorders of the nervous system.

7.23 Describe nervous system disorders and relate abnormal function to the pathology.

word roots & combining forms

cephal/o: head

cerebell/o: cerebellum

cerebr/o: cerebrum

dur/o: tough

encephal/o: brain

gangli/o: ganglion

gli/o: glue

medull/o: medulla

mening/o: meninges

myel/o: spinal cord

neur/o: nerve

poli/o: gray matter

Neuron

Figure 7.1 shows the anatomy of a myelinated multipolar neuron. Color the box next to each term. Use the same color for the corresponding structures in the figure.

- ☐ **Dendrites**(A)
- ☐ **Cell body**(B)
- ☐ **Axon**(C)
- ☐ **Myelin sheath**(D)
- ☐ **Terminal arborization**(E)
- ☐ **Synaptic knobs**(F)
- ☐ **Trigger zone**(G)

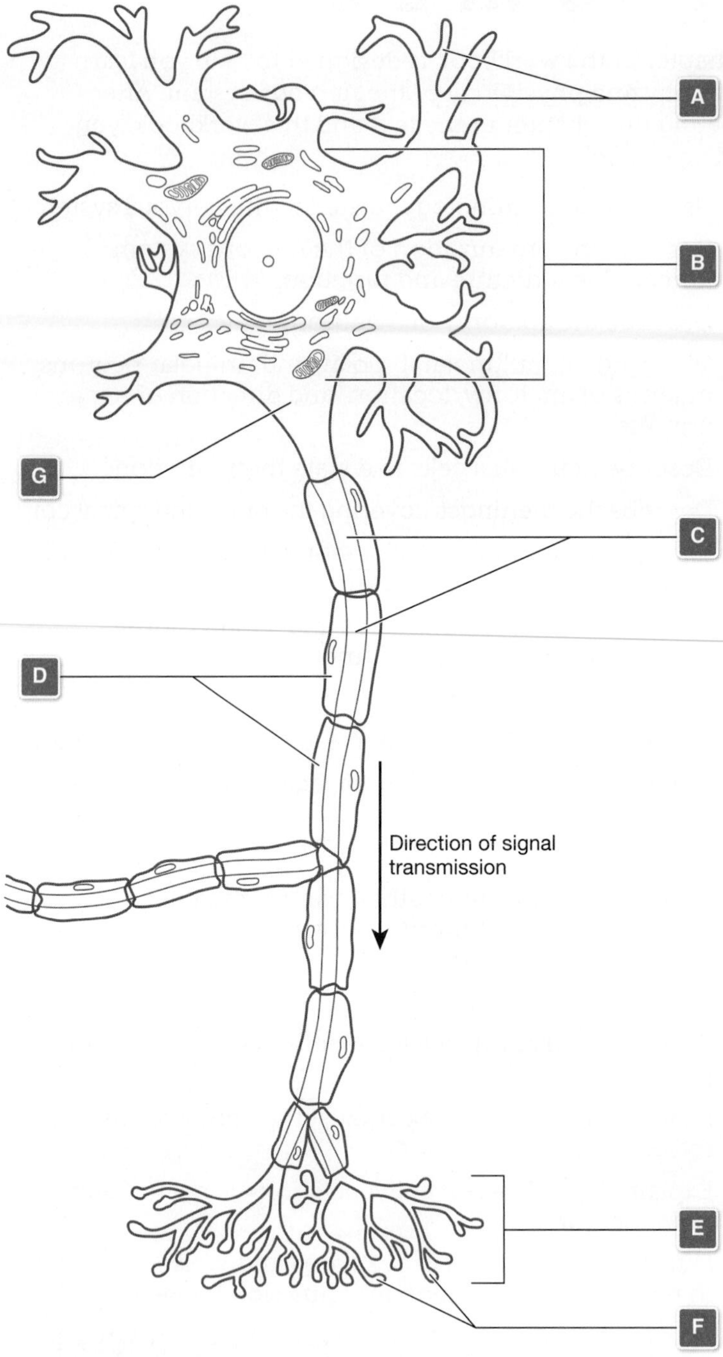
A
B
G
C
D
Direction of signal transmission
E
F

FIGURE 7.1 Myelinated multipolar neuron.

COLORING BOOK

Brain

Figures 7.2 and 7.3 show the anatomy of a brain. Color the box next to each term. Use the same color for the corresponding structures in the figures.

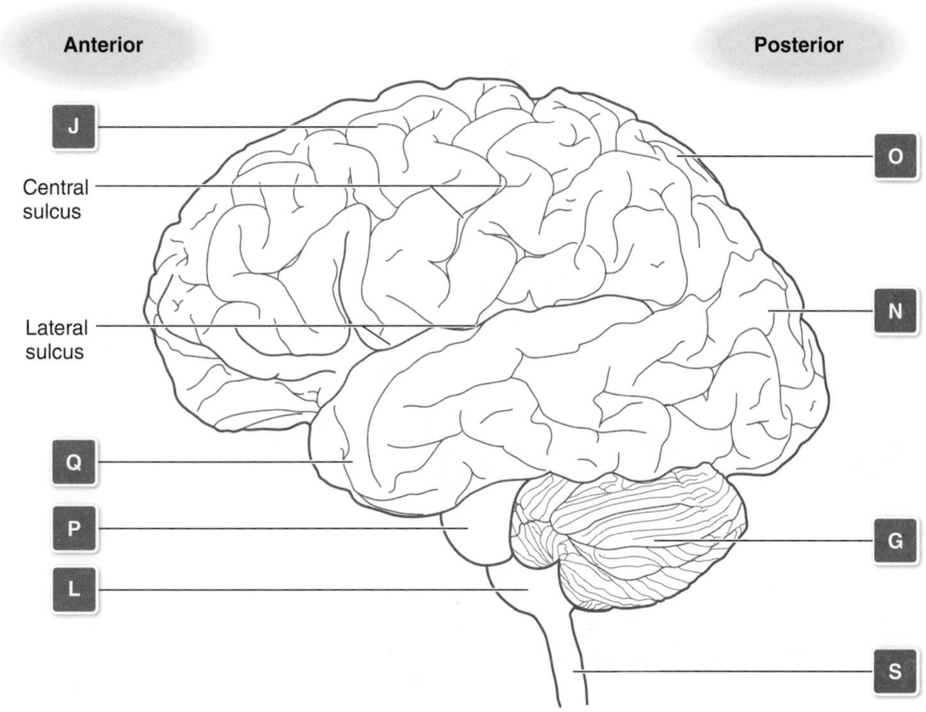

FIGURE 7.2 The brain, lateral view.

- ☐ **Cerebellum**(G)
- ☐ **Colliculi**(H)
- ☐ **Corpus callosum**(I)
- ☐ **Frontal lobe**(J)
- ☐ **Hypothalamus**(K)
- ☐ **Medulla oblongata**(L)
- ☐ **Midbrain**(M)
- ☐ **Occipital lobe**(N)
- ☐ **Parietal lobe**(O)
- ☐ **Pons**(P)
- ☐ **Temporal lobe**(Q)
- ☐ **Thalamus**(R)
- ☐ **Spinal cord**(S)

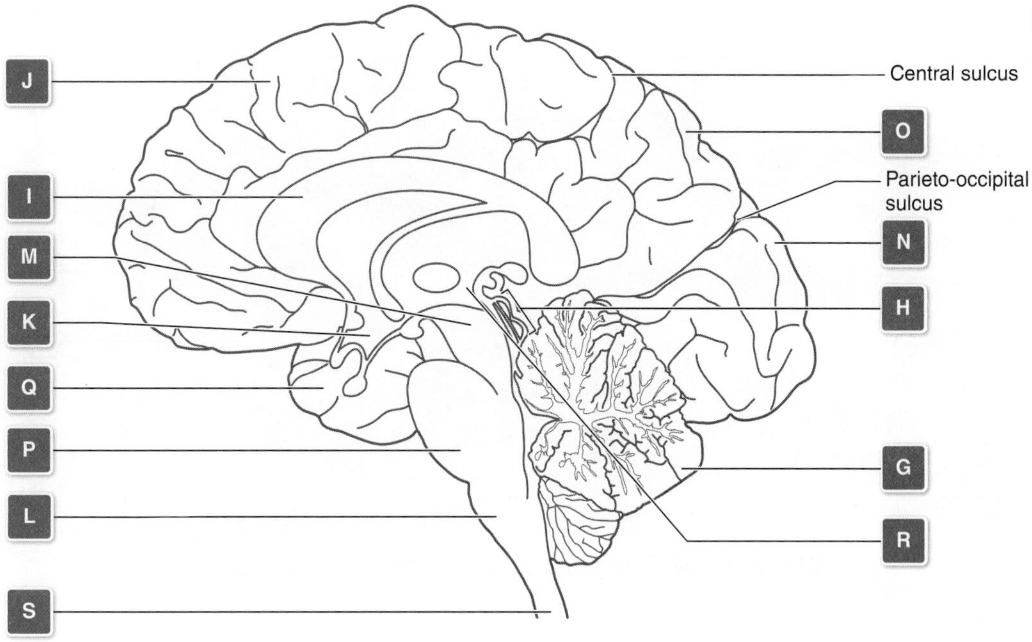

FIGURE 7.3 The brain, midsagittal view.

COLORING BOOK

Spinal Cord and Spinal Nerves

Figure 7.4 shows the anatomy of the spinal cord and the spinal nerves. Color the box next to each term. Use the same color for the corresponding structures in the figure.

- ☐ **Spinal cord** (S)
- ☐ **Cauda equina** (T)
- ☐ **Cervical spinal nerves** (U)
- ☐ **Thoracic spinal nerves** (V)
- ☐ **Lumbar spinal nerves** (W)
- ☐ **Sacral spinal nerves** (X)
- ☐ **Coccygeal spinal nerves** (Y)

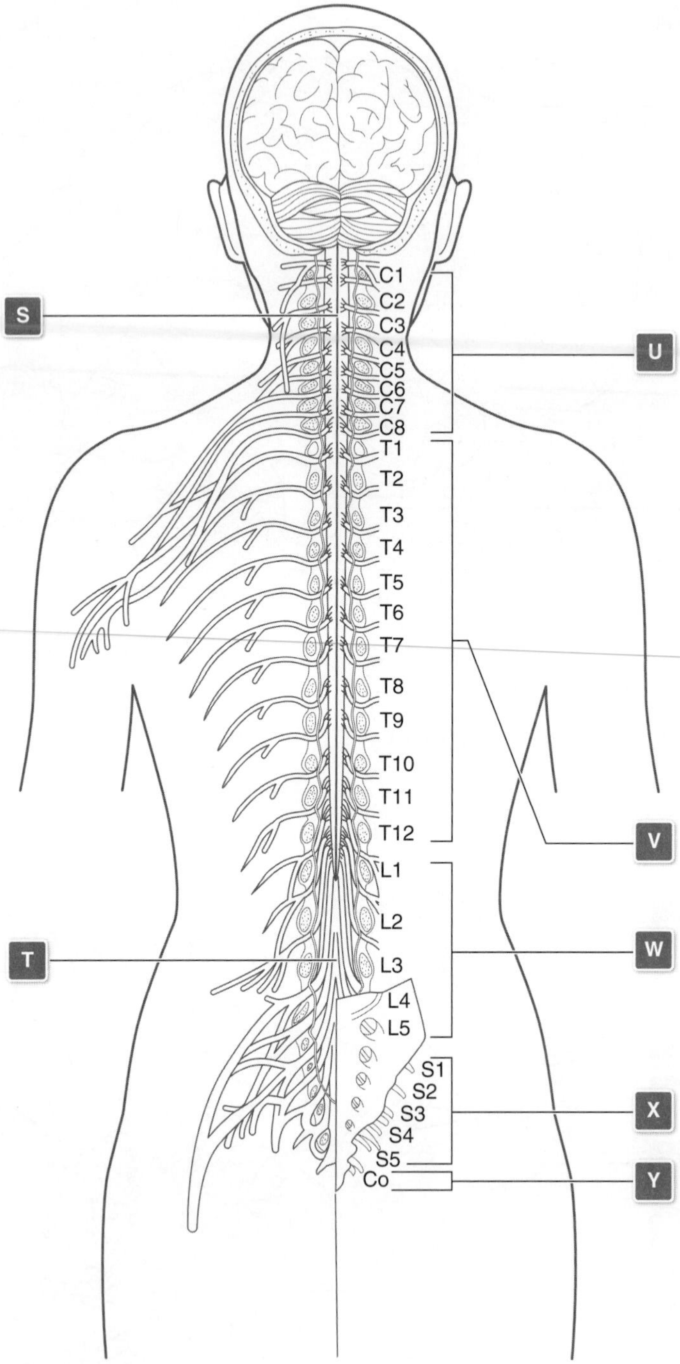

FIGURE 7.4 The spinal cord and spinal nerves.

COLORING BOOK

Spinal Cord and Meninges

Figure 7.5 shows the anatomy of the spinal cord with the meninges. Color the box next to each term. Use the same color for the corresponding structures in the figure.

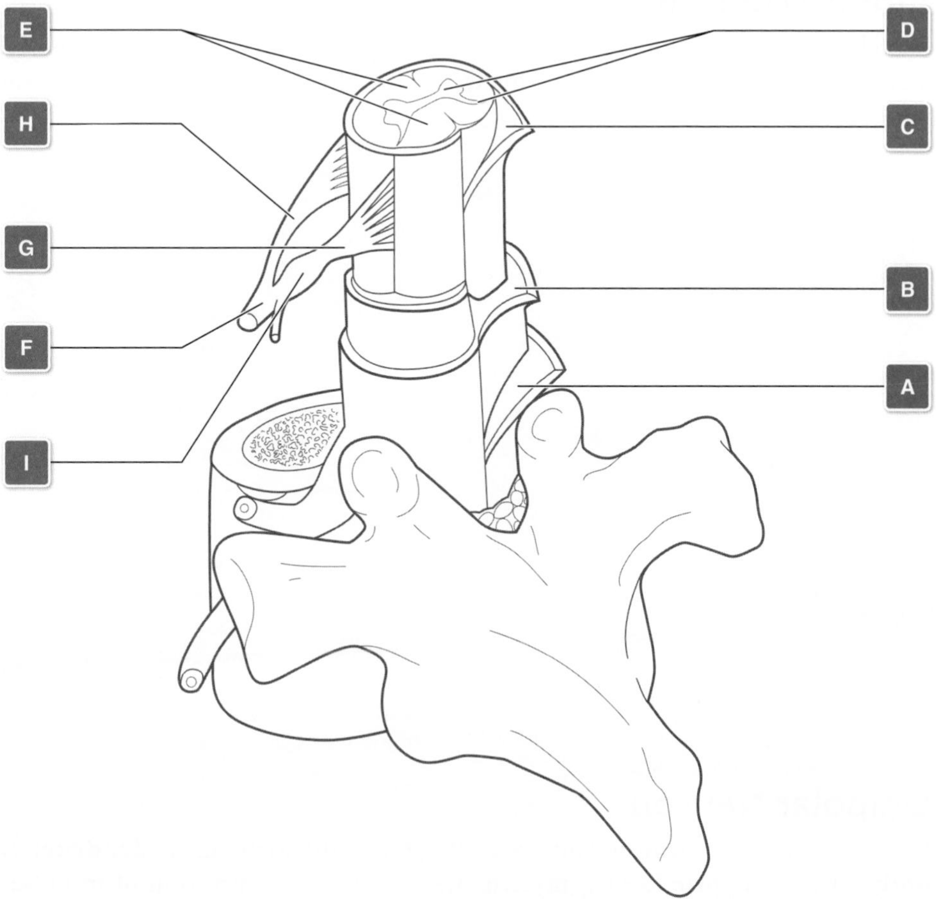

- ☐ **Dura mater(A)**
- ☐ **Arachnoid mater(B)**
- ☐ **Pia mater(C)**
- ☐ **Horns(D)**
- ☐ **Columns(E)**
- ☐ **Spinal nerve(F)**
- ☐ **Dorsal nerve root(G)**
- ☐ **Ventral nerve root(H)**
- ☐ **Dorsal root ganglion(I)**

FIGURE 7.5 Cross section of the spinal cord with meninges.

COLORING BOOK

Types of Neurons

Figure 7.1 in the Coloring Book section shows a myelinated multipolar neuron. In this activity you are to sketch the other two types of neurons, with labels, in the space provided.

Bipolar Neuron

Sketch a bipolar neuron and label the following structures: **dendrite(s), body, axon, synaptic knobs, myelin sheath.** Show the direction of impulses on your sketch.

Unipolar Neuron

Sketch a unipolar neuron and label the following structures: **dendrite(s), body, axon, synaptic knobs, myelin sheath.** Show the direction of impulses on your sketch.

Nerve Model

In the muscular system chapter you created a muscle model using spaghetti and three different types of wraps to show the connective tissues. The connective tissues extended past the ends of the muscle fibers to represent a tendon. Nerves have similar structure (see text Outcome 7.10, "Anatomy of a Nerve").

In this activity you are to create a nerve model using household supplies. You may be as creative as you like. Be able to explain how the structures are represented in your model.

Your model is to include:	This is represented in my model by:
• Two multipolar neurons	
• Two unipolar neurons	
• Myelin sheath on all neurons	
• Two fascicles	
• Endoneurium	
• Perineurium	
• Epineurium	
• Two nerve roots	
• One ganglion	

Reflexes

Reflexes are predictable, involuntary motor responses to a stimulus. They happen in reflex arcs involving a receptor, an afferent neuron, an integration center, an efferent neuron, and an effector. For this exercise you are to perform the given reflex and then respond to the following questions.

Pupillary Light Reflex

You will need a partner, a room with dim light, and a flashlight. Turn on the flashlight and hold it off to the side. Have your partner stare directly at you. Bring the light from the side and quickly shine it in your partner's right eye (only for a second). You should observe what happens to the size of the pupils (black opening in the center of the eye) in both eyes.

1. What happened to the size of the pupil in the right eye?

____

2. What happened to the size of the pupil in the left eye?

____

This is a good example of a reflex in the parasympathetic division. It involves several structures. You are to put the structures in order and label them according to what part of a reflex arc they represent. Some parts of the reflex arc may be used more than once. One answer is provided.

Order	Structure	Part of reflex arc
______	Brain	______
______	Dendrite of a bipolar neuron in the eye	______
______	Preganglionic neuron	______
______	Bipolar neuron in the optic nerve	______
______	Postganglionic neuron	______
______	Iris of the eye (smooth muscle)	Effector

3. Since the effector in this reflex is smooth muscle, is this a somatomotor or autonomic reflex?

____

4. What do you think happens in the brain to have the effect you described for both pupils?

____

____

____

Pathway

Pathways demonstrate the complexity and efficiency of communication in the nervous system. Complete the following pathway for a blind person reading Braille (raised dots on a page that represent letters) aloud. Start with sensing the raised dot on the paper and end with skeletal messages being sent out from the brain.

Unipolar neuron → ______________ → medulla oblongata → ______________ → ______________ → thalamus → ______________ lobe (general sensory area → ______________ area) → temporal lobe (______________ area) → frontal lobe (______________ area → premotor area → ______________ area).

LABORATORY EXERCISES AND ACTIVITIES

Key Words

The following terms are defined in the glossary of the textbook.

action potential
afferent
autonomic
axon
bipolar
cerebrovascular accident (CVA)
cerebrospinal fluid (CSF)
decremental
dementia
dendrite
depolarize
efferent
hydrocephalus
multipolar
myelin
neuroglia
paralysis
parasympathetic
reflex
repolarize
resting membrane potential
sympathetic
synapse
unipolar

Concept Maps

Use key words and other bold words from the chapter to complete the following concept maps (Figures 7.6 to 7.12).

Neuroglia

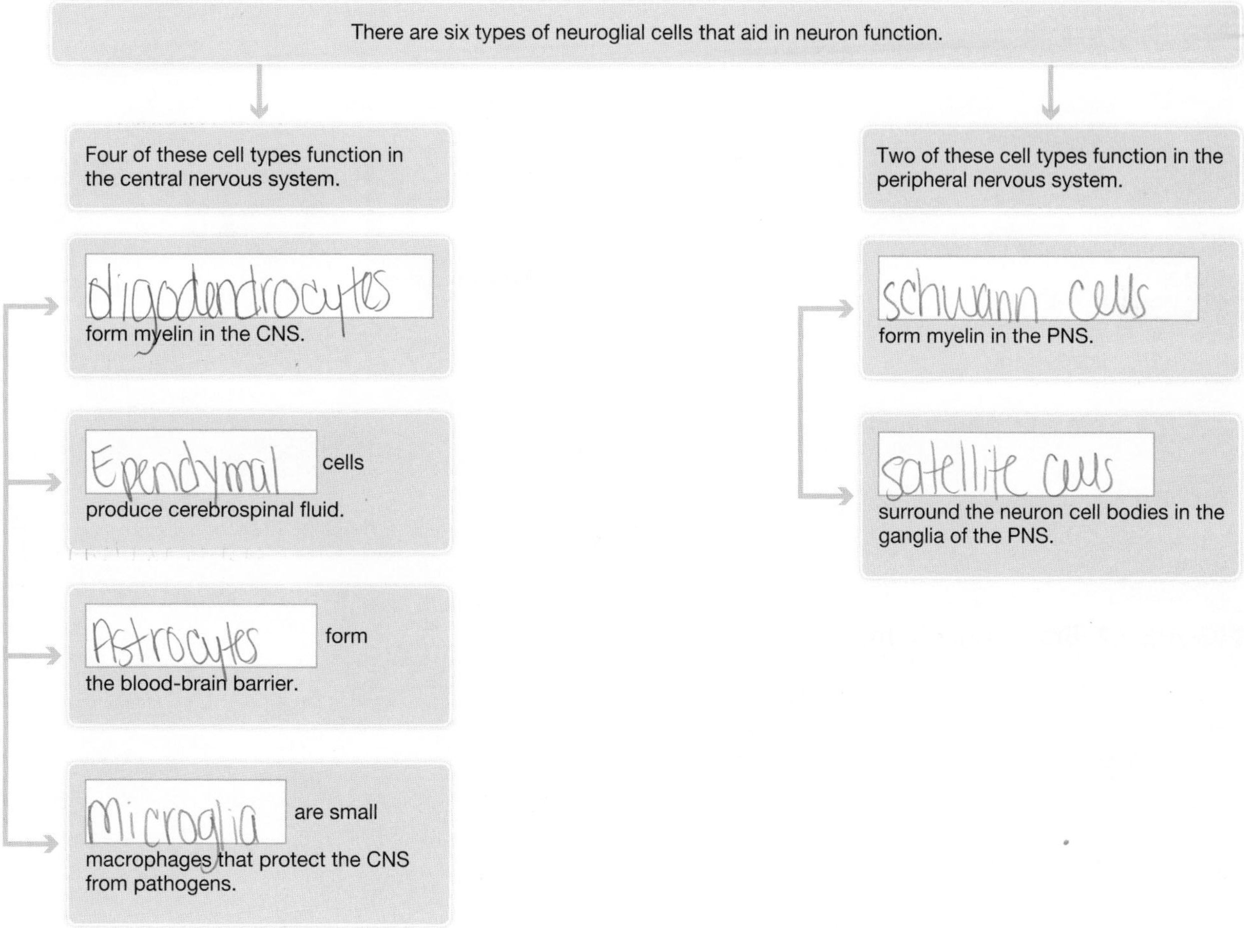

There are six types of neuroglial cells that aid in neuron function.
Four of these cell types function in the central nervous system.
Two of these cell types function in the peripheral nervous system.
oligodendrocytes
form myelin in the CNS.
Ependymal cells
produce cerebrospinal fluid.
Astrocytes form
the blood-brain barrier.
Microglia are small
macrophages that protect the CNS from pathogens.
schwann cells
form myelin in the PNS.
satellite cells
surround the neuron cell bodies in the ganglia of the PNS.

FIGURE 7.6 Neuroglia concept map.

Brain

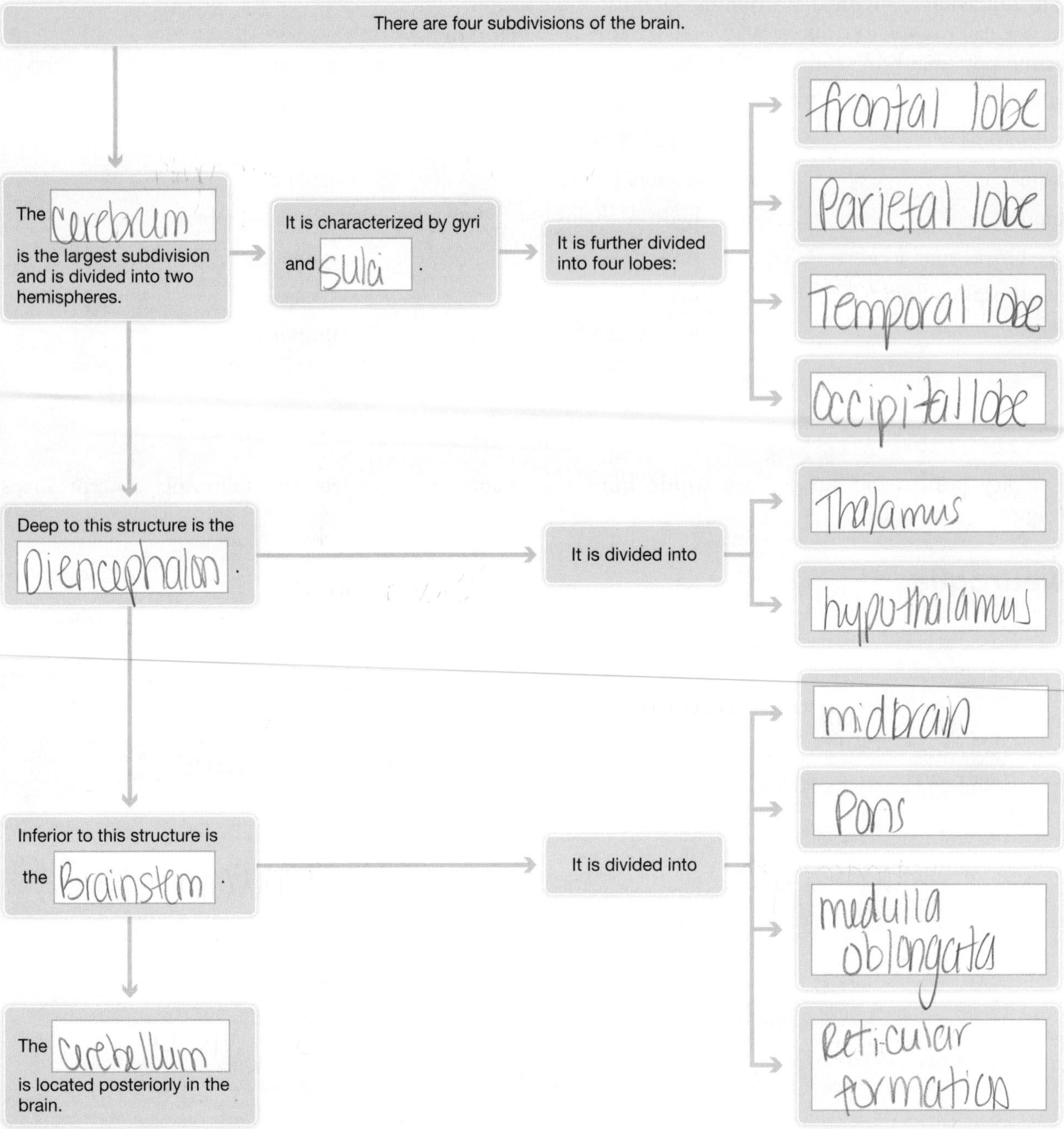
There are four subdivisions of the brain.
The Cerebrum is the largest subdivision and is divided into two hemispheres.
It is characterized by gyri and Sulci.
It is further divided into four lobes:
frontal lobe
Parietal lobe
Temporal lobe
Occipital lobe
Deep to this structure is the Diencephalon.
It is divided into
Thalamus
hypothalamus
Inferior to this structure is the Brainstem.
It is divided into
midbrain
Pons
medulla oblongata
Reticular formation
The Cerebellum is located posteriorly in the brain.

FIGURE 7.7 Brain concept map.

Meninges

The brain and spinal cord are covered with three membranes.

- The dura mater is the most superficial layer.
- The arachnoid mater is the middle layer. It is very delicate and characterized by a weblike appearance.
- The pia mater is tight to the brain and spinal cord.

FIGURE 7.8 Meninges concept map.

Nerves

There are two categories of nerves in the PNS.

- Cranial nerves
  - Twelve pairs are attached directly to the brain.
  - They are classified as sensory, motor, or both.
- Spinal nerves
  - 31 pairs are attached directly to the spinal cord.
  - They are all sensory and motor neurons that split into a(n)
    - dorsal root, which is afferent.
    - ventral root, which is efferent.

FIGURE 7.9 Nerves concept map.

KEY WORD CONCEPT MAPS

Resting Membrane Potential

Resting membrane potential is achieved when the outside of the neuron has a(n) positive charge (due to high concentrations of Na^+) and the inside of the neuron has a(n) negative charge.

Polarization is the difference in these charges on either side of the neuron's cellular membrane.

↓

repolarizing will occur when the charges of the membrane change due to the opening of Na^+ channels allowing Na^+ to flow into the cell by facilitated diffusion.

↓

The sodium-potassium pump will restore a(n) resting membrane potential by pumping the Na^+ out of the cell.

↓

The result is nerve impulse.

FIGURE 7.10 Resting membrane potential concept map.

KEY WORD CONCEPT MAPS

Local and Action Potentials

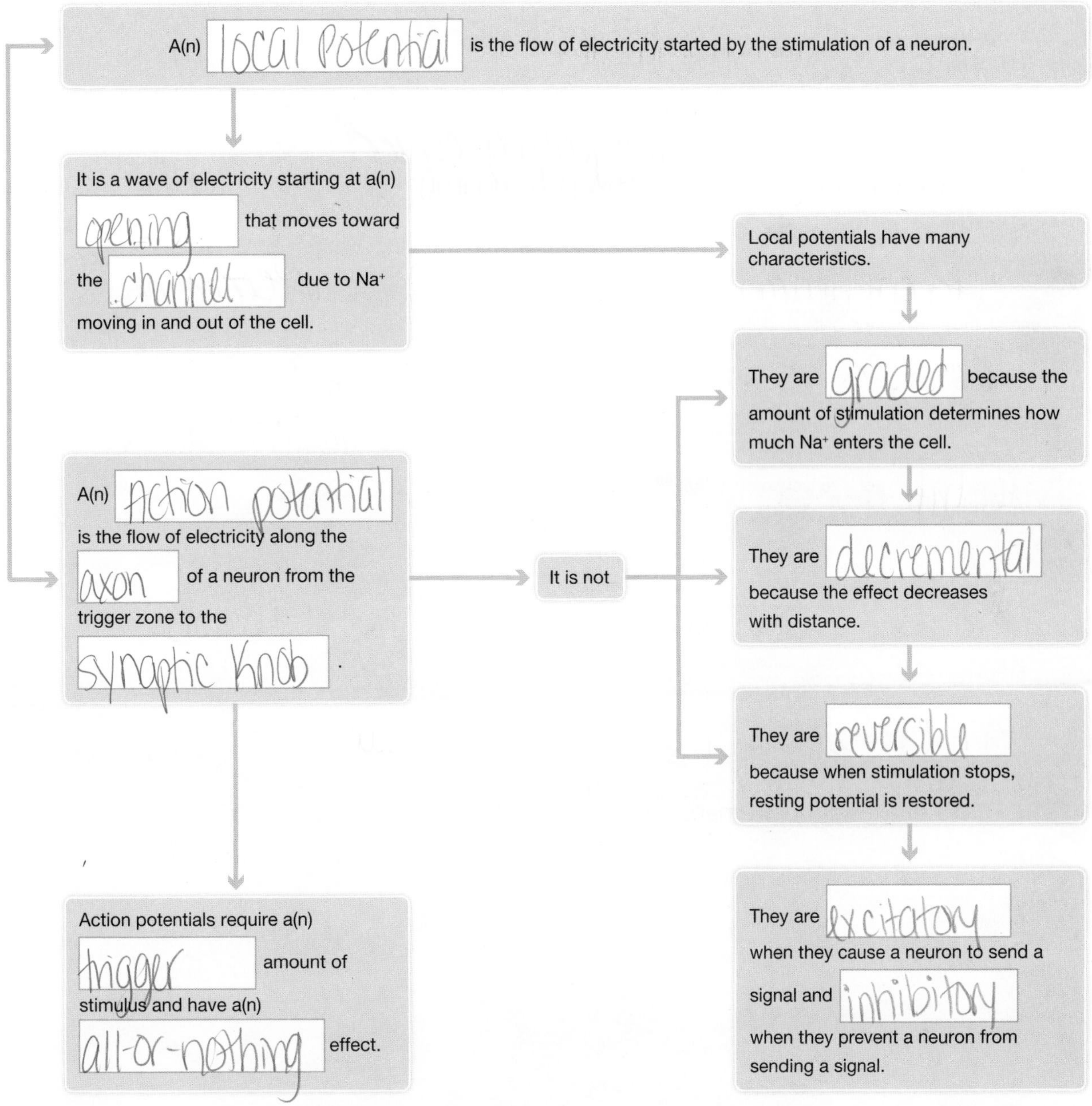
A(n) local Potential is the flow of electricity started by the stimulation of a neuron.
It is a wave of electricity starting at a(n) opening that moves toward the channel due to Na+ moving in and out of the cell.
Local potentials have many characteristics.
They are graded because the amount of stimulation determines how much Na+ enters the cell.
A(n) Action potential is the flow of electricity along the axon of a neuron from the trigger zone to the synaptic knob.
It is not
They are decremental because the effect decreases with distance.
They are reversible because when stimulation stops, resting potential is restored.
Action potentials require a(n) trigger amount of stimulus and have a(n) all-or-nothing effect.
They are excitatory when they cause a neuron to send a signal and inhibitory when they prevent a neuron from sending a signal.

FIGURE 7.11 Local and action potentials concept map.

Reflexes

Reflexes are involuntary, predictable motor responses to stimulus without conscious thought.

1. They begin with a receptor on the dentrite of a sensory neuron that receives the stimulus.

2. The afferent neuron carries the signal to the CNS.

3. The integrating center receives the signal and conducts it to the

4. motor or efferent neuron, which carries the signal away from the CNS to

5. the effector, the structure causing the effect.

If it is skeletal muscle, it is considered a(n) somatic reflex.

If it is smooth muscle or a gland, it is considered a(n) autonomic reflex.

FIGURE 7.12 Reflexes concept map.

Word Deconstruction: *In the textbook, you built words to fit a definition using combining forms, prefixes, and suffixes. Here you are to break down the term into its parts (prefixes, roots, and suffixes) and give a definition. Prefixes and suffixes can be found inside the back cover of the textbook.*

FOR EXAMPLE Dermatitis: **dermat/itis—inflammation of the skin**

1. Poliomyelitis: ____________________
2. Gangliectomy: ____________________
3. Neurodynia: ____________________
4. Encephalitis: ____________________
5. Cephalocele: ____________________

Multiple Select: *Select the correct choices for each statement. The choices may be all correct, all incorrect, or any combination of correct and incorrect.*

1. How is the nervous system organized?
   - **a.** The nervous system is divided into the central nervous system and the autonomic nervous system.
   - **b.** The cerebrum is composed of three lobes.
   - **c.** The hypothalamus is part of the diencephalon.
   - **d.** The peripheral nervous system is composed of afferent and efferent neurons.
   - **e.** The autonomic division is composed of afferent neurons only.
2. How are neurons classified?
   - **a.** Bipolar neurons are sensory.
   - **b.** Multipolar neurons are sensory.
   - **c.** Unipolar neurons are efferent.
   - **d.** Unipolar neurons are motor.
   - **e.** Bipolar neurons are afferent.
3. What is (are) the function(s) of neuroglial cells?
   - **a.** Astrocytes fight pathogens.
   - **b.** Ependymal cells prevent medications from reaching the brain.
   - **c.** Schwann cells form myelin in the PNS.
   - **d.** Satellite cells regulate the composition of the CSF.
   - **e.** Microglia regulate the environment of ganglia in the PNS.
4. Agnes is suspected of having meningitis. Her physician performed a lumbar puncture. Why is this a good idea?
   - **a.** Cerebrospinal fluid may contain the pathogen.
   - **b.** The lumbar area contains an enlargement of the cord, so it will be easier to hit.
   - **c.** The cauda equina is located in the lumbar region.
   - **d.** Cerebrospinal fluid can be found in the subdural space.
   - **e.** Cerebrospinal fluid circulates over the entire brain and spinal cord, so it will likely pick up a pathogen if it is in the CNS.
5. How is the anatomy of a nerve organized?
   - **a.** Epineurium surrounds a neuron.
   - **b.** Neuron axons are bundled in fascicles.
   - **c.** Endoneurium surrounds a fascicle.
   - **d.** Perineurium surrounds a nerve.
   - **e.** Epineurium surrounds a nerve.
6. How does the sympathetic division compare to the parasympathetic division?
   - **a.** The preganglionic neuron is longer in the sympathetic division than in the parasympathetic division.

b. The postganglionic neuron is shorter in the parasympathetic division than in the sympathetic division.
c. They both have neurons coming off the cord in the same place.
d. They both are part of the autonomic nervous system.
e. They both involve afferent neurons.

7. How do Broca's area and Wernicke's area compare?
   a. They are both in the frontal lobe.
   b. They are both in the temporal lobe.
   c. They are both in the cerebrum.
   d. Problems in either area are called *aphasia.*
   e. They both function for language.

8. What can you expect to typically see concerning the nervous system in an 80-year-old?
   a. Short-term memory is diminished.
   b. Reaction times are slow.
   c. Verbal skills are diminished.
   d. Long-term memory is absent.
   e. The same effects as those in a 70-year-old.

9. What happens at the spinal cord?
   a. Bipolar neurons enter the cord through the dorsal root.
   b. Preganglionic neurons exit the cord through the ventral root.
   c. Postganglionic neurons have no connection to the cord.
   d. Unipolar neuron cell bodies are found in the gray matter of the spinal cord.
   e. Multipolar neuron cell bodies are found in the dorsal root ganglia.

10. What happens in Alzheimer's disease?
    a. Insoluble proteins called *tangles* form in neurons in the brain.
    b. Groups of dead cells called *plaques* form in the brain.
    c. The onset of symptoms may be years after the disease process began.
    d. There is a loss of cognitive function, called *dementia.*
    e. A definite diagnosis can be made only after death.

Matching: *For each of the following cranial nerves, pick one answer from each column to match the function, the name, and the type (whether it is sensory, motor, or both). For example (using the choices below), the answers for CN I would be c, i, o.*

		Function	Name	Type
______	1. CN VII	a. Eye movement	h. Oculomotor	o. Sensory
______	2. CN III	b. Vision	i. Olfactory	p. Motor
______	3. CN IX	c. Smell	j. Facial	q. Both
______	4. CN V	d. Chewing	k. Optic	
______	5. CN VIII	e. Hearing and equilibrium	l. Trigeminal	
		f. Taste and swallowing	m. Glossopharyngeal	
		g. Taste and facial expression	n. Auditory	

Matching: *Match the part of the brainstem to the function. Some answers may be used more than once.*

______	6. Is responsible for sleep-wake cycle	a. Medulla oblongata
______	7. Regulates heart rate	b. Midbrain

_______ **8.** Serves as a bridge to the cerebellum

_______ **9.** Has colliculi for vision and hearing

_______ **10.** Has pyramids where motor messages cross

c. Pons

d. Reticular formation

Completion: *Fill in the blanks to complete the following statements.*

1. In a resting membrane potential, _______________ ions are on the inside of the cell and _______________ ions are on the outside of the cell.
2. When sodium rushes inside the cell, the membrane becomes _______________.
3. The _______________ restores a resting membrane potential.
4. Acetylcholine is carried down an axon to the _______________, where it is released.
5. _______________ is needed in the diet of children for proper myelination of developing neurons.

Critical Thinking

1. A woman enters the room. You are standing next to your friend. You comment on the woman's perfume. Your friend says, "What perfume? I don't smell anything." Explain how one person can smell the perfume and the other person cannot in terms of local and action potentials.

2. Considering what you have learned about how memory works at the cellular and molecular levels, what would be the best method of studying for your A&P final exam? Explain.

3. Why might a surgeon hesitate before performing surgery on a patient who has recently come out of a coma? Explain in terms of anatomy and physiology.

CHAPTER 7 REVIEW QUESTIONS

CHAPTER 7 REVIEW QUESTIONS

Case Study

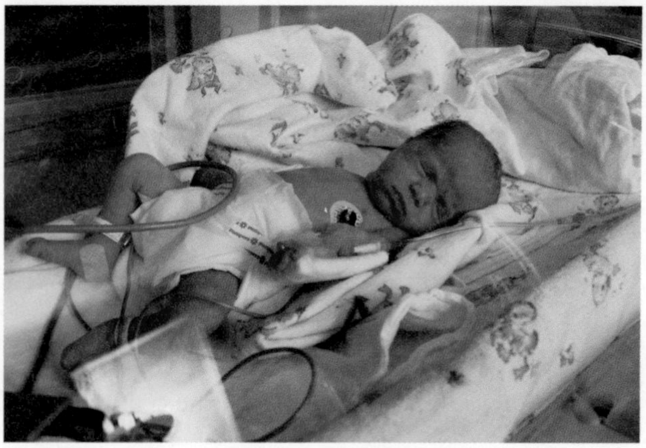

Bill and Nora just gave birth to a baby girl. During the birth, the baby was deprived of oxygen for an extended period of time due to various complications. As a result, physicians are concerned about the baby's health status and will continue to monitor her for any evidence of developmental problems associated with the difficult birth.

1. Could Bill and Nora's newborn daughter be diagnosed with cerebral palsy? Explain your answer.

2. What other conditions (covered in this chapter) could accompany that diagnosis?

3. If the baby is diagnosed with cerebral palsy, will the baby's symptoms progress as she develops?

This section of the chapter is designed to help you find where each outcome is covered in the textbook and workbook.

Learning Outcomes		Text Readings, Figures, and Tables	Workbook Activities	Text and Workbook Assessments
7.1	Use medical terminology related to the nervous system.	Word roots & combining forms, pronunciation key: p. 231	Word roots & combining forms	*Text:* Review: 7.1 *Workbook:* Word Deconstruction: 1–5
7.2	Describe the organization of the nervous system in regard to structure and function.	Overview: pp. 232–233 Figure 7.2		*Text:* Review: 7.2 *Workbook:* Multiple Select: 1
7.3	Describe the anatomy of a neuron.	Anatomy of a neuron: pp. 233–235 Figure 7.3	*Coloring book:* Neuron Figure 7.1	*Text:* Spot Check: 6 Review: 7.3 *Workbook:* Completion: 4
7.4	Differentiate multipolar, bipolar, and unipolar neurons in terms of anatomy, location, and direction of nerve impulses.	Types of neurons: pp. 235–236 Table 7.1	*Lab exercises and activities:* Types of neurons Figure 7.1	*Text:* Spot Check: 1 Review: 7.4 *Workbook:* Multiple Select: 2
7.5	Describe neuroglial cells and state their function.	Neuroglia: pp. 236–239 Figures 7.4–7.5 Table 7.2	*Concept maps:* Neuroglia Figure 7.6	*Text:* Spot Check: 2 Review: 7.5 *Workbook:* Multiple Select: 3
7.6	Describe the meninges covering the brain and spinal cord.	Meninges: pp. 239–240 Figure 7.6	*Coloring book:* Spinal cord and meninges Figure 7.5 *Concept maps:* Meninges Figure 7.8	*Text:* Review: 7.6
7.7	Explain the importance of cerebrospinal fluid, including its production, circulation, and function.	Cerebrospinal fluid: pp. 240–242 Figure 7.7		*Text:* Spot Check: 3 Review: 7.7 *Workbook:* Multiple Select: 4
7.8	Describe the major landmarks and subdivisions of the brain and state their functions.	Brain: pp. 242–250 Figures 7.8–7.12 Table 7.3	*Coloring book:* Brain Figures 7.2–7.3 *Concept maps:* Brain Figure 7.7	*Text:* Review: 7.8 *Workbook:* Matching: 6–10 Critical Thinking: 3

CHAPTER 7 MAPPING

Learning Outcomes		Text Readings, Figures, and Tables	Workbook Activities	Text and Workbook Assessments
7.9	Describe the spinal cord.	Spinal cord: pp. 250–253 Figures 7.13–7.15	*Coloring book:* Spinal cord and spinal nerves; Spinal cord and meninges Figures 7.4–7.5	*Text:* Review: 7.9
7.10	Describe the anatomy of a nerve and its connective tissues.	Anatomy of a nerve: pp. 253–254 Figure 7.16	*Lab exercises and activities:* Nerve model	*Text:* Review: 7.10 *Workbook:* Multiple Select: 5
7.11	List the cranial nerves in order, stating their function and whether they are sensory, motor, or both.	Cranial nerves: pp. 254–255 Figure 7.17 Table 7.4		*Text:* Spot Check: 4–5, 9 Review: 7.11 *Workbook:* Matching: 1–5
7.12	Describe the attachment of nerves to the spinal cord.	Spinal nerves: pp. 255–257 Figures 7.18–7.19	*Coloring book:* Spinal cord and meninges Figure 7.5 *Concept maps:* Nerves Figure 7.9	*Text:* Review: 7.12 *Workbook:* Multiple Select: 9
7.13	Compare the parasympathetic and sympathetic divisions of the autonomic nervous system in terms of anatomy and function.	Autonomic nervous system: pp. 257–260 Figures 7.20–7.21		*Text:* Spot Check: 6, 7 Review: 7.13 *Workbook:* Multiple Select: 6
7.14	Describe a resting membrane potential.	Nerve impulses: pp. 260–261 Figure 7.22	*Concept maps:* Resting membrane potential Figure 7.10	*Text:* Review: 7.14 *Workbook:* Completion: 1–3
7.15	Compare and contrast a local potential and an action potential.	Local potential: pp. 261–262 Figure 7.23 Action potential: pp. 262–264 Figure 7.24	*Concept maps:* Local and action potentials Figure 7.11	*Text:* Spot Check: 8 Review: 7.15 *Workbook:* Critical Thinking: 1
7.16	Describe a specific reflex and list the components of its reflex arc.	Reflexes: pp. 264–265 Figure 7.25	*Lab exercises and activities:* Reflexes *Concept maps:* Reflexes Figure 7.12	*Text:* Spot Check: 9 Review: 7.16

Learning Outcomes		Text Readings, Figures, and Tables	Workbook Activities	Text and Workbook Assessments
7.17	Explain the difference between short-term and long-term memory.	Memory: p. 266		*Text:* Spot Check: 10 Review: 7.17 *Workbook:* Critical Thinking: 2
7.18	Differentiate between Broca's area and Wernicke's area in regard to their location and function in speech.	Language: p. 266		*Text:* Review: 7.18 *Workbook:* Multiple Select: 7
7.19	Explain the function of the nervous system by writing a pathway for a sensory message sent to the brain to be processed for a motor response.	Function of the nervous system: p. 267 Figure 7.26	*Lab exercises and activities:* Pathway	*Text:* Spot Check: 11 Review: 7.19
7.20	Explain the nutritional requirements of the nervous system.	Nutritional requirements of the nervous system: p. 268		*Text:* Review: 7.20 *Workbook:* Completion: 5
7.21	Explain the effects of aging on the nervous system.	Effects of aging on the nervous system: p. 268		*Text:* Review: 7.21 *Workbook:* Multiple Select: 8
7.22	Describe common diagnostic tests used to diagnose disorders of the nervous system.	Diagnostic tests for nervous system disorders: p. 268 Table 7.5		*Text:* Spot Check: 12 Review: 7.22
7.23	Describe nervous system disorders and relate abnormal function to the pathology.	Nervous system disorders: pp. 268–272 Figures 7.27–7.28 Table 7.6		*Text:* Spot Check: 13 Review: 7.23 *Workbook:* Multiple Select: 10 Case Study: 1–3

8 The Nervous System—Senses

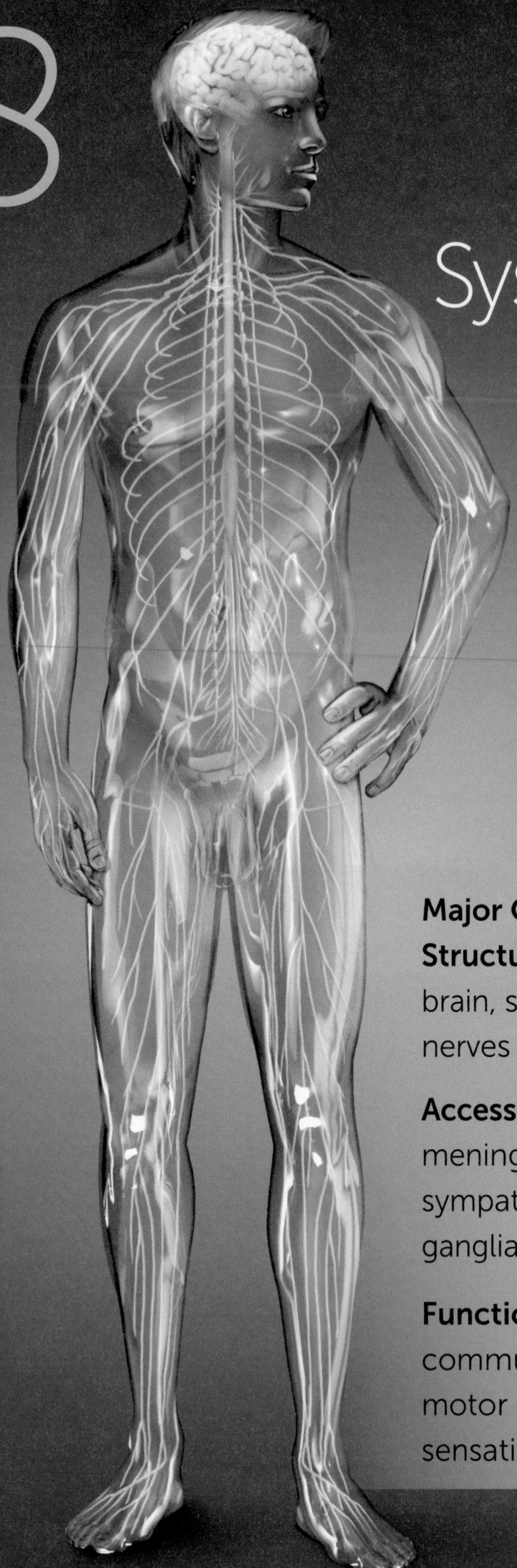

Major Organs and Structures:
brain, spinal cord, nerves

Accessory Structures:
meninges, sympathetic chain of ganglia

Functions:
communication, motor control, sensation

learning outcomes

This chapter of the workbook is designed to help you learn the anatomy and physiology of the senses of the nervous system. After completing this chapter in the text and this workbook, you should be able to:

- **8.1** Use medical terminology related to the senses of the nervous system.
- **8.2** Classify the senses in terms of what is sensed and where the receptors are located.
- **8.3** Describe the sensory receptors for the general senses in the skin.
- **8.4** Explain the types of information transmitted by sensory receptors in the skin.
- **8.5** Describe the pathway for pain.
- **8.6** Describe the sensory receptors for taste.
- **8.7** Describe the different tastes and explain how flavor is perceived.
- **8.8** Describe the pathway for taste.
- **8.9** Describe the sensory receptors for smell.
- **8.10** Explain how odors are perceived.
- **8.11** Describe the pathway for smell.
- **8.12** Describe the anatomy of the ear.
- **8.13** Explain how sound is perceived.
- **8.14** Describe the pathway for hearing.
- **8.15** Describe the anatomy of the vestibular apparatus.
- **8.16** Explain how equilibrium is perceived.
- **8.17** Describe the pathway for equilibrium.
- **8.18** Describe the anatomy of the eye.
- **8.19** Explain how vision is perceived.
- **8.20** Describe the pathway for vision.
- **8.21** Describe the effects of aging on the senses.
- **8.22** Describe common diagnostic tests used to diagnose disorders of the senses.
- **8.23** Describe disorders of the senses and relate abnormal function to the pathology.

word roots & combining forms

audi/o: hearing

aur/o: ear

cochle/o: cochlea

corne/o: cornea

lacrim/o: tears

lith/o: stone

ocul/o: eye

ophthalm/o: eye

opt/o: eye, vision

ot/o: ear

presby/o: old age

propri/o: own

retin/o: retina

scler/o: sclera

tympan/o: eardrum

The Ear

Figure 8.1 shows the structures of the ear. Color the box next to each term. Use the same color for the corresponding structures in the figure.

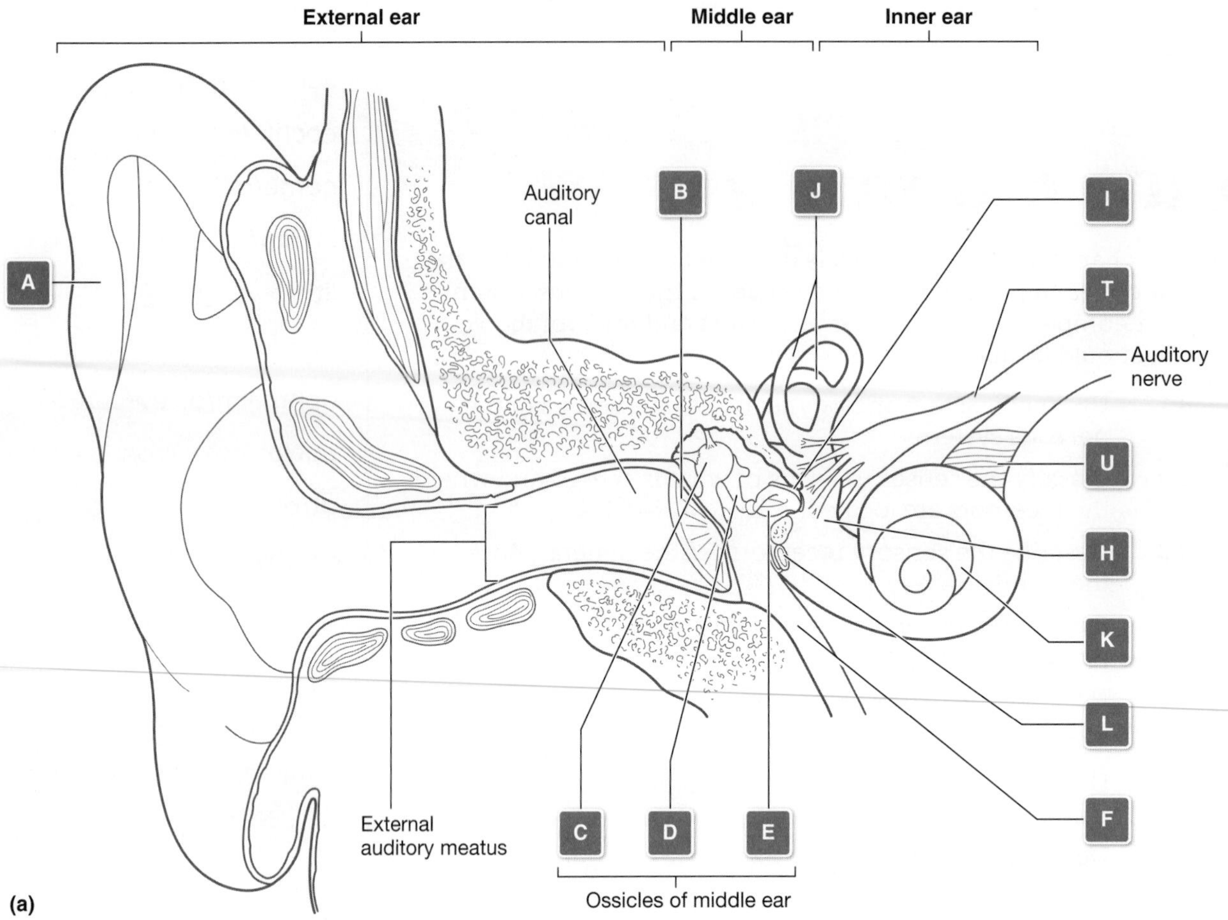

External ear
Middle ear
Inner ear
A
Auditory canal
B
J
I
T
Auditory nerve
U
H
K
L
F
External auditory meatus
C
D
E
Ossicles of middle ear
(a)

FIGURE 8.1 The ear: (a) major structures of the ear, (b) inner ear, (c) cochlear section, (d) tectorial membrane and organ of Corti, (e) hair cell.

Outer Ear

- ☐ **Pinna** (A)
- ☐ **Tympanic membrane** (B)

Middle Ear

- ☐ **Malleus** (C)
- ☐ **Incus** (D)
- ☐ **Stapes** (E)
- ☐ **Auditory tube** (F)

Inner Ear

- ☐ **Vestibule** (H)
- ☐ **Oval window** (I)
- ☐ **Semicircular canals** (J)
- ☐ **Cochlea** (K)
- ☐ **Round window** (L)
- ☐ **Saccule** (M)
- ☐ **Utricle** (N)
- ☐ **Cochlear duct** (O)
  - ☐ **Tectorial membrane** (P)
  - ☐ **Organ of Corti** (Q)
  - ☐ **Hair cell** (R)
  - ☐ **Basilar membrane** (S)

Nerves

- ☐ **Vestibular nerve** (T)
- ☐ **Cochlear nerve** (U)

Fluids

- ☐ **Perilymph** (V)
- ☐ **Endolymph** (W)

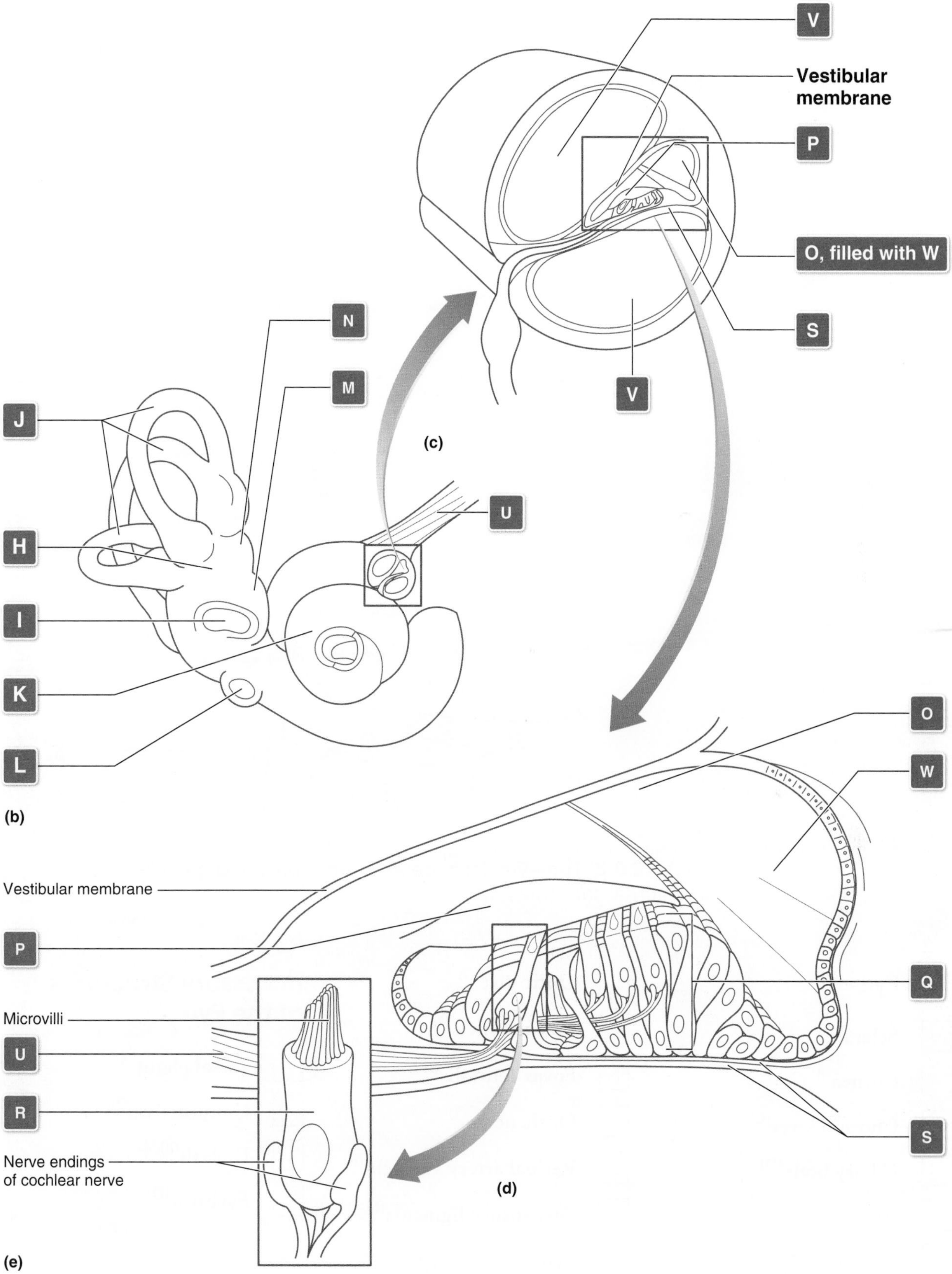
V
Vestibular membrane
P
O, filled with W
S
V
N
M
J
(c)
U
H
I
K
L
(b)
O
W
Vestibular membrane
P
Q
Microvilli
U
R
Nerve endings of cochlear nerve
S
(d)
(e)

FIGURE 8.1 concluded

The Eye

Figure 8.2 shows the structures of the eye. Color the box next to each term. Use the same color for the corresponding structures in the figures.

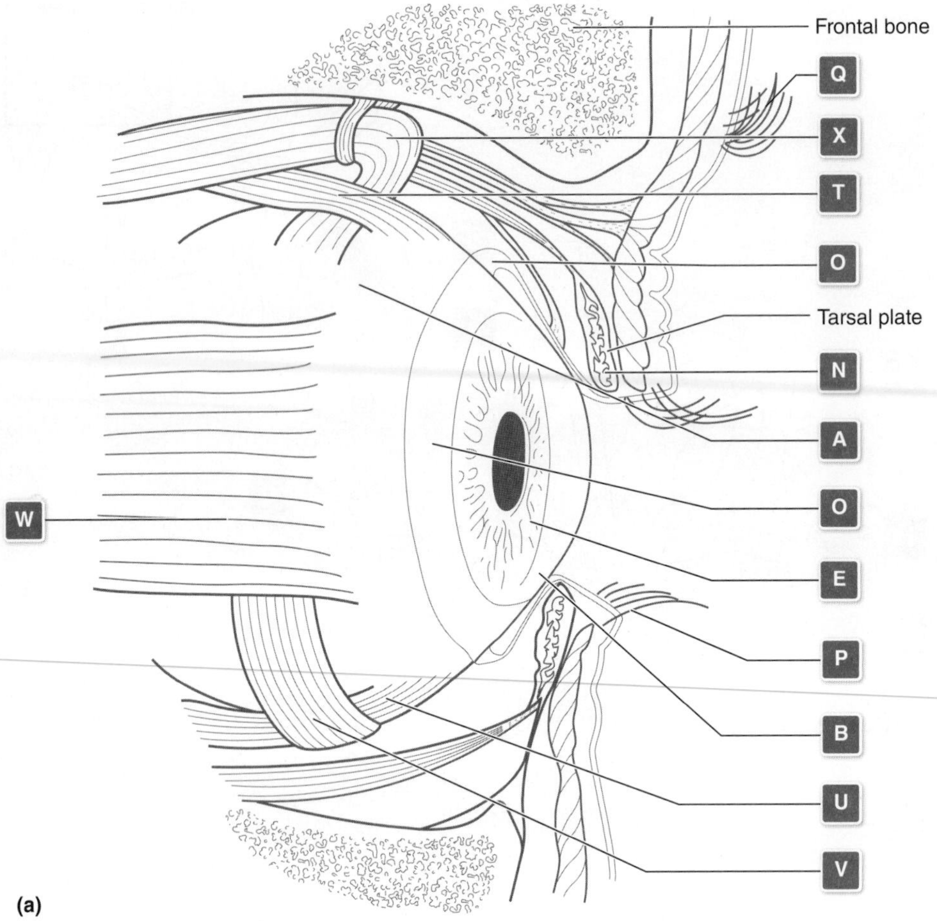

Frontal bone
Q
X
T
O
Tarsal plate
N
A
O
E
P
B
U
V
W
(a)

FIGURE 8.2 The eye: (a) accessory structures and muscles, (b) internal structures, (c) retina.

The Eye

- ☐ **Sclera**(A)
- ☐ **Cornea**(B)
- ☐ **Choroid layer**(C)
- ☐ **Ciliary body**(D)
- ☐ **Iris**(E)
- ☐ **Retina**(F)
- ☐ **Optic disc**(G)
- ☐ **Macula lutea**(H)
- ☐ **Fovea centralis**(I)
- ☐ **Optic nerve**(J)
- ☐ **Retinal artery and vein**(K)
- ☐ **Suspensory ligments**(L)
- ☐ **Lens**(M)

Accessory Structures of the Eye

- ☐ **Tarsal gland**(N)
- ☐ **Conjunctiva**(O)
- ☐ **Eyelash**(P)
- ☐ **Eyebrow**(Q)

COLORING BOOK

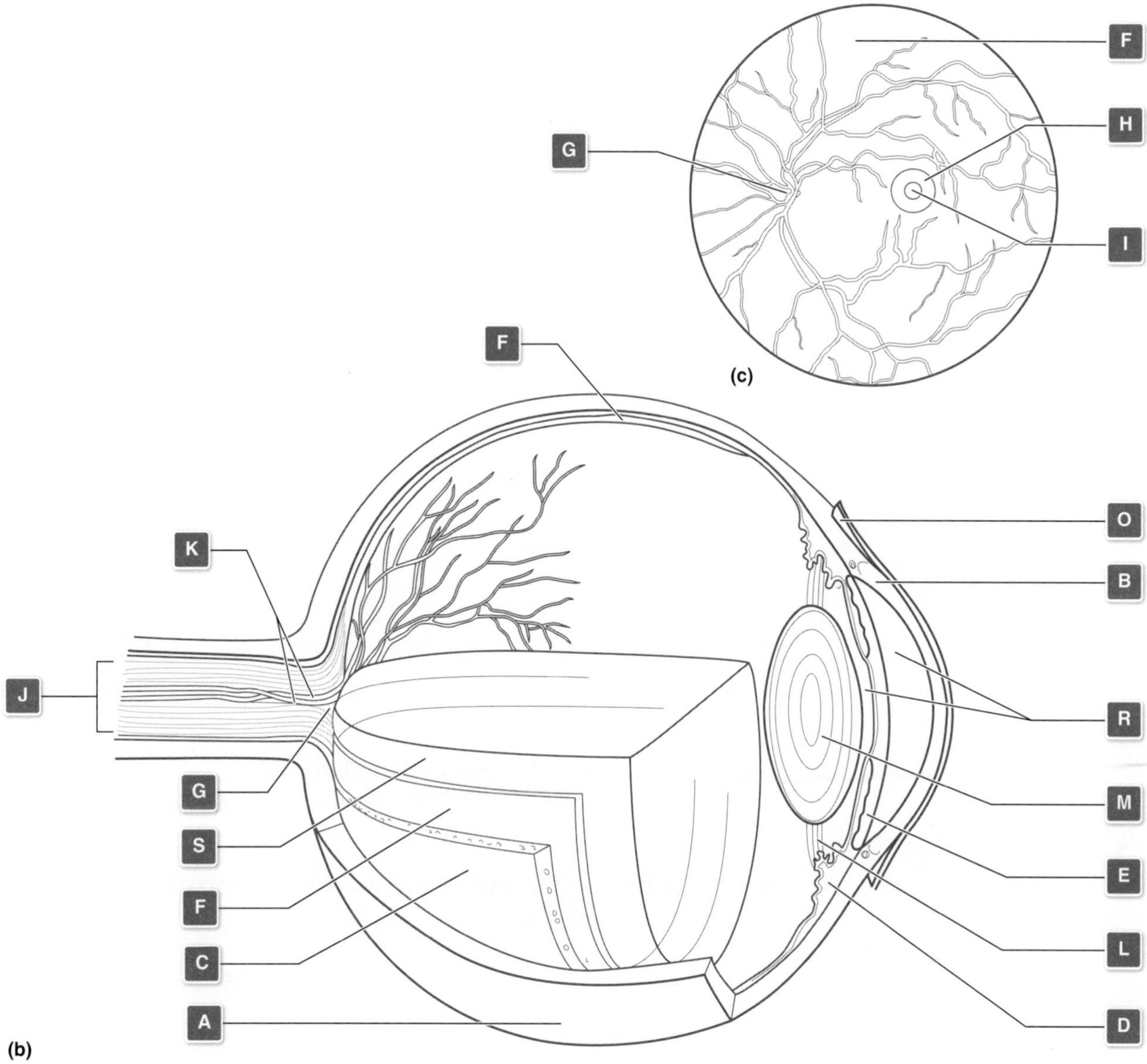

FIGURE 8.2 concluded

Fluids of the Eye

- ☐ **Aqueous humor**(R)
- ☐ **Vitreous humor**(S)

Muscles of the Eye

- ☐ **Superior rectus muscle**(T)
- ☐ **Inferior rectus muscle**(U)
- ☐ **Inferior oblique muscle**(V)
- ☐ **Lateral rectus muscle**(W)
- ☐ **Superior oblique muscle**(X)

Paths of the Stimulus

The pathways for action potentials from the sensory receptors to the areas of the brain were covered in Chapters 7 and 8. In this exercise you are to trace the path for sound in the ear and light in the eye from where they enter the organ to where they initiate a local potential. Then answer the questions that follow.

Sound in the Ear

List in order the structures and fluids that sound waves/mechanical vibrations/fluid waves would travel through to initiate a local potential. You may want to look at **Figure 8.1** as a guide for this activity.

Pinna → auditory canal → ______________________________

____

____

____

1. The sound entering the ear was in the form of waves of air molecules. At what structure does that change to mechanical vibrations? ______________________________

2. What happens to the vibrations in the middle ear? ______________________________

   ____

3. At what point are the mechanical vibrations converted to waves of fluid? ______________________________

   ____

   ____

4. What causes hair cells to bend? ______________________________

   ____

5. Where is a local potential initiated? ______________________________

   ____

6. Assuming the local potential is a threshold stimulus, what nerve will carry the action potential? ______________________________

   ____

Light in the Eye

List in order the structures and fluids that light would travel through to initiate a local potential. You may want to look at **Figure 8.2** as a guide for this activity.

Light → cornea → aqueous humor → ____________________

____

____

____

7. What happens to light as it passes through the cornea? ____________________

____

8. What smooth muscle regulates how much light will pass through an opening within the eye? ____________________

____

9. What is that opening called? ____________________

10. What happens to the light as it passes through the lens? ____________________

____

11. What structures work together to adjust the thickness of the lens?

____

12. What is the term for adjusting the thickness of the lens? ____________________

13. By what age is that ability usually significantly diminished? ____________________

14. Where on the retina should an object be focused for the best vision?

____

15. What type of photoreceptors are at that location?

____

16. Where will a local potential be initiated? ____________________

17. Assuming the local potential is a threshold stimulus, what nerve will carry the action potential? ____________________

Blind Spot

In this activity you will focus an image on the optic disc of your retina to see what happens. It is at the optic disc that all the axons from the neurons leave the retina to form the optic nerve. There are no rods or cones to detect light at the optic disc, so it is a functional blind spot.

To conduct the test, you will need to hold the workbook page with **Figure 8.3** in front of you. (If you are using an eBook, you will have to print the page or move your head instead of the book in the following instructions.)

FIGURE 8.3 Blind spot test.

1. Close your right eye.
2. Hold the page with **Figure 8.3** so that the figure is 30 cm (approx. 1 ft) in front of you.
3. Stare at the X on the figure with your left eye. Can you see the red dot even while staring at the X? ____________________
4. While continuing to stare at the X, slowly move the figure toward your eye or slightly right or left until the red dot disappears.
5. Record what you see in place of the red dot. ____________________
6. Where on the retina is the X focused if you are staring right at it? __________
____
7. Where on the retina was the red dot focused when you were staring at the X at the start of this exercise? ____________________
____
8. What type of photoreceptors are located on that part of the retina?
____
9. How do you know? ______________________________
10. When the red dot disappeared, what took its place? ____________________
____
11. Filling in the missing image with what is surrounding it is called *visual filling*. Why do you suppose the brain does visual filling?
____
____
____

Key Words

The following terms are defined in the glossary of the textbook.

accommodation
astigmatism
cataract
conjunctivitis
dynamic equilibrium
glaucoma
gustation
humor
hyperopia
lacrimal
lingual
myopia
nociceptor
olfactory
otoliths
perilymph
presbyopia
receptive field
refraction
sensorineural
static equilibrium
umami
vestibular apparatus
visual acuity

Concept Maps

Use key words and other bold words from the chapter to complete the following concept maps (**Figures 8.4** to **8.10**).

General Senses

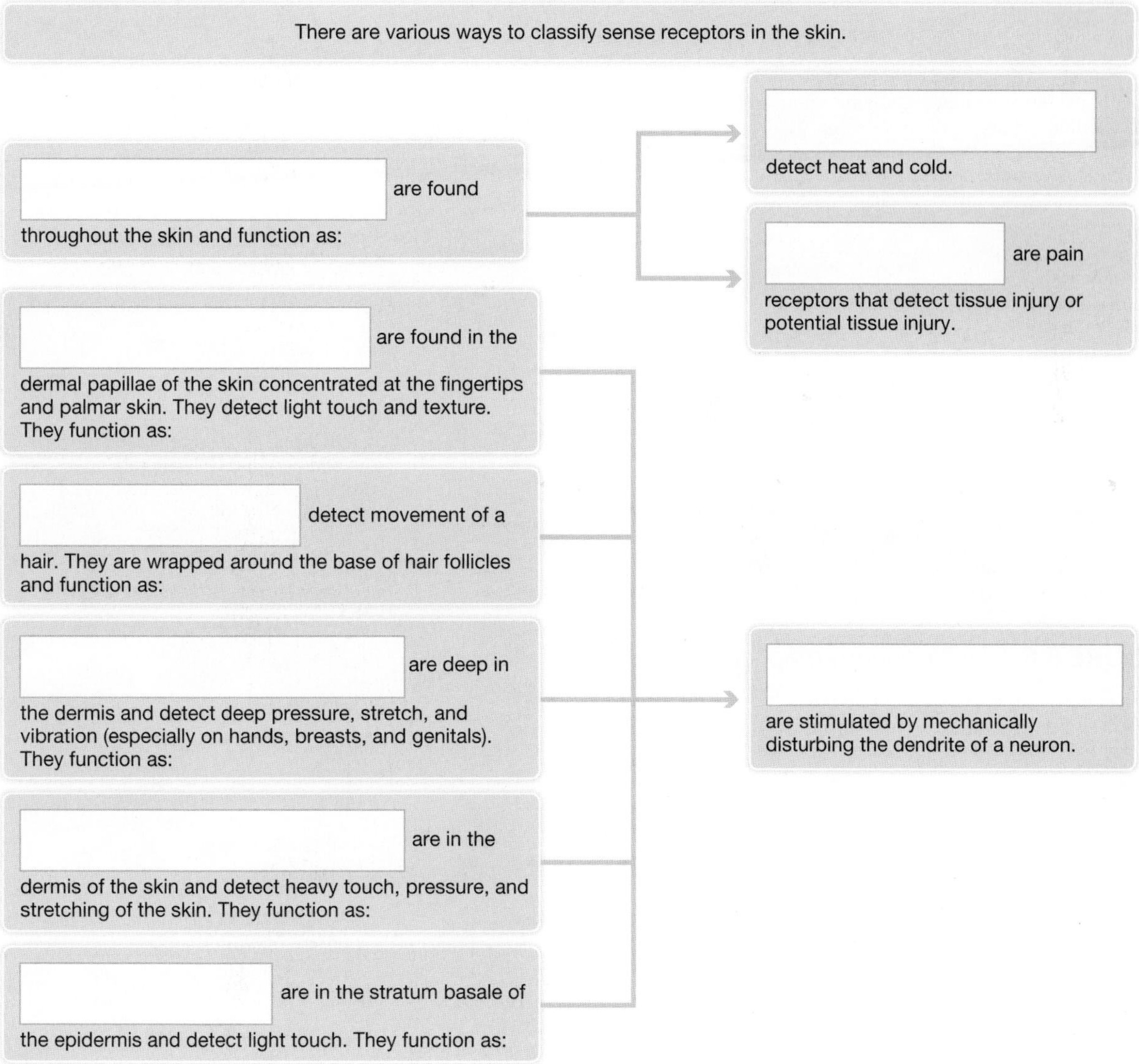
There are various ways to classify sense receptors in the skin.
are found throughout the skin and function as:
detect heat and cold.
are pain receptors that detect tissue injury or potential tissue injury.
are found in the dermal papillae of the skin concentrated at the fingertips and palmar skin. They detect light touch and texture. They function as:
detect movement of a hair. They are wrapped around the base of hair follicles and function as:
are deep in the dermis and detect deep pressure, stretch, and vibration (especially on hands, breasts, and genitals). They function as:
are stimulated by mechanically disturbing the dendrite of a neuron.
are in the dermis of the skin and detect heavy touch, pressure, and stretching of the skin. They function as:
are in the stratum basale of the epidermis and detect light touch. They function as:

FIGURE 8.4 General senses in the skin concept map.

Taste

Molecules taken into the mouth dissolve in saliva. Receptors for the five primary tastes are located all over the tongue and oral cavity but are concentrated on certain areas of the tongue.

- Lateral edges of the tongue are the most sensitive areas for ____ sensation. It is detected when ions from salts bind to receptors.
- The tip of the tongue is most sensitive to ____ taste sensations. They are detected when sugars bind to receptors.
- ____ is a meaty taste derived from some amino acids binding to the taste hairs.
- ____ taste is associated with alkaloids such as caffeine, quinine, and nicotine. It is most concentrated at the back of the tongue.
- The ____ taste sensation is associated with acids. It is most concentrated on the lateral edges of the tongue.

FIGURE 8.5 Taste concept map.

Smell

____ are bipolar neurons that pass from the olfactory bulb of the olfactory nerve (CN I) through the cribriform plate.

→ Their ____ spread out in the olfactory mucosa of the roof of the nasal cavity.

→ They serve as chemoreceptors.

→ Each one has one type of receptor to detect one particular odor.

→ The binding of an odor molecule to the receptor initiates a(n) ____.

→ A threshold stimulus at the trigger zone of the olfactory cell will generate an action potential that will be passed on to a synapse within the ____ of CN I.

FIGURE 8.6 Smell concept map.

Hearing

Hearing involves converting sound waves outside the head to action potentials in the bipolar neurons of the cochlear nerve.

This process begins when external sound waves are directed into the auditory canal by the __________ .

↓

The waves of air molecules hit the __________ and cause it to vibrate.

↓

The __________ is connected and also begins to vibrate.

↓

That causes the __________ and then the __________ to vibrate as well.

↓

Vibrations are transferred to the __________ of the inner ear.

↓

Vibrations create waves within the __________ of the cochlear tube, eventually causing the __________ to bulge out.

↓

If the frequency is slow enough, the wave can make it all the way around the cochlear tube.

If the frequency is fast, the wave pushes on the __________ , causing the hair cells to vibrate.

↓

The hair cells are pushed against the __________ and bend each time there is a vibration.

↓

Hair cells are mechanoreceptors; bending them releases a neurotransmitter to the bipolar neuron at its base to start a local potential.

FIGURE 8.7 Hearing concept map.

Equilibrium

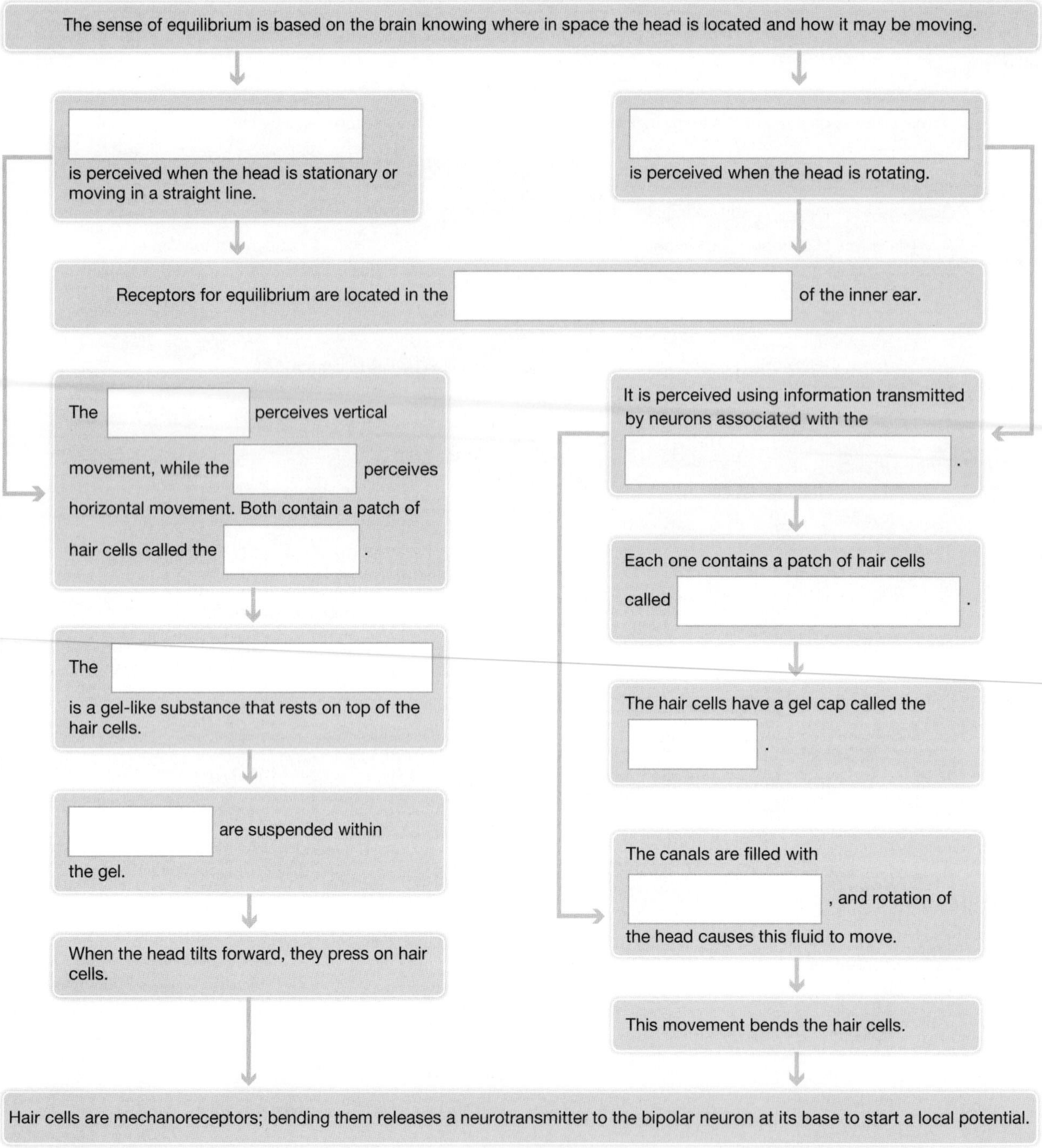
The sense of equilibrium is based on the brain knowing where in space the head is located and how it may be moving.
is perceived when the head is stationary or moving in a straight line.
is perceived when the head is rotating.
Receptors for equilibrium are located in the
of the inner ear.
The
perceives vertical movement, while the
perceives horizontal movement. Both contain a patch of hair cells called the
.
The
is a gel-like substance that rests on top of the hair cells.
are suspended within the gel.
When the head tilts forward, they press on hair cells.
It is perceived using information transmitted by neurons associated with the
.
Each one contains a patch of hair cells called
.
The hair cells have a gel cap called the
.
The canals are filled with
, and rotation of the head causes this fluid to move.
This movement bends the hair cells.
Hair cells are mechanoreceptors; bending them releases a neurotransmitter to the bipolar neuron at its base to start a local potential.

FIGURE 8.8 Equilibrium concept map.

Retina

There are two types of photoreceptors in the retina of the eye.

______ are bipolar photoreceptors used for noncolor vision and low-light conditions.

They contain a chemical called ______ that reacts to light to initiate a local potential.

______ are bipolar photoreceptors used for color vision and are responsible for best ______.

They contain the chemical ______ that reacts to light to initiate a local potential.

Photoreceptors are deep in the retina, close to the ______.

They synapse with bipolar connecting neurons, which synapse with still another layer of neurons called ganglion cells.

The axons of these cells form the ______.

FIGURE 8.9 Retina concept map.

Vision

Light reflected off an object strikes the ______.

↓

The light then passes through the ______

↓

and passes through the ______.

↓

The light then passes through the ______ and is projected on the retina upside down.

↓

In this case the light was refracted to perfectly focus the image on the retina.

↓

If the image is focused ahead of the retina, the condition is called ______ (nearsightedness).

If the image is focused behind the retina, the condition is called ______ (farsightedness).

↓

Either condition can be corrected with artificial lenses or refractive surgery.

For someone to see objects up close and objects far away, the eye must accommodate for distant and near vision.

↓

For distant vision, the ring of ______ around the lens is relaxed.

↓

That creates tension in the ______ that hold the lens in place.

↓

The tension pulls on the lens from all sides, causing it to become ______.

In ______ vision, the ciliary muscles around the lens contract.

↓

That takes tension off the ______ that hold the lens in place.

↓

The lack of tension on the lens allows it to become ______.

↓

Thus the eye has been able to change its focus so that the image of the object is perfectly focused on the retina whether it is near or far.

FIGURE 8.10 Vision concept map.

KEY WORD CONCEPT MAPS

Word Deconstruction: *In the textbook, you built words to fit a definition using combining forms, prefixes, and suffixes. Here you are to break down the term into its parts (prefixes, roots, and suffixes) and give a definition. Prefixes and suffixes can be found inside the back cover of the textbook.*

FOR EXAMPLE Dermatitis: dermat/itis—inflammation of the skin

1. Proprioceptors: ______
2. Retinopathy: ______
3. Corneal: ______
4. Otoscopic: ______
5. Intraocular: ______

Multiple Select: *Select the correct choices for each statement. The choices may be all correct, all incorrect, or any combination of correct and incorrect.*

1. How are senses classified?
   a. Special senses include taste, touch, smell, vision, and hearing.
   b. General senses include pain.
   c. Special senses are detected by organs in the head.
   d. General senses are detected throughout the body.
   e. Equilibrium is a general sense.
2. What is the pathway for pain?
   a. Damaged tissues release chemicals to stimulate local potentials on the dendrites of nociceptors.
   b. If the pain is in the head, the sensory message is carried on cranial nerves.
   c. If the pain is below the neck, the sensory messages go to the spinal cord.
   d. The sensory messages for pain may go to the hypothalamus and amygdala.
   e. The pathway for pain begins with a bipolar neuron.
3. How do the receptors for taste work?
   a. Taste cells have taste hairs that have receptors for chemicals.
   b. The chemicals detected are dissolved in saliva.
   c. There are six primary tastes.
   d. Salt taste sensations are concentrated at the back of the tongue.
   e. Flavor is derived from multiple sensory inputs.
4. What happens to our senses as we age?
   a. The number of receptors in olfaction and gustation decrease.
   b. It gets harder to accommodate.
   c. The iris is not as responsive to changes in the amount of light.
   d. Pain sensitivity diminishes starting about age 50.
   e. Light touch is usually increased in the elderly.
5. Which of the following is (are) true about the forms of hearing loss?
   a. Sensorineural loss is a problem of the organ of Corti or the vestibular nerve.
   b. Conductive hearing loss can be caused by a ruptured tympanic membrane.
   c. Sensorineural hearing loss can be caused by arthritis in the ossicles of the middle ear.

d. The type of hearing loss can be determined by comparing air conduction and bone conduction using a tuning fork.
e. Tinnitus is often associated with mild hearing loss.

6. How is vision perceived?
   a. Rhodopsin and iodopsin react to light to initiate a local potential on a bipolar neuron.
   b. Images are projected upside down on the retina.
   c. The cornea and lens reflect light.
   d. The choroid layer is dark, so more light is reflected within the eye.
   e. Depth perception allows objects to be located in space.

7. How can the anatomy of the eye be described?
   a. The cornea is perfectly smooth to properly refract light.
   b. The uvea is the outer layer of the back of the eye.
   c. The ciliary body is smooth muscle.
   d. The diaphragm is smooth muscle.
   e. The canal of Schlemm drains aqueous humor.

8. How are odors perceived?
   a. Chemicals bind to hair cells.
   b. Odors are detected in the mucosa of the nasal cavity.
   c. Bipolar neurons are used.
   d. Messages travel through the sella turcica to the olfactory bulb.
   e. Adaptation diminishes the duration of the sense of smell.

9. Which of the following statements is (are) true concerning receptors for general senses in the skin?
   a. Thermoreceptors detect heat and cold.
   b. Receptors are distributed evenly over the skin.
   c. Mechanoreceptors detect touch, vibration, stretch, and pressure.
   d. Free nerve endings detect heat, cold, and pain.
   e. Tactile disks are the only general sense receptor located in the epidermis.

10. What type(s) of information is (are) transmitted to the brain from receptors?
    a. The location, based on receptive fields.
    b. The intensity, based on the sensitivity of the receptors.
    c. The type of sensation, based on the pathway.
    d. The intensity of the stimulus, based on the frequency of nerve impulses.
    e. The duration of the stimulus, based on how much the neuron adapts the frequency of action potentials.

Matching: *Match the description to the structure. Some answers may be used more than once. Some descriptions may fit more than one structure.*

_______ 1. Contains the saccule and utricle
_______ 2. Contains the organ of Corti and endolymph
_______ 3. Contains perilymph
_______ 4. Is found in the bony labyrinth
_______ 5. Connects the middle ear to the nasopharynx

a. Vestibular apparatus
b. Semicircular canals
c. Cochlear duct
d. Ossicles
e. Auditory canal
f. Auditory tube

CHAPTER 8 REVIEW QUESTIONS

Matching: *Match the final destination in a pathway to the sense. Some answers may be used more than once. Some senses may have more than one destination.*

_______	**6.** Pain	**a.** Occipital lobe
_______	**7.** Vision	**b.** Temporal lobe
_______	**8.** Taste	**c.** Frontal lobe
_______	**9.** Equilibrium	**d.** Parietal lobe
_______	**10.** Hearing	**e.** Hypothalamus and amygdala

Completion: *Fill in the blanks to complete the following statements.*

1. Each neuron in the skin is responsible for detecting a stimulus in a given area called a(n) _______________.

2. Taste cells are found in taste buds located in bumps on the tongue called _______________.

3. _______________ is a meaty taste.

4. The _______________ of sound is measured in hertz.

5. The volume of sound is measured in units called _______________.

Critical Thinking

1. Explain in terms of the physiology of the vestibular apparatus why astronauts may have difficulty with equilibrium in a weightless environment.

____

____

____

2. In terms of anatomy and physiology, why might the smell of your mother's home cooking produce a visceral and an emotional response?

____

____

____

____

3. Write a pathway for singing along with the radio, starting where sound waves enter the ear and ending where skeletal messages are sent.

____

____

____

____

Case Study

Michael has worked in construction for the last 20 years. He is only in his early 40s but has been experiencing trouble with his hearing for the last few years. He has finally decided to go and get his hearing checked.

1. What are the two types of hearing loss? What type is Michael most likely suffering from?

2. What diagnostic test could be used to help determine his diagnosis? How does the test work to determine hearing loss?

3. Is there anything Michael could have done to prevent damaging his hearing?

4. What, if anything, can be done to improve Michael's hearing?

CHAPTER 8 REVIEW QUESTIONS

This section of the chapter is designed to help you find where each outcome is covered in the textbook and workbook.

Learning Outcomes		Text Readings, Figures, and Tables	Workbook Activities	Text and Workbook Assessments
8.1	Use medical terminology related to the senses of the nervous system.	Word roots & combining forms, pronunciation key: p. 279	Word roots & combining forms	*Text:* Review: 8.1 *Workbook:* Word Deconstruction: 1–5
8.2	Classify the senses in terms of what is sensed and where the receptors are located.	Overview: p. 280		*Text:* Review: 8.2 *Workbook:* Multiple Select: 1
8.3	Describe the sensory receptors for the general senses in the skin.	General senses: pp. 280–281 Figure 8.2 Table 8.1	*Concept maps:* General senses in the skin Figure 8.4	*Text:* Spot Check: 1, 2 Review: 8.3 *Workbook:* Multiple Select: 9
8.4	Explain the types of information transmitted by sensory receptors in the skin.	Physiology of general senses in the skin: pp. 281–283 Figure 8.3	*Concept maps:* General senses in the skin Figure 8.4	*Text:* Review: 8.4 *Workbook:* Multiple Select: 10 Completion: 1
8.5	Describe the pathway for pain.	Pathway for pain: pp. 283–284 Figure 8.4		*Text:* Review: 8.5 *Workbook:* Multiple Select: 2 Matching: 6–10
8.6	Describe the sensory receptors for taste.	Taste: pp. 284–285 Figure 8.5		*Text:* Review: 8.6 *Workbook:* Multiple Select: 3 Completion: 2
8.7	Describe the different tastes and explain how flavor is perceived.	Physiology of taste: p. 286	*Concept maps:* Taste Figure 8.5	*Text:* Spot Check: 3 Review: 8.7 *Workbook:* Multiple Select: 3 Completion: 3
8.8	Describe the pathway for taste.	Pathway for taste: p. 286		*Text:* Review: 8.8 *Workbook:* Matching: 8
8.9	Describe the sensory receptors for smell.	Smell: pp. 286–287 Figure 8.6	*Concept maps:* Smell Figure 8.6	*Text:* Review: 8.9 *Workbook:* Multiple Select: 8

CHAPTER 8 MAPPING

Learning Outcomes		Text Readings, Figures, and Tables	Workbook Activities	Text and Workbook Assessments
8.10	Explain how odors are perceived.	Physiology of smell: p. 287		*Text:* Spot Check: 4 Review: 8.10 *Workbook:* Multiple Select: 8
8.11	Describe the pathway for smell.	Pathway for olfaction: p. 288		*Text:* Review: 8.11 *Workbook:* Critical Thinking: 2
8.12	Describe the anatomy of the ear.	Hearing: pp. 288–292 Figures 8.7–8.11	*Coloring book:* The ear Figure 8.1 *Lab activities and exercises:* Paths of the stimulus—sound in the ear Figure 8.1 *Concept maps:* Hearing Figure 8.7	*Text:* Review: 8.12 *Workbook:* Matching: 1–5
8.13	Explain how sound is perceived.	Physiology of hearing: pp. 292–293 Figures 8.12–8.13	*Concept maps:* Hearing Figure 8.7	*Text:* Spot Check: 5 Review: 8.13 *Workbook:* Completion: 4, 5
8.14	Describe the pathway for hearing.	Pathway for hearing: p. 293		*Text:* Review: 8.14 *Workbook:* Critical Thinking: 3 Matching: 10
8.15	Describe the anatomy of the vestibular apparatus.	Equilibrium: pp. 293–295 Figures 8.14–8.17	*Coloring book:* The ear Figure 8.1	*Text:* Review: 8.15 *Workbook:* Matching: 1–5
8.16	Explain how equilibrium is perceived.	Physiology of equilibrium: pp. 295–297		*Text:* Spot Check: 6 Review: 8.16 *Workbook:* Critical Thinking: 1
8.17	Describe the pathway for equilibrium.	Pathway for equilibrium: p. 297		*Text:* Review: 8.17 *Workbook:* Matching: 9

Learning Outcomes		Text Readings, Figures, and Tables	Workbook Activities	Text and Workbook Assessments
8.18	Describe the anatomy of the eye.	Vision: pp. 297–303 Figures 8.18–8.24	*Coloring book:* The eye Figure 8.2 *Lab exercises and activities:* Paths of the stimulus—light in the eye Figure 8.2 *Concept maps:* Retina Figure 8.9	*Text:* Spot Check: 7–10 Review: 8.18 *Workbook:* Multiple Select: 7
8.19	Explain how vision is perceived.	Physiology of vision: pp. 303–306 Figure 8.25	*Lab exercises and activities:* Blind spot Figure 8.3 *Concept maps:* Retina; Vision Figures 8.9–8.10	*Text:* Spot Check: 11 Review: 8.19 *Workbook:* Multiple Select: 6
8.20	Describe the pathway for vision.	Pathway for vision: p. 306 Figure 8.26		*Text:* Review: 8.20 *Workbook:* Matching: 7
8.21	Describe the effects of aging on the senses.	Effects of aging on the senses: pp. 306–307		*Text:* Review: 8.21 *Workbook:* Multiple Select: 4
8.22	Describe common diagnostic tests used to diagnose disorders of the senses.	Diagnostic tests for disorders of the senses: p. 307 Table 8.2		*Text:* Spot Check: 12 Review: 8.22 *Workbook:* Case Study: 2
8.23	Describe disorders of the senses and relate abnormal function to the pathology.	Disorders of the senses: pp. 308–309 Figure 8.27 Table 8.3		*Text:* Spot Check: 12, 13 Review: 8.23 *Workbook:* Multiple Select: 5 Case Study: 1, 3–4

9 The Endocrine System

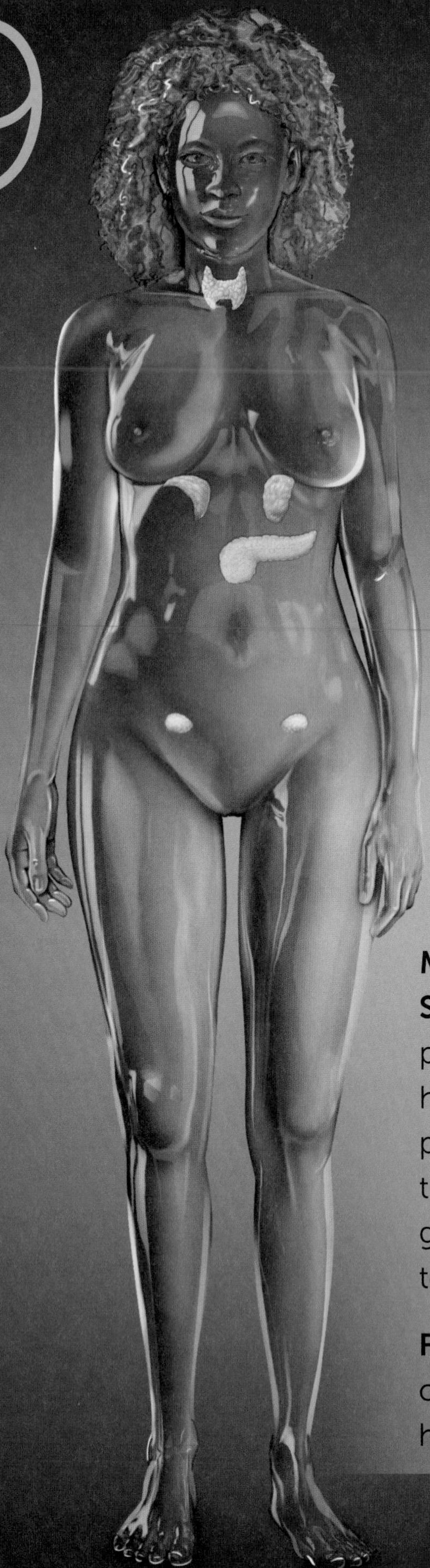

Major Organs and Structures:
pineal gland, hypothalamus, pituitary gland, thyroid gland, adrenal glands, pancreas, testes, ovaries

Functions:
communication, hormone production

learning outcomes

This chapter of the workbook is designed to help you learn the anatomy and physiology of the endocrine system. After completing this chapter in the text and this workbook, you should be able to:

9.1 Use medical terminology related to the endocrine system.

9.2 Compare and contrast the endocrine and nervous systems in terms of type, specificity, speed, and duration of communication.

9.3 Define *gland, hormone,* and *target tissue.*

9.4 List the major hormones, along with their target tissues and functions, of each of the endocrine system glands.

9.5 Locate and identify endocrine system glands.

9.6 Describe the chemical makeup of hormones, using estrogen, insulin, and epinephrine as examples.

9.7 Compare the location of receptors for protein hormones with that of receptors for steroid hormones.

9.8 Differentiate autocrine, paracrine, endocrine, and pheromone chemical signals in terms of the proximity of the target tissue.

9.9 Explain the regulation of hormone secretion and its distribution.

9.10 Explain how the number of receptors can be changed.

9.11 Explain how hormones are eliminated from the body.

9.12 Explain the function of hormones by showing how they interact to maintain homeostasis.

9.13 Explain the effects of aging on the endocrine system.

9.14 Describe common diagnostic tests used to diagnose endocrine system disorders.

9.15 Describe endocrine system disorders and relate abnormal function to pathology.

word **roots** & combining **forms**

aden/o: gland

adren/o: adrenal glands

adrenal/o: adrenal glands

andr/o: male

cortic/o: cortex

crin/o: secrete

dips/o: thirst

gluc/o: sugar

glyc/o: sugar

gonad/o: sex glands

hormon/o: hormone

pancreat/o: pancreas

ster/o: steroid

thyr/o: thyroid gland

Endocrine System Glands

Figure 9.1 shows the endocrine system glands. Color the box next to each term. Use the same color for the corresponding structures in the figure.

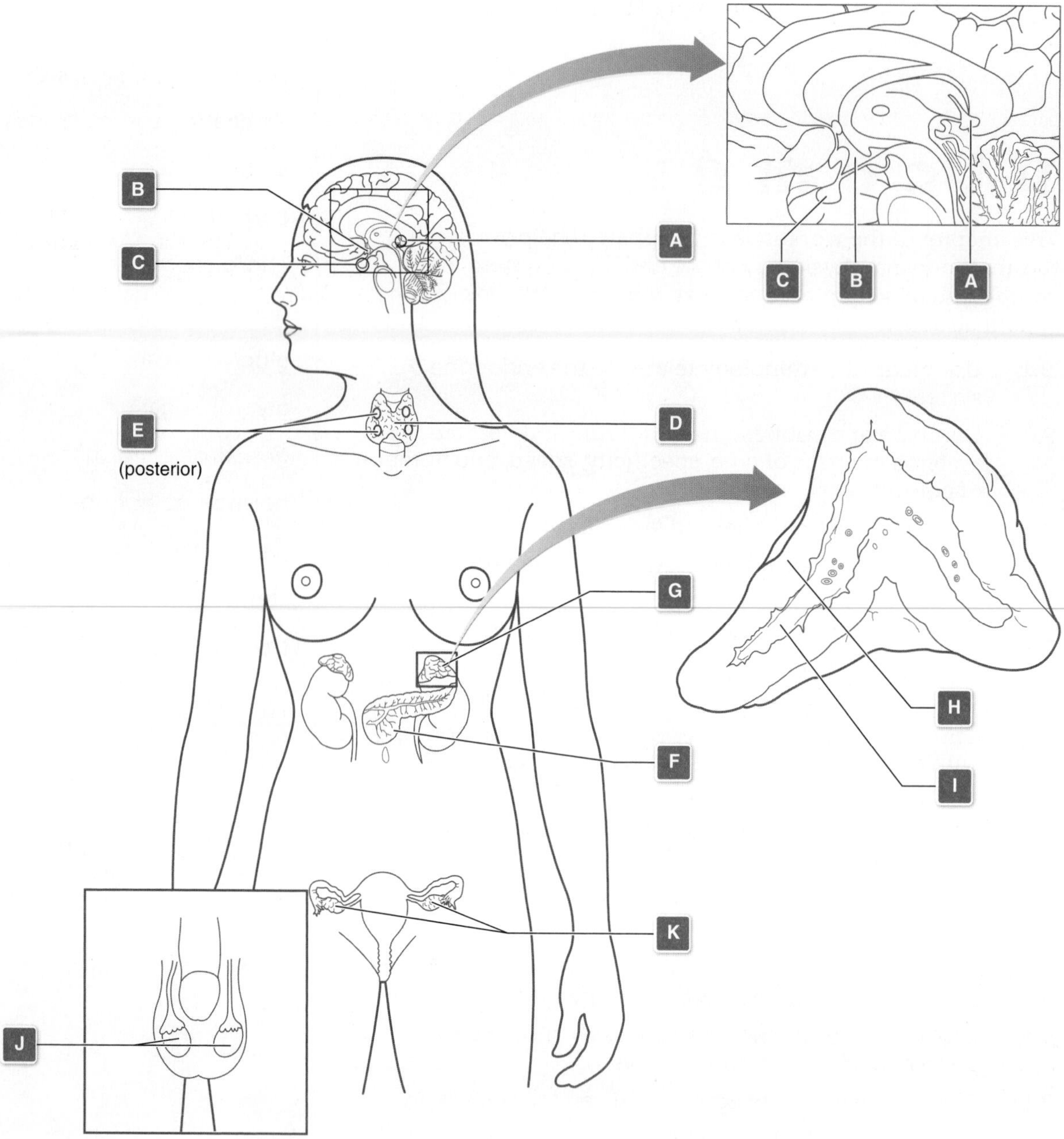

FIGURE 9.1 Endocrine system glands.

- ☐ **Pineal gland** (A)
- ☐ **Hypothalamus** (B)
- ☐ **Pituitary gland** (C)
- ☐ **Thyroid gland** (D)
- ☐ **Parathyroid glands** (E)
- ☐ **Pancreas** (F)
- ☐ **Adrenal gland** (G)
  - ☐ **Adrenal cortex** (H)
  - ☐ **Adrenal medulla** (I)
- ☐ **Testes** (J)
- ☐ **Ovaries** (K)

Comparison of the Endocrine and Nervous Systems

The endocrine system and the nervous system are used for communication. Complete Table 9.1, comparing the communication of the two systems.

TABLE 9.1 Comparison of endocrine and nervous system communication

System	Type of communication	Specificity of communication	Speed of communication	Duration of communication (include how the communication is stopped)
Endocrine system				
Nervous system				

Graphing an Endocrine Problem

One way to see whether you grasp the fundamental concepts of the endocrine system is to critique other students' work. Six groups of anatomy and physiology students were given an endocrine problem to graph. The four students in each group discussed the problem and together put their graph on the board for all to see. After all the graphs were displayed, the students in each group were given time to discuss, within their group, all of the graphs. Each group was then given the opportunity to amend their graph. Group 4 amended their graph once. Group 5 amended their graph twice.

The following instructions were given to the six groups of students.

ENDOCRINE PROBLEM GIVEN TO SIX GROUPS OF A&P STUDENTS

Consider hormone A. For hormone A to work, it must be modified by the liver before traveling to its target tissue. Hormones are usually produced when there is a need and are eliminated when the need is met. Figure 9.2 shows a graph of the normal amount of hormone A produced over time to meet a need. The level of hormone A increases to a point at which the need is met. The hormone is then quickly excreted by the kidney.

Using the graph in Figure 9.2, you are to add the following to the graph: (1) what you think will happen to the level of the hormone if the liver is diseased and can no longer modify the hormone and (2) what you think will happen to the level of the hormone if the liver is functioning to modify the hormone but the kidney is diseased and can no longer eliminate the hormone.

Your task is to work through the endocrine problem and either (a) select one of the groups' graphs as being correct and justify your choice (see **Figure 9.3**) or (b) produce your own graph using **Figure 9.2** and give a justification for why your graph is correct.

Your answer:

Either add to the graph in **Figure 9.2** or select one of the student graphs in **Figure 9.3**. On the line below, indicate which graph is correct.

____

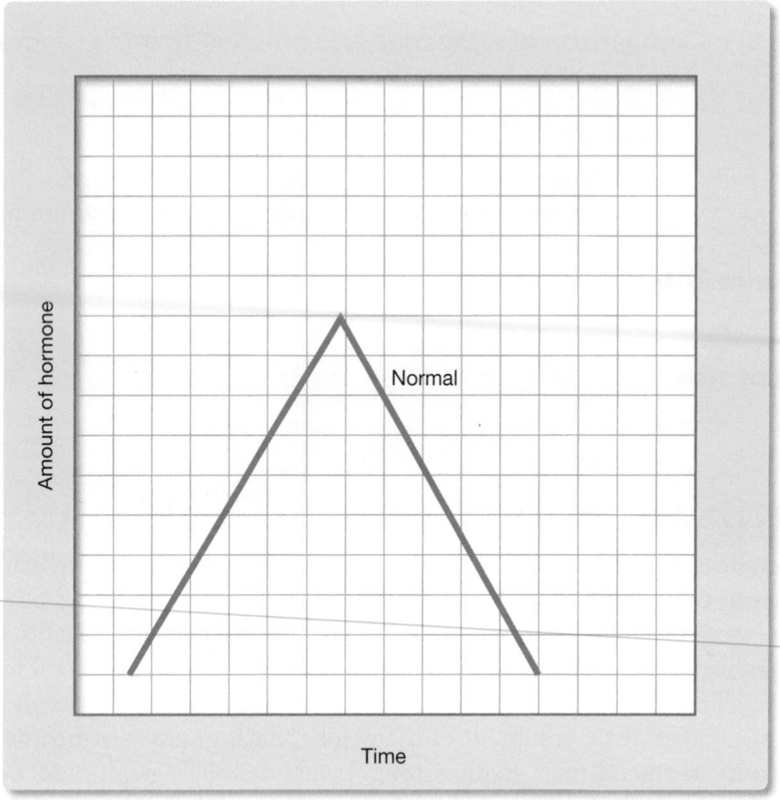

FIGURE 9.2 Graph of normal response to a need for hormone A.

Justification for your answer:

____

____

____

____

____

____

____

____

____

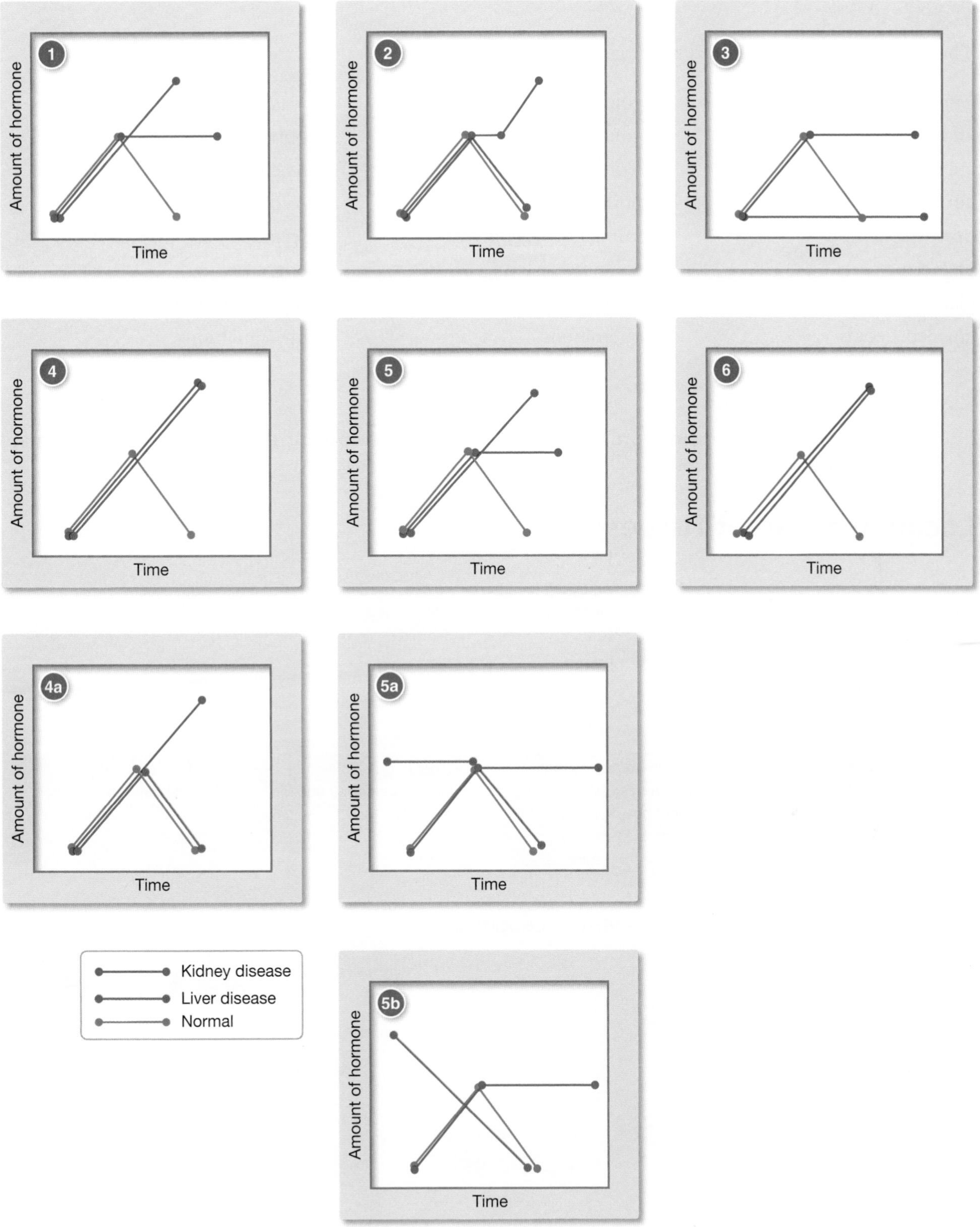

FIGURE 9.3 The graphs of the six groups of A&P students.

Key Words

The following terms are defined in the glossary of the textbook.

acromegaly
androgens
autoantibody
autocrine
down-regulation
endocrine
gland
glucocorticoids
goiter
gonads
half-life
hormone
mineralocorticoids
pancreatic islets
paracrine
pheromone
plasma protein
polydipsia
receptor
second messenger
secondary sex characteristics
target tissue
thyroid hormone
up-regulation

Concept Maps

Use key words and other bold words from the chapter to complete the following concept maps (Figures 9.4 to 9.6).

Location of Target Tissues

Hormone function can be classified by the location of the cells producing the hormone relative to the location of the hormone's target.

- ______ refers to the secretion of a hormone by the cells of the same tissue type that it targets.
- ______ refers to hormones that work on neighboring cells without having to go through the blood to get to the target tissue.
- ______ refers to chemicals that cause a response outside the body in another individual.
- ______ refers to hormones that travel through the blood to get to their target tissue.

FIGURE 9.4 Location of target tissues concept map.

Regulation of Hormone Secretion

Hormones are not secreted at a constant rate.

↓

They are secreted when there is a need. Their secretion is usually regulated by ________ mechanisms.

↓

Hormone secretion can be initiated in three ways:

________ A neuron stimulating a gland.	Another ________ stimulating a gland.	A(n) ________ stimulating a gland.
An example of this type of stimulation occurs when sympathetic neurons stimulate the adrenal medulla to secrete epinephrine in times of pain, fear, and/or stress.	An example of this can be seen when the hormones released by the hypothalamus stimulate the secretion of hormones from the anterior pituitary.	An example of this can be seen as the pancreas monitors blood glucose levels. Glucose is not a hormone, but if glucose levels are high, the pancreas responds to that stimulation by secreting insulin.

FIGURE 9.5 Regulation of hormone secretion concept map.

Hormone Elimination

The effects of a hormone will continue until the hormone is eliminated from the system. Hormones can be eliminated in four ways:

________	________	________	________
For example, the kidneys can remove a hormone from the blood and excrete it in urine, or the liver can remove a hormone from the blood and excrete it in bile.	For example, the breakdown of protein hormones results in amino acids that can be used for protein synthesis.	For example, a hormone like epinephrine can be taken back up by a cell so that it can be recycled and released at another time.	For example, the liver can bind water-soluble molecules to a hormone so that it will be excreted at a faster rate.

↓

________ is the length of time it takes for one-half of a substance to be eliminated from the circulatory system.

FIGURE 9.6 Hormone elimination concept map.

Word Deconstruction: *In the textbook, you built words to fit a definition using combining forms, prefixes, and suffixes. Here you are to break down the term into its parts (prefixes, roots, and suffixes) and give a definition. Prefixes and suffixes can be found inside the back cover of the textbook.*

FOR EXAMPLE Dermatitis: dermat/itis—inflammation of the skin

1. Polydipsia: ______________________________
2. Hyperglycemia: ______________________________
3. Gonadotropin: ______________________________
4. Adenoma: ______________________________
5. Corticosteroid: ______________________________

Multiple Select: *Select the correct choices for each statement. The choices may be all correct, all incorrect, or any combination of correct and incorrect.*

1. Diana comes into the clinic complaining of being thirsty all the time and having to use the bathroom often. Which of the following statements seem(s) consistent with this scenario?
   - **a.** She is describing polyuria and polydipsia.
   - **b.** She may have a problem with ACTH.
   - **c.** She may have a problem with ADH.
   - **d.** She may have a problem with receptors for insulin.
   - **e.** These symptoms are consistent with diabetes mellitus and diabetes insipidus.
2. What is important to understand about the chemical composition of a hormone?
   - **a.** Protein hormones can be delivered medically using a patch on the skin.
   - **b.** Steroid hormones are broken down in the stomach, so they cannot be given orally.
   - **c.** Protein hormones bind to plasma proteins.
   - **d.** Steroid hormones require a second messenger.
   - **e.** The chemical composition matters in the clinical delivery of a hormone.
3. How do the endocrine and nervous systems compare?
   - **a.** The nervous system is faster.
   - **b.** The endocrine system is more specific.
   - **c.** The effects of the endocrine system last longer.
   - **d.** The endocrine system uses chemicals and the nervous system does not.
   - **e.** All communication originates in the brain for both systems.
4. Emily is pregnant, and the fetus growing inside her is actively building bone. The calcium for the fetal bone deposition is taken from Emily's blood, lowering her normal blood calcium level. What is (are) the hormonal consequence(s) for Emily?
   - **a.** Emily's thyroid gland will produce calcitonin to bring her blood calcium level back to normal.
   - **b.** Emily's thyroid gland will produce PTH to bring her calcium level back to homeostasis.
   - **c.** ACTH is involved.
   - **d.** TSH is involved.
   - **e.** There are no hormonal consequences.
5. What is (are) the hormonal consequence(s) of a consistent diet overly high in carbohydrates?
   - **a.** The receptors for glucose will remain the same.
   - **b.** The receptors for insulin will likely be up-regulated.
   - **c.** The receptors for glucagon will likely be down-regulated.
   - **d.** The target tissues will likely become less sensitive to insulin.
   - **e.** The target tissues will likely become less sensitive to glucagon.

6. Cells in the heart produce a chemical that travels through the blood to the kidney to regulate urine production. Which of the following statements is (are) true given the definitions of *gland, hormone,* and *target tissue?*
   a. The cells in the heart qualify as an endocrine gland.
   b. The chemical the heart produces qualifies as a hormone.
   c. The kidney is the target tissue.
   d. The kidney must have receptors for the chemical.
   e. The chemical must fit in the receptors in the kidney like a key in a lock.
7. Which of the following is (are) accurate concerning target tissues in relation to the gland that produces the hormone?
   a. *Pheromone* refers to hormones that affect another individual as the target tissue.
   b. *Autocrine* refers to hormones that affect the cells that produced them.
   c. *Endocrine* refers to hormones that affect neighboring cells without going through the bloodstream.
   d. Ceruminous glands are endocrine glands.
   e. *Paracrine* refers to hormones that travel through the bloodstream.
8. Which of the following statements is (are) accurate concerning insulin and estrogen?
   a. The receptors for insulin are on a cell membrane.
   b. The receptors for insulin are inside the cell.
   c. The receptors for estrogen must be on the cell membrane.
   d. The receptors for estrogen can be anywhere in the cell.
   e. The locations of the receptors for estrogen and insulin are the same since both hormones target most tissues.
9. Why is the liver important to the endocrine system?
   a. It is a target tissue for glucagon.
   b. It produces plasma proteins.
   c. It produces cholesterol.
   d. It can excrete hormones in bile.
   e. It can conjugate hormones.
10. Why is the kidney important to the endocrine system?
   a. It is a target tissue for PTH.
   b. It produces plasma proteins.
   c. It produces cholesterol.
   d. It can excrete hormones in urine.
   e. It produces aldosterone.

Matching: *Match the hormone to its function. Some answers may be used more than once.*

_______ **1.** Tells cells to take in glucose, lowering blood glucose levels
_______ **2.** Tells the liver to convert glycogen to glucose, raising blood glucose levels
_______ **3.** Tells the kidney to reabsorb Na^+ and excrete K^+
_______ **4.** Increases heart and respiration rate and increases blood glucose levels
_______ **5.** Stimulates appetite

a. Glucagon
b. Epinephrine
c. Thyroid hormone
d. Insulin
e. Aldosterone
f. Cortisol

CHAPTER 9 REVIEW QUESTIONS

Matching: *Match the hormone to its chemical composition. Some answers may be used more than once.*

______	**6.** Insulin	**a.**	Protein
______	**7.** Epinephrine	**b.**	Amino acid derivative
______	**8.** Estrogen	**c.**	Steroid
______	**9.** Thyroid hormone		
______	**10.** Cortisol		

Completion: *Fill in the blanks to complete the following statements.*

1. ____________________ is the amount of time it takes for half of a substance to be removed from the system.

2. The adrenal ____________________ is in the middle of the adrenal gland and produces the hormone____________________.

3. The ____________________ is a stalk that connects the hypothalamus and pituitary glands.

4. The ____________________ sits in the sella turcica.

5. The ____________________ is located inferior to the posterior corpus callosum.

Critical Thinking

1. How could the effects of aging on the endocrine system be reduced? Explain one example.

____

____

____

____

____

2. The liver produces plasma proteins to extend the half-lives of some hormones. What will be the effect on the half-life of insulin if the liver is damaged and cannot produce plasma proteins? Explain.

____

____

____

____

3. Iodine is needed in the diet for the thyroid to produce functional thyroid hormone. What will happen to levels of TRH and TSH if there is insufficient iodine in the diet? Explain.

____

____

____

____

Case Study

Kenny is a 16-year-old who has visited his physician complaining of frequent urination, excessive thirst, and rapid weight loss. When he arrived, the medical assistant took a medical history, vital signs, height, and weight. Kenny's vital signs indicated that his blood pressure was slightly elevated, he is considered obese with a weight of 210 pounds and a body mass index over 30, and his patient history revealed diabetes on both sides of the family. When the physician interviewed Kenny, he discovered that some of Kenny's favorite things to do are play video games and peruse the Internet.

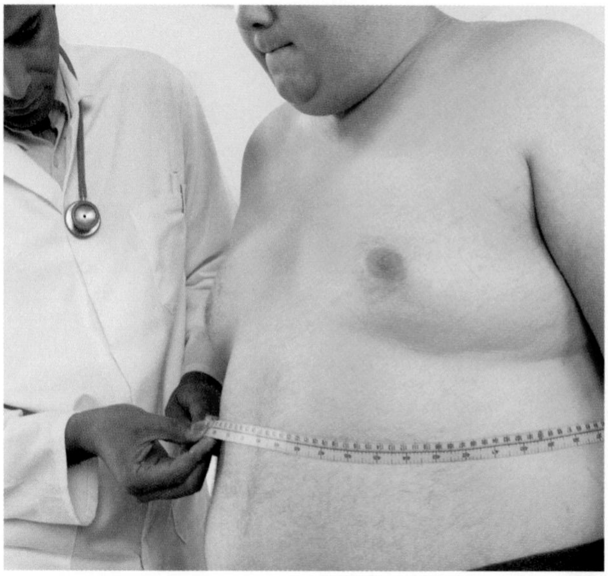

1. With the symptoms listed, what endocrine disorder might Kenny have?

____

2. Are there multiple types of this disease (your answer for number 1)? If so, which types might Kenny have? Explain how you came to this conclusion.

____

____

3. What additional diagnostic tests should be performed to determine what Kenny is suffering from? What might these tests show that would indicate a problem?

____

____

____

CHAPTER 9 MAPPING

This section of the chapter is designed to help you find where each outcome is covered in the textbook and workbook.

Learning Outcomes		Text Readings, Figures, and Tables	Workbook Activities	Text and Workbook Assessments
9.1	Use medical terminology related to the endocrine system.	Word roots & combining forms, pronunciation key: p. 315	Word roots & combining forms	*Text:* Review: 9.1 *Workbook:* Word Deconstruction: 1–5
9.2	Compare and contrast the endocrine and nervous systems in terms of type, specificity, speed, and duration of communication.	Overview: p. 316 Figure 9.2	*Lab exercises and activities:* Comparison of the endocrine and nervous systems Table 9.1	*Text:* Spot Check: 2 Review: 9.2 *Workbook:* Multiple Select: 3
9.3	Define *gland, hormone,* and *target tissue.*	Anatomy of the endocrine system: p. 317		*Text:* Spot Check: 1 Review: 9.3 *Workbook:* Multiple Select: 6
9.4	List the major hormones, along with their target tissues and functions, of each of the endocrine system glands.	Glands: pp. 317–319 Figure 9.3 Table 9.1		*Text:* Spot Check: 3, 4 Review: 9.4 *Workbook:* Multiple Select: 4, 9, 10 Matching: 1–5 Completion: 2, 4, 5
9.5	Locate and identify endocrine system glands.	Glands: pp. 319–324 Figures 9.3–9.8	*Coloring Book:* Endocrine system glands Figure 9.1	*Text:* Review: 9.5 *Workbook:* Completion: 2, 3
9.6	Describe the chemical makeup of hormones, using estrogen, insulin, and epinephrine as examples.	Hormones: pp. 324–325 Figure 9.9		*Text:* Spot Check: 5 Review: 9.6 *Workbook:* Multiple Select: 2 Matching: 6–10
9.7	Compare the location of receptors for protein hormones with that of receptors for steroid hormones.	Target tissues: p. 325 Figures 9.10–9.11		*Text:* Review: 9.7 *Workbook:* Multiple Select: 8
9.8	Differentiate autocrine, paracrine, endocrine, and pheromone chemical signals in terms of the proximity of the target tissue.	Location of target tissues: pp. 325–327	*Concept maps:* Location of target tissues Figure 9.4	*Text:* Review: 9.8 *Workbook:* Multiple Select: 7

Learning Outcomes		Text Readings, Figures, and Tables	Workbook Activities	Text and Workbook Assessments
9.9	Explain the regulation of hormone secretion and its distribution.	Regulation of hormone secretion and distribution: pp. 327–329 Figures 9.12–9.13	*Concept maps:* Regulation of hormone secretion Figure 9.5	*Text:* Spot Check: 6, 8 Review: 9.9 *Workbook:* Multiple Select: 9, 10 Critical Thinking: 2
9.10	Explain how the number of receptors can be changed.	Receptor regulation: pp. 329–330 Figure 9.14		*Text:* Review: 9.10 *Workbook:* Multiple Select: 5
9.11	Explain how hormones are eliminated from the body.	Hormone elimination: pp. 330–331	*Concept maps:* Hormone elimination Figure 9.6	*Text:* Spot Check: 7 Review: 9.11 *Workbook:* Multiple Select: 9, 10 Completion: 1
9.12	Explain the function of hormones by showing how they interact to maintain homeostasis.	Function: 4 scenarios: pp. 331–335 Figure 9.15 Tables 9.2–9.6	*Lab exercises and activities:* Graphing an endocrine problem Figures 9.2–9.3	*Text:* Spot Check: 9–17 Review: 9.12 *Workbook:* Critical Thinking: 3
9.13	Explain the effects of aging on the endocrine system.	Effects of aging on the endocrine system: p. 335		*Text:* Review: 9.13 *Workbook:* Critical Thinking: 1
9.14	Describe common diagnostic tests used to diagnose endocrine system disorders.	Diagnostic test for endocrine system disorders: pp. 335–336 Table 9.7		*Text:* Spot Check: 18 *Review: 9.14* Workbook: Case Study: 3
9.15	Describe endocrine system disorders and relate abnormal function to pathology.	Endocrine system disorders: pp. 336–343 Figures 9.16–9.20 Table 9.8		*Text:* Spot Check: 19 Review: 9.15 *Workbook:* Multiple Select: 1 Case Study: 1, 2

10 The Cardiovascular System—Blood

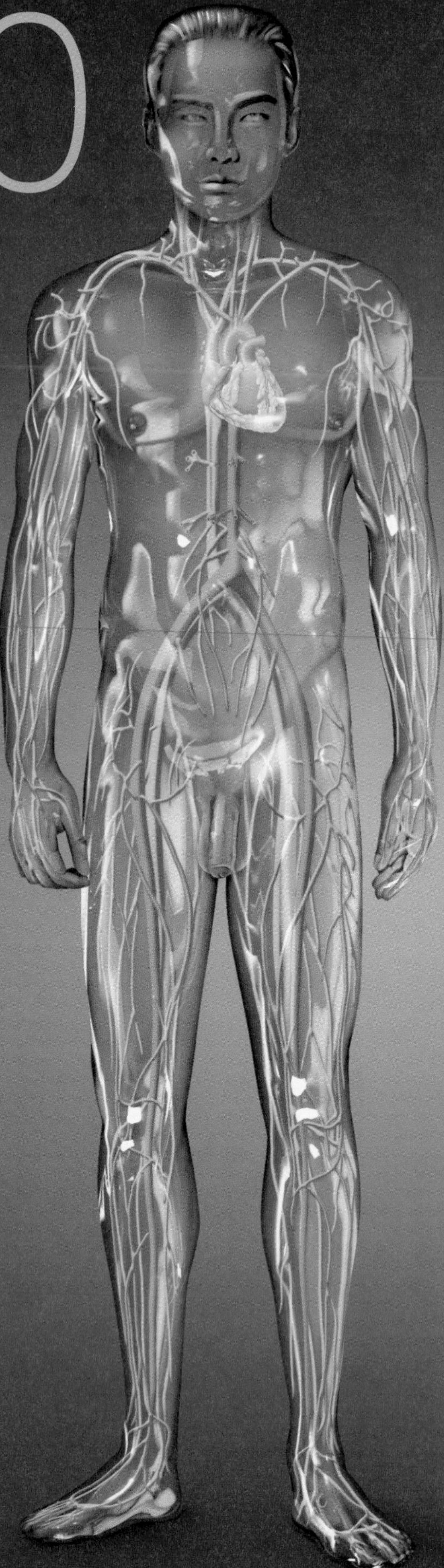

Major Organs and Structures:
heart, aorta, superior and inferior venae cavae

Accessory Structures:
arteries, veins, capillaries

Functions:
transportation, protection by fighting foreign invaders and clotting to prevent its own loss, acid-base balance, fluid and electrolyte balance, temperature regulation

learning outcomes

This chapter of the workbook is designed to help you learn the anatomy and physiology of the cardiovascular system concerning blood. After completing this chapter in the text and this workbook, you should be able to:

10.1 Use medical terminology related to the cardiovascular system.

10.2 Identify the components of blood.

10.3 List the constituents of plasma and their functions.

10.4 Identify the formed elements and list their functions.

10.5 Compare the various forms of hemopoiesis in terms of starting cell, factors influencing production, location, and final product.

10.6 Describe the structure and function of hemoglobin.

10.7 Summarize the nutritional requirements of red blood cell production.

10.8 Describe the life cycle of a red blood cell from its formation to removal.

10.9 Describe the body's mechanisms for controlling bleeding.

10.10 Describe two pathways for blood clotting in terms of what starts each, their relative speed, and the clotting factors involved.

10.11 Describe what happens to blood clots when they are no longer needed.

10.12 Explain what keeps blood from clotting in the absence of injury.

10.13 Explain what determines ABO and Rh blood types.

10.14 Explain how a blood type relates to transfusion compatibility.

10.15 Determine, from a blood type, the antigens and antibodies present and the transfusion compatibility.

10.16 Predict the compatibility between mother and fetus given Rh blood types for both and describe the possible effects.

10.17 Summarize the functions of blood by giving an example or explanation of each.

10.18 Describe common diagnostic blood tests and explain what can be learned from them.

10.19 Describe disorders of the cardiovascular system concerning blood and relate abnormal function to pathology.

word **roots** & combining **forms**

agglutin/o: clumping

blast/o: primitive cell

coagul/o: clot

cyt/o: cell

erythr/o: red

granul/o: granules

hem/o: blood

hemat/o: blood

leuk/o: white

phag/o: eat, swallow

thromb/o: clot

COLORING BOOK

Formed Elements

Color each of the formed elements in **Figure 10.1** as they would appear on a prepared slide.

FIGURE 10.1 Formed elements.

- [ ] **Erythrocyte**(A)
- [ ] **Neutrophil**(B)
- [ ] **Basophil**(C)
- [ ] **Eosinophil**(D)
- [ ] **Monocyte**(E)
- [ ] **Lymphocyte**(F)
- [ ] **Thrombocyte**(G)

Blood Cell Identification

Identify the formed element indicated by the arrow on each of the figures of prepared blood slides (**Figures** 10.2 to 10.5).

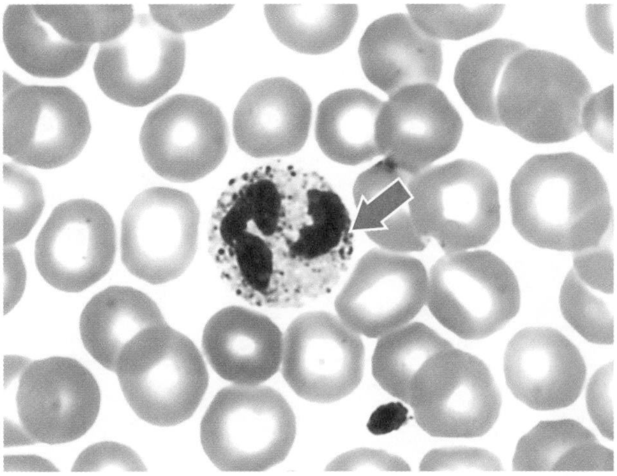

FIGURE 10.2

1. Identify the cell indicated by the arrow in **Figures** 10.2.

    ____

2. What is the function of this formed element?

    ____

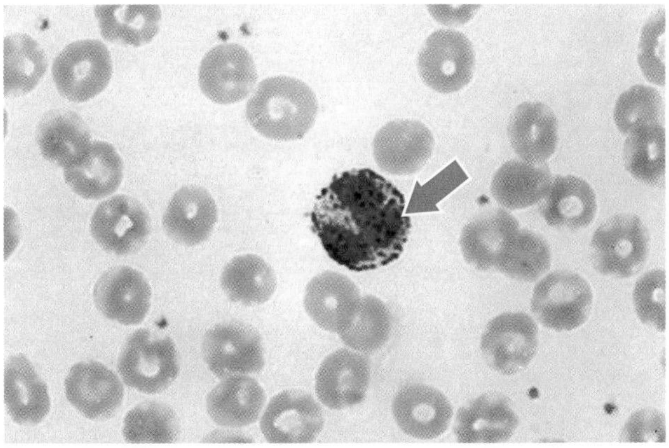

FIGURE 10.3

3. Identify the cell indicated by the arrow in **Figures** 10.3.

    ____

4. What is the function of this formed element?

    ____

LABORATORY EXERCISES AND ACTIVITIES

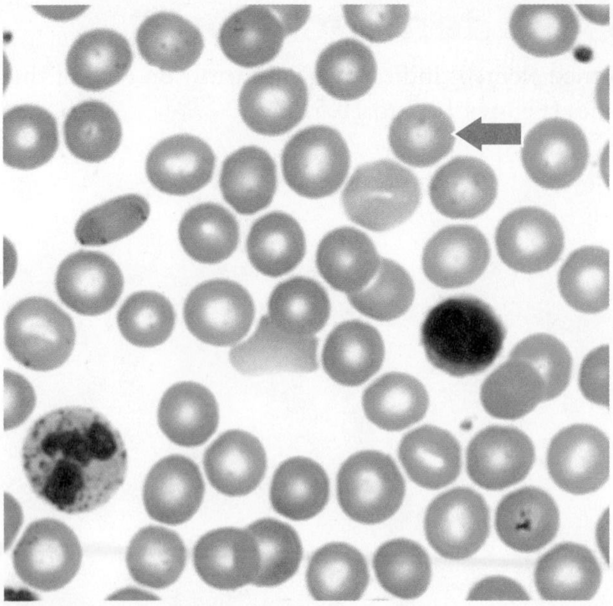

FIGURE 10.4

5. Identify the cell indicated by the arrow in **Figures** 10.4.

____

6. What is the function of this formed element?

____

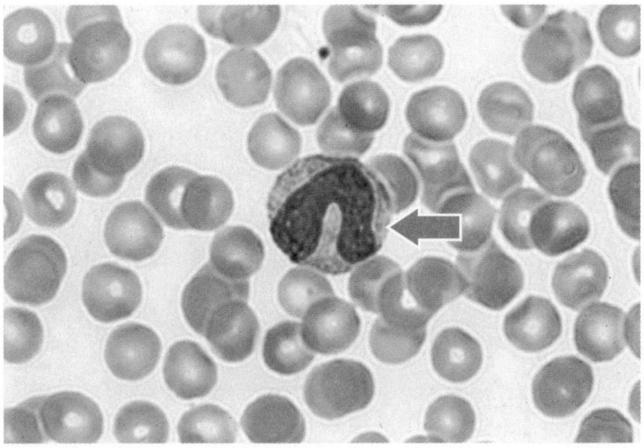

FIGURE 10.5

7. Identify the cell indicated by the arrow in **Figures** 10.5.

____

8. What is the function of this formed element?

____

Blood Doping

Blood doping is a process tried by some athletes to increase the number of their red blood cells. Many athletic organizations screen athletes to make sure the practice is not used. In the process, whole blood is withdrawn and stored. This is similar to a blood donation at a blood bank. Two weeks later, the stored blood is administered back into the athlete's bloodstream before an athletic event. Answer the following questions about this practice:

1. What happens to the oxygen level in the blood immediately after the blood is withdrawn?

2. What organ(s) in the body notice(s) the changed oxygen level?

3. How will the organ(s) respond?

4. What happens to the blood in the body during the two weeks between the time the blood is withdrawn and the administration of the stored blood? Explain.

5. What advantage does this process provide an athlete?

6. What blood condition is produced when the stored blood is administered back into the body?

7. Is there any physiological disadvantage to this practice? Explain.

LABORATORY EXERCISES AND ACTIVITIES

8. What could be administered to achieve the same effect without withdrawing blood? Explain.

____

____

____

Blood Typing

Complete Table 10.1 concerning blood types and transfusion compatibility. Two of the rows have been completed for you. Here are some points to remember:

- Blood type is determined by the antigens on the cells.
- Antibodies, dissolved in plasma, function to seek out and agglutinate foreign antigens.
- Rh antibodies are not present in Rh− blood unless the person has been exposed to the Rh antigen. An Rh− person should not be exposed to the Rh antigen.
- The donor's antigens must survive the recipient's antibodies in a transfusion.

TABLE 10.1 Blood typing and transfusion compatibility

Type	Antigens	Antibodies	To whom this type can donate blood	From whom this type can receive blood
A+	A and Rh	Anti-B	A+, AB+	A+, A−, O+, O−
A−				
B+				
B−				
AB+				
AB−	A and B	None, except Rh if previously exposed to the Rh antigen	AB− and AB+	A−, B−, AB−, and O−
O+				
O−				

Blood Tests

Compare the following blood test results for a 39-year-old male with the normal blood values listed in Table 10.3 of the textbook.

- Hematocrit: 58%
- Hemoglobin: 16 g/dL of blood
- Red blood cell count: 6.8 million/mm^3 of blood
- White blood cell count: 12,000/mm^3 of blood
- White blood cell differential:
  - Neutrophils: 74%
  - Basophils: 1%
  - Eosinophils: 1%
  - Monocytes: 4%
  - Lymphocytes: 20%
- Platelet count: 300,000/mm^3 of blood

1. Which tests are abnormal?

____

____

____

2. What questions might a physician ask to understand the reason(s) for the abnormal results?

____

____

____

3. What disorders might be indicated by these results?

____

____

____

LABORATORY EXERCISES AND ACTIVITIES

KEY WORD CONCEPT MAPS

Key Words

The following terms are defined in the glossary of the textbook.

agglutination
clotting factor
coagulation
erythrocyte
erythropoietin (EPO)
fibrin
fibrinogen
formed elements
granulocyte
hematocrit
hemocytoblast
hemoglobin
hemopoiesis
hemostasis
leukocyte
leukocytosis
leukopenia
lymphoid
myeloid
pluripotent
polycythemia
septicemia
serum
thrombocyte

Concept Maps

Use key words and other bold words from the chapter to complete the following concept maps (**Figures 10.6 to 10.10**).

c. Sara is type AB−.
d. Sara is O+.
e. Sara does not have any ABO or Rh antigens on her cells.

7. What is the composition of blood?
   a. Blood is composed mostly of formed elements.
   b. Blood is composed mostly of plasma.
   c. The buff-color layer in a hematocrit shows that less than 1 percent of blood is composed of leukocytes and platelets.
   d. Blood is a connective tissue composed of cells in a fluid matrix.
   e. Blood contains a solution called *plasma.*
8. Which of the following statements describe(s) a function of a formed element?
   a. Erythrocytes carry oxygen.
   b. Basophils secrete histamine.
   c. Eosinophils attack worms.
   d. Monocytes become macrophages.
   e. Platelets secrete growth factors to stimulate mitosis in broken blood vessel walls.
9. Which of the following statements describe(s) a form of hemopoiesis?
   a. Myeloid hemopoiesis produces any of the formed elements in the red bone marrow.
   b. Lymphoid hemopoiesis produces any of the formed elements in lymphoid tissue, such as the lymph nodes and spleen.
   c. Thrombopoiesis is started by a chemical from the liver and spleen.
   d. Leukopoiesis is started by chemicals secreted by macrophages and lymphocytes.
   e. Erythropoiesis is started by a chemical produced by the kidneys.
10. Which of the following statements is (are) true concerning hemoglobin?
   a. Hemoglobin is composed of four amino acid chains and a heme group.
   b. Hemoglobin is a complex protein.
   c. Hemoglobin contains iron.
   d. Hemoglobin is found in leukocytes.
   e. Hemoglobin is a solute in the cytoplasm of erythrocytes.

Matching: *Match each example with the type of solute found in plasma. Some answers may be used more than once.*

_______	1. Bilirubin	a. Albumins
_______	2. Oxygen	b. Globulins
_______	3. Glucose	c. Waste products
_______	4. Transport protein	d. Nutrients
_______	5. Anti-A antibodies	e. Gases

Matching: *Match the nutrient with its function in blood. Some answers may be used more than once.*

_______	6. Copper	a. Needed for clotting
_______	7. Iron	b. Needed for cell division
_______	8. Folic acid	c. Needed to carry oxygen
_______	9. Vitamin B_{12}	d. Needed for hemoglobin
_______	10. Calcium	e. Needed for enzymes to form hemoglobin

Completion: *Fill in the blanks to complete the following statements.*

1. ______________________________ is a disorder caused by a blood-type incompatibility between a mother and her fetus.
2. ______________________________ is the process of dissolving a clot when it is no longer useful.
3. ______________________________ can result from overproduction of red blood cells or dehydration.
4. In a transfusion, the donor's ______________________________ must survive the recipient's ______________________________.
5. In blood typing, ______________________________ seek out and attack foreign ______________________________.

Critical Thinking

1. Explain four reasons why blood clotting might be slower than normal.

____

____

____

____

____

____

2. George has tapeworms. His only symptom is mild abdominal discomfort. His physician ordered routine blood tests. What do you predict the results will show? Explain.

____

____

____

____

3. How would the white blood count and white blood cell differential differ for a person with leukemia and a person with leukocytosis? Why is there a difference?

____

____

____

____

Case Study

MaryAnn is pregnant with her second child. MaryAnn is Rh− and her first child is Rh+. Answer the following questions about MaryAnn and her children.

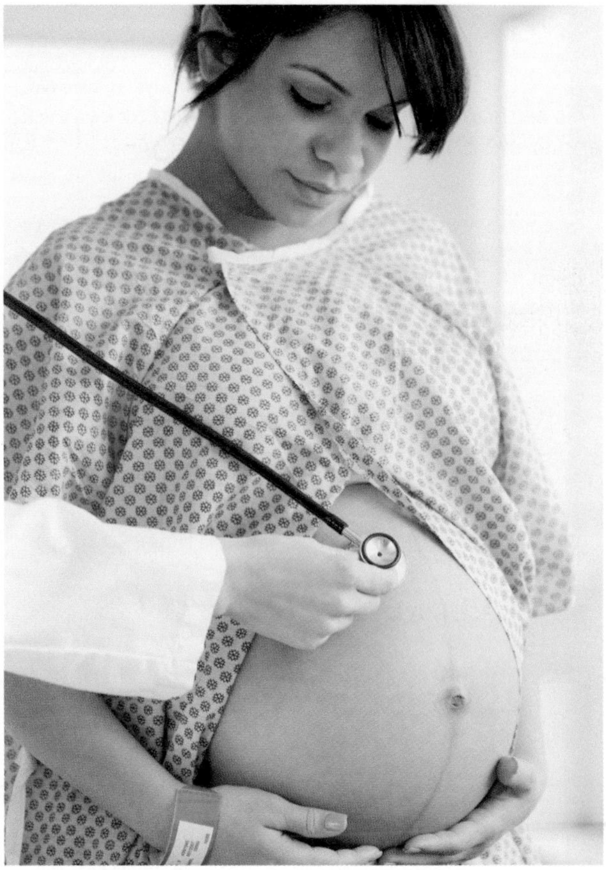

1. Does MaryAnn have anti-Rh antibodies? Explain your answer.

____

____

2. What could possibly happen to the second child if his or her blood type is Rh−? What might happen if the second child is Rh+?

____

____

____

____

3. What can result if this incompatibility of blood types occurs between the mother and the fetus and how can this incompatibility be avoided?

____

____

____

____

CHAPTER 10 MAPPING

This section of the chapter is designed to help you find where each outcome is covered in the textbook and workbook.

Learning Outcomes		Text Readings, Figures, and Tables	Workbook Activities	Text and Workbook Assessments
10.1	Use medical terminology related to the cardiovascular system.	Word roots & combining forms, pronunciation key: pp. 349–350	Word roots & combining forms	*Text:* Review: 10.1 *Workbook:* Word Deconstruction: 1–5
10.2	Identify the components of blood.	Anatomy of blood: p. 350 Figure 10.2	*Concept maps:* Blood composition Figure 10.6	*Text:* Spot Check: 1, 3 Review: 10.2 *Workbook:* Multiple Select: 7
10.3	List the constituents of plasma and their functions.	Plasma: pp. 350–352 Figure 10.2		*Text:* Spot Check: 2 Review: 10.3 *Workbook:* Matching: 1–5
10.4	Identify the formed elements and list their functions.	Formed elements: pp. 352–356 Table 10.1	*Coloring book:* Formed elements Figure 10.1 *Lab exercises and activities:* Blood cell identification Figures 10.2–10.5	*Text:* Spot Check: 4, 5 Review: 10.4 *Workbook:* Multiple Select: 8
10.5	Compare the various forms of hemopoiesis in terms of starting cell, factors influencing production, location, and final product.	Hemopoiesis: pp. 356–359 Figures 10.3–10.4 Table 10.2	*Lab exercises and activities:* Blood doping *Concept maps:* Hemopoiesis Figure 10.7	*Text:* Spot Check: 6, 7 Review: 10.5 *Workbook:* Multiple Select: 9
10.6	Describe the structure and function of hemoglobin.	Hemoglobin: pp. 359–360 Figures 10.5–10.6		*Text:* Review: 10.6 *Workbook:* Multiple Select: 10
10.7	Summarize the nutritional requirements of red blood cell production.	Nutritional requirements for erythropoiesis: p. 361		*Text:* Review: 10.7 *Workbook:* Matching: 6–9
10.8	Describe the life cycle of a red blood cell from its formation to removal.	Life cycle of a red blood cell: pp. 361–362 Figure 10.7		*Text:* Spot Check: 8 Review: 10.8 *Workbook:* Multiple Select: 3
10.9	Describe the body's mechanisms for controlling bleeding.	Hemostasis: pp. 362–363 Figure 10.8	*Concept maps:* Hemostasis Figure 10.8	*Text:* Spot Check: 9 Review: 10.9 *Workbook:* Multiple Select: 1

Learning Outcomes	Text Readings, Figures, and Tables	Workbook Activities	Text and Workbook Assessments
10.10 Describe two pathways for blood clotting in terms of what starts each, their relative speed, and the clotting factors involved.	Pathways of blood clotting: pp. 363–365 Figures 10.9–10.10		*Text:* Review: 10.10 *Workbook:* Matching: 10 Critical Thinking: 1
10.11 Describe what happens to blood clots when they are no longer needed.	Elimination of blood clots: p. 365		*Text:* Review: 10.11 *Workbook:* Completion: 2
10.12 Explain what keeps blood from clotting in the absence of injury.	Preventing inappropriate clotting: pp. 365–366		*Text:* Review: 10.12 *Workbook:* Multiple Select: 4
10.13 Explain what determines ABO and Rh blood types.	Blood typing: pp. 366–367 Figures 10.11–10.12		*Text:* Spot Check: 10, 11 Review: 10.13 *Workbook:* Multiple Select: 2
10.14 Explain how a blood type relates to transfusion compatibility.	Determining a blood type and transfusion compatibility: pp. 367–368 Figure 10.13		*Text:* Review: 10.14 *Workbook:* Completion: 4, 5
10.15 Determine, from a blood type, the antigens and antibodies present and the transfusion compatibility.	Determining a blood type and transfusion compatibility: p. 368	*Lab exercises and activities:* Blood typing Table 10.1 *Concept maps:* Blood typing Figure 10.9	*Text:* Spot Check: 12 Review: 10.15 *Workbook:* Multiple Select: 6
10.16 Predict the compatibility between mother and fetus given Rh blood types for both and describe the possible effects.	Mother-fetus blood-type compatibility: pp. 368–369 Figure 10.14		*Text:* Review: 10.16 *Workbook:* Completion: 1 Case Study: 1, 2
10.17 Summarize the functions of blood by giving an example or explanation of each.	Functions of blood: pp. 369–370 Figure 10.15		*Text:* Review: 10.17 *Workbook:* Multiple Select: 5
10.18 Describe common diagnostic blood tests and explain what can be learned from them.	Diagnostic tests for blood disorders: pp. 371–372 Figure 10.16 Table 10.3	*Lab exercises and activities:* Blood tests	*Text:* Spot Check: 13 Review: 10.18 *Workbook:* Critical Thinking: 2
10.19 Describe disorders of the cardiovascular system concerning blood and relate abnormal function to pathology.	Blood disorders: pp. 372–377 Figures 10.17–10.18 Table 10.4	*Lab exercises and activities:* Blood doping *Concept maps:* Anemia Figure 10.10	*Text:* Spot Check: 14 Review: 10.19 *Workbook:* Completion: 3 Case Study: 3

11 The Cardiovascular System—Heart and Vessels

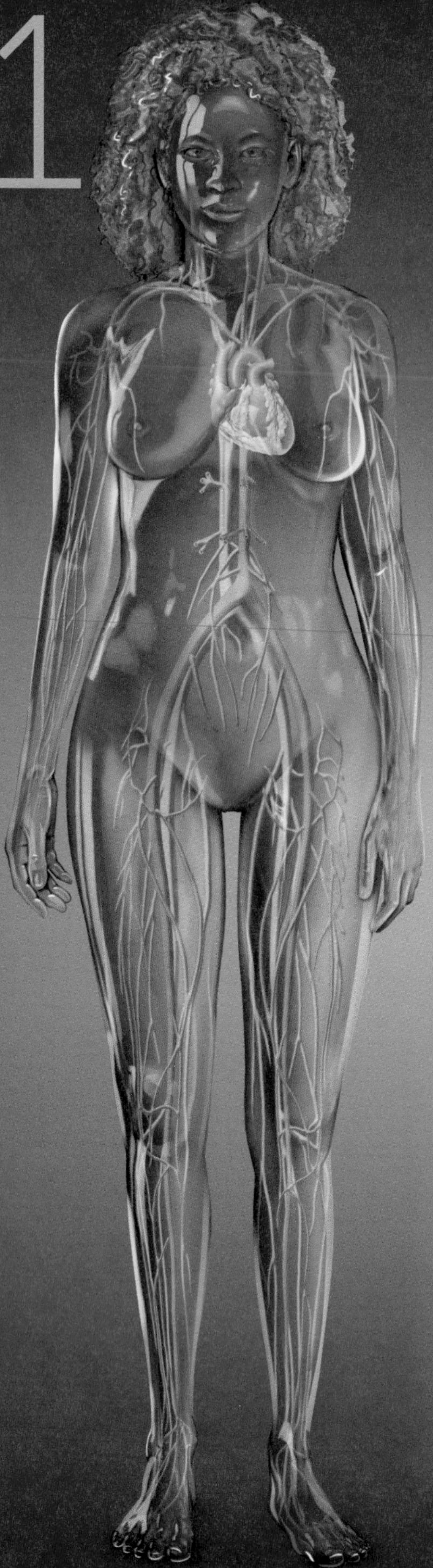

Major Organs and Structures:
heart, aorta, superior and inferior venae cavae

Accessory Structures:
arteries, veins, capillaries

Functions:
transportation, protection by fighting foreign invaders and clotting to prevent its own loss, acid-base balance, fluid and electrolyte balance, temperature regulation

learning outcomes

This chapter of the workbook is designed to help you learn the anatomy and physiology of the cardiovascular system. After completing this chapter in the text and this workbook, you should be able to:

11.1 Use medical terminology related to the cardiovascular system.

11.2 Identify the chambers, valves, and features of the heart.

11.3 Relate the structure of cardiac muscle to its function.

11.4 Explain why the heart does not fatigue.

11.5 Trace blood flow through the heart.

11.6 Describe the heart's electrical conduction system.

11.7 Describe the events that produce the heart's cycle of contraction and relaxation.

11.8 Interpret a normal EKG, explaining what is happening electrically in the heart.

11.9 Calculate cardiac output given heart rate and stroke volume.

11.10 Explain the factors that govern cardiac output.

11.11 Summarize nervous and chemical factors that alter heart rate, stroke volume, and cardiac output.

11.12 Locate and identify the major arteries and veins of the body.

11.13 Compare the anatomy of the three types of blood vessels.

11.14 Describe coronary and systemic circulatory routes.

11.15 Explain how blood in veins is returned to the heart.

11.16 Explain the relationship between blood pressure, resistance, and flow.

11.17 Describe how blood pressure is expressed and how mean arterial pressure and pulse pressure are calculated.

11.18 Explain how blood pressure and flow are regulated.

11.19 Explain the effect of exercise on cardiac output.

11.20 Summarize the effects of aging on the cardiovascular system.

11.21 Describe common diagnostic tests used to diagnose heart and vessel disorders.

11.22 Describe heart and vessel disorders and relate abnormal function to pathology.

word **roots** & combining **forms**

arter/o, arteri/o: artery

ather/o: fatty substance

atri/o: atrium

brady/: slow

cardi/o: heart

coron/o: heart

pericardi/o: pericardium

rhythm/o: rhythm

sphygm/o: pulse

steth/o: chest

tachy/: rapid

vas/o: vessel

vascul/o: vessel

ven/o, ven/i: vein

ventricul/o: ventricle

Heart

Figure 11.1 shows the external anatomy of the heart and the coronary vessels. Color the box next to each term. Use the same color for the corresponding structure in the figure.

- ☐ Right auricle[(A)]
- ☐ Left auricle[(B)]
- ☐ Left atrium[(C)]
- ☐ Right ventricle[(D)]
- ☐ Left ventricle[(E)]
- ☐ Left coronary artery[(F)]
- ☐ Right coronary artery[(G)]
- ☐ Coronary sinus[(H)]
- ☐ Great cardiac vein[(I)]
- ☐ Anterior interventricular artery[(J)]
- ☐ Right atrium[(K)]

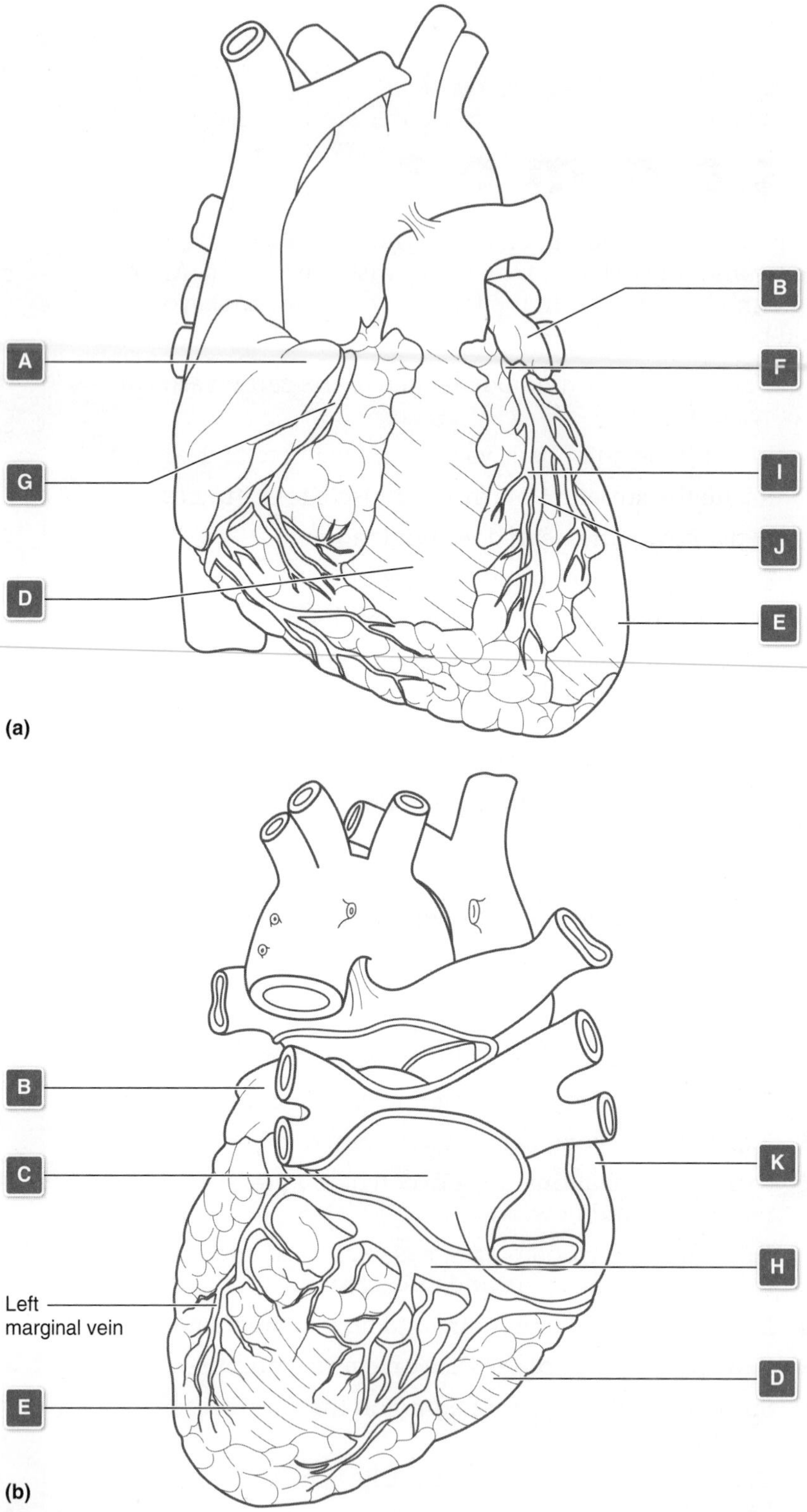

FIGURE 11.1 External anatomy of the heart: (a) anterior view, (b) posterior view.

Pericardial Sac

Figure 11.2 shows the pericardial membrane in relation to the heart wall. Color the box next to each term. Use the same color for the corresponding structure in the figure.

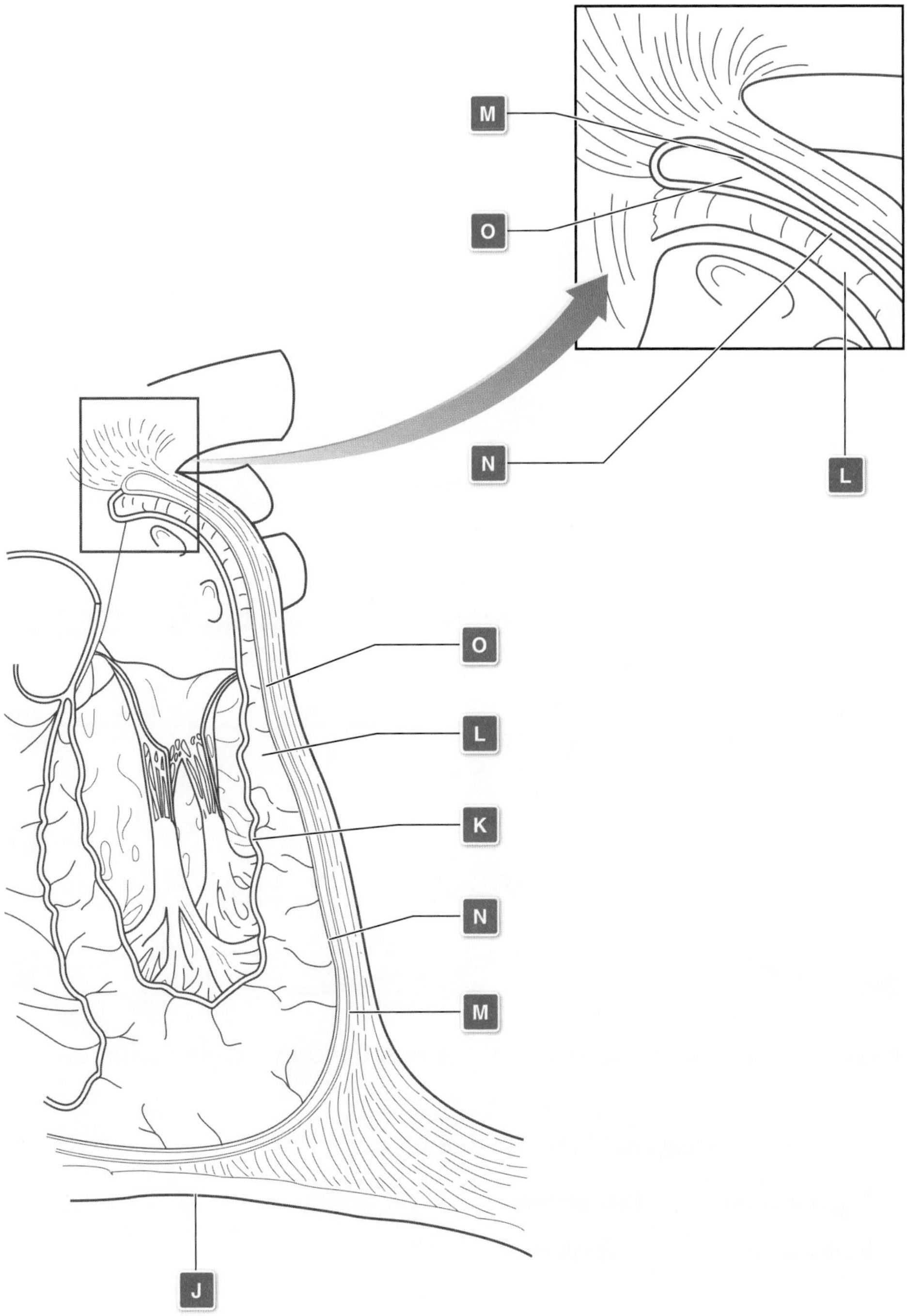

- [ ] Diaphragm(J)
- [ ] Endocardium(K)
- [ ] Myocardium(L)
- [ ] Parietal pericardium(M)
- [ ] Visceral pericardium (epicardium)(N)
- [ ] Pericardial cavity with serous fluid(O)

FIGURE 11.2 Pericardial sac.

COLORING BOOK

Internal Heart Anatomy

Figure 11.3 shows the internal anatomy of the heart and associated structures. Color the box next to each term. Use the same color for the corresponding structures in the figure.

- ☐ Aorta(P)
- ☐ Pulmonary artery(Q)
- ☐ Superior and inferior venae cavae(R)
- ☐ Pulmonary trunk(S)
- ☐ Pulmonary veins(T)
- ☐ Interatrial septum(V)
- ☐ Interventricular septum(W)
- ☐ Tricuspid valve(X)
- ☐ Bicuspid valve(Y)
- ☐ Aortic valve(Z)
- ☐ Tendinous cords(A)
- ☐ Papillary muscle(B)
- ☐ Epicardium(C)
- ☐ Endocardium(D)

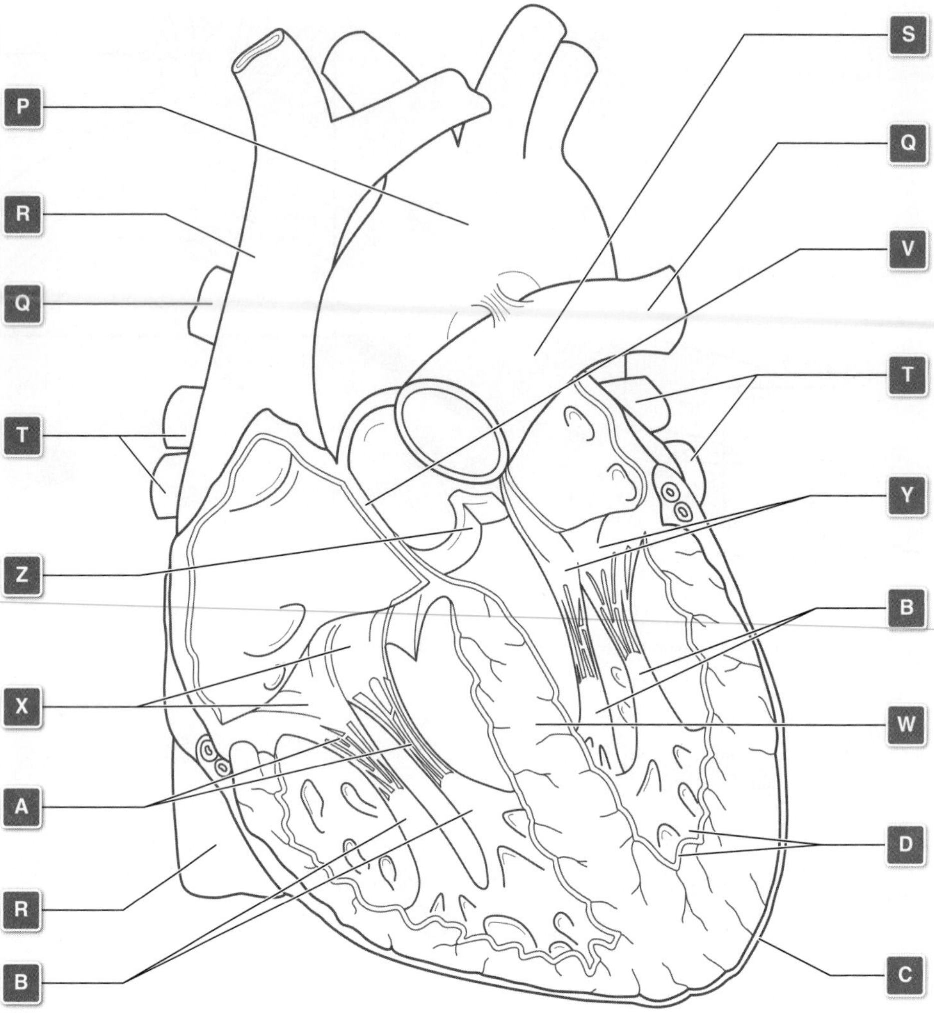

FIGURE 11.3 Internal anatomy of the heart and associated structures.

Label the following chambers in **Figure 11.3**:

Right atrium	**Left atrium**
Right ventricle	**Left ventricle**

Blood Flow through the Heart

Put the following steps in order to trace blood flow through the heart to the lungs and back from the lungs through the heart to go out to the body. The first and last steps have been provided.

Blood travels:	
To the left atrium	1. From the superior and inferior vena cavae
Through the pulmonary valve	2. ______
Through the tricuspid valve	3. ______
To the aorta	4. ______
To the left ventricle	5. ______
To the pulmonary trunk	6. ______
From the superior and inferior venae cavae	7. ______
To the right atrium	8. ______
Through the aortic valve	9. ______
Through the mitral valve	10. ______
To the pulmonary veins	11. ______
To the right ventricle	12. ______
To the lungs	13. ______
From the lungs	14. ______
To the pulmonary arteries	15. ______
To go out to the body	16. To go out to the body

Labeling Vessels

Choose from the list of arteries to label the vessels in **Figure 11.4**.

Subclavian artery
Femoral artery
Renal artery
Axillary artery
Common iliac artery
Aorta
Brachial artery
External iliac artery
Common carotid artery

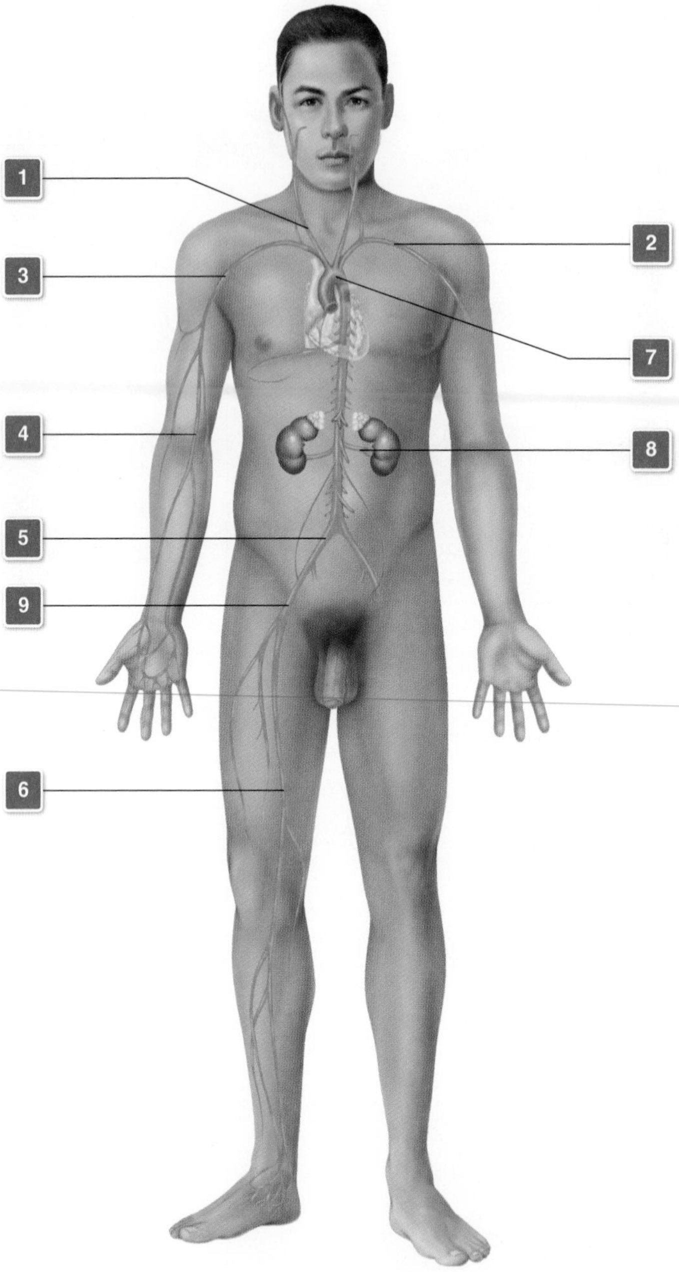

FIGURE 11.4 Major arteries.

1. ______________________________

2. ______________________________

3. ______________________________

4. ____________________

5. ____________________

6. ____________________

7. ____________________

8. ____________________

9. ____________________

Choose from the list of veins to label the vessels in **Figure 11.5**.

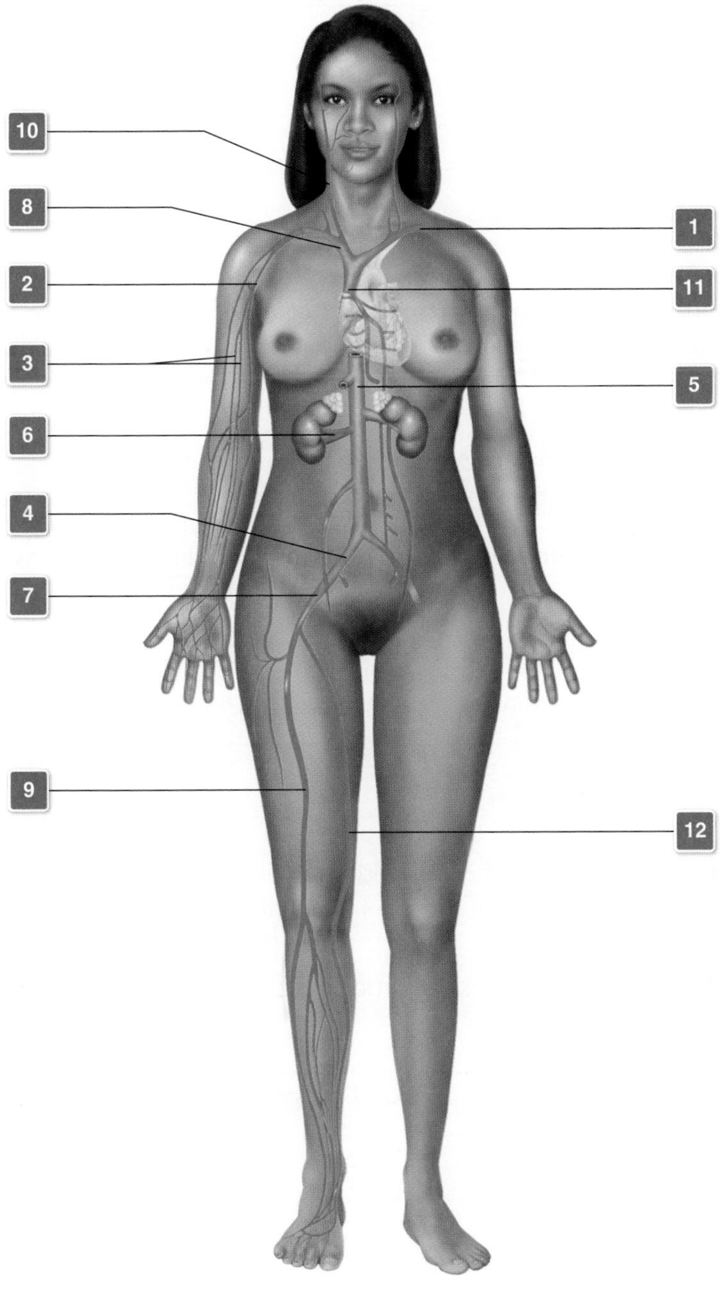

Internal jugular vein
Renal vein
Great saphenous vein
Axillary vein
Superior vena cava
Brachial veins
External iliac vein
Subclavian vein
Inferior vena cava
Femoral vein
Common iliac vein
Brachiocephalic vein

FIGURE 11.5 Major veins.

LABORATORY EXERCISES AND ACTIVITIES

1. ______________________________
2. ______________________________
3. ______________________________
4. ______________________________
5. ______________________________
6. ______________________________
7. ______________________________
8. ______________________________
9. ______________________________
10. ______________________________
11. ______________________________
12. ______________________________

BLOOD PRESSURES

Five students had their blood pressure data collected in a lab. Three of the students had their blood pressure data collected while at rest. Complete their data in Table 11.1.

TABLE 11.1 Blood pressures for students 1 to 3 at rest

At rest	Heart rate	Blood pressure	Pulse pressure	Mean arterial pressure
Student 1	72 beats/min	118/62 mmHg		
Student 2	64 beats/min	120/60 mmHg		
Student 3	69 beats/min	116/78 mmHg		

The same three students then exercised for five minutes and again had their blood pressure data collected in the lab. Complete their data in Table 11.2.

TABLE 11.2 Blood pressures for students 1 to 3 after exercise

After exercise	Heart rate	Blood pressure	Pulse pressure	Mean arterial pressure
Student 1	104 beats/min	128/76 mmHg		
Student 2	116 beats/min	140/58 mmHg		
Student 3	88 beats/min	140/74 mmHg		

1. With what you know about heart rate and blood pressure, what accounts for the difference between the at-rest and after-exercise data? Explain in terms of physiology.

___

___

___

___

___

___

___

___

___

___

___

The other two students had their blood pressure data collected while they were in a supine position on the lab table. Complete their data in Table 11.3.

TABLE 11.3 Blood pressures for students 4 and 5 while supine

While supine	Heart rate	Blood pressure	Pulse pressure	Mean arterial pressure
Student 4	60 beats/min	94/68 mmHg		
Student 5	56 beats/min	110/70 mmHg		

These students then had their blood pressure data collected immediately upon standing. Complete their data in Table 11.4.

TABLE 11.4 Blood pressures for students 4 and 5 immediately upon standing

Immediately upon standing	Heart rate	Blood pressure	Pulse pressure	Mean arterial pressure
Student 4	76 beats/min	108/84 mmHg		
Student 5	84 beats/min	120/80 mmHg		

LABORATORY EXERCISES AND ACTIVITIES

Students 4 and 5 had their blood pressure data collected one more time after standing for two minutes. Complete their data in Table 11.5.

TABLE 11.5 Blood pressures for students 4 and 5 after standing two minutes

After standing for 2 minutes	Heart rate	Blood pressure	Pulse pressure	Mean arterial pressure
Student 4	68 beats/min	98/76 mmHg		
Student 5	83 beats/min	120/68 mmHg		

2. With what you now know about heart rate and blood pressure, what accounts for the differences due to body position? Explain in terms of physiology.

6. What is the relationship between blood pressure, resistance, and flow?
   - **a.** Blood pressure is dependent on cardiac output, blood volume, and resistance.
   - **b.** Resistance is dependent on blood viscosity, vessel length, and blood pressure.
   - **c.** Vessel length can be changed by the vasomotor center.
   - **d.** Blood flow is the amount of blood flowing to an area in a given amount of time.
   - **e.** A hematocrit is a good indicator of blood viscosity.

7. How is blood returned to the heart?
   - **a.** Through cardiac suction.
   - **b.** Through gravity.
   - **c.** Through the skeletal muscle pump.
   - **d.** From pressure generated in the heart.
   - **e.** Through the thoracic pump.

8. What happens during a cardiac cycle?
   - **a.** AV valves close, causing afterload.
   - **b.** All chambers fill during ventricular diastole.
   - **c.** Papillary muscles contract to keep AV valves closed.
   - **d.** Atrial systole comes before atrial diastole.
   - **e.** The heart rests just before atrial diastole.

9. Which of the following statements is (are) accurate concerning an electrocardiogram.
   - **a.** It can also be called an EEG.
   - **b.** It measures the force of heart contractions.
   - **c.** It shows the electrical activity of the heart.
   - **d.** A normal electrocardiogram shows three electrical events and a rest.
   - **e.** An atrial depolarization is not shown on a normal electrocardiogram.

10. Bill is an 80-year-old who has not exercised much in the last 20 years and does not watch his diet. What would you expect concerning his cardiovascular health?
   - **a.** His vessels have stiffened.
   - **b.** He has more collateral circulation in his heart.
   - **c.** His vascular resistance has decreased.
   - **d.** He is probably hypertensive if he has a lot of sodium in his diet.
   - **e.** His stroke volume has increased.

Matching: *Match each description with the type of vessel. Some answers may be used more than once.*

______	**1.** Receive blood from capillaries	**a.** Large veins
______	**2.** Deliver blood to capillaries	**b.** Arterioles
______	**3.** Have three tunics	**c.** Capillaries
______	**4.** Allow for the exchange of materials	**d.** Venules
______	**5.** Have large amounts of elastic fibers to expand and contract with each heartbeat	**e.** Conducting arteries

Matching: *Match the value with its unit. Some answers may be used more than once.*

______	**6.** Cardiac output	**a.** beats/min
______	**7.** MAP	**b.** mL/beat
______	**8.** Pulse pressure	**c.** mmHg
______	**9.** Stroke volume	**d.** mL/min
______	**10.** Heart rate	

Completion: *Fill in the blanks to complete the following statements.*

1. Cardiac muscle tissue contains ______________ to store oxygen.
2. Cardiac muscle tissue contains ______________, a storage molecule for glucose.
3. A(n) ______________ rhythm is produced if the SA node is the pacemaker.
4. A(n) ______________ rhythm is produced if the AV node is the ectopic focus.
5. The cardiac output for someone with a heart rate of 68 beats/min and a stroke volume of 75 mL/beat is ______________.

Critical Thinking

1. Compare the composition of the blood (as far as oxygen and nutrients) and the direction of blood flow (in regard to the liver) for the hepatic artery, the hepatic vein, and the hepatic portal vein.

____

____

____

____

2. Paramedics arrived on the scene of a car accident involving a driver with a deep cut on his leg that was bleeding a great deal. A preliminary assessment determined that the driver's blood pressure was falling and his heart rate was above normal. Explain the mechanisms that produce this result from severe bleeding. How did the body know what to do?

____

____

____

____

3. Atherosclerosis can happen in arteries other than coronary arteries. What could be the possible consequences of atherosclerosis in the carotid arteries? Explain.

____

____

____

____

Case Study

Betty is a 60-year-old woman who has been diagnosed with atherosclerosis.

1. Explain what effect this disorder can have on Betty's coronary arteries? What condition can atherosclerosis lead to?

2. If Betty's coronary arteries are affected by atherosclerosis, what can potentially happen to Betty's heart if her condition is not treated? List two to three disorders that can result and explain how they affect the heart's ability to function.

3. What diagnostic procedures can be performed to assess the health of Betty's heart?

4. What are potential treatment options for Betty?

CHAPTER 11 MAPPING

This section of the chapter is designed to help you find where each outcome is covered in the textbook and workbook.

Learning Outcomes		Text Readings, Figures, and Tables	Workbook Activities	Text and Workbook Assessments
11.1	Use medical terminology related to the cardiovascular system.	Word roots & combining forms, pronunciation key: pp. 383–384	Word roots & combining forms	*Text:* Review: 11.1 *Workbook:* Word Deconstruction: 1–5
11.2	Identify the chambers, valves, and features of the heart.	Heart anatomy: pp. 384–391 Figures 11.2–11.7	*Coloring book:* Heart; Pericardial sac; Internal heart anatomy Figures 11.1–11.3	*Text:* Spot Check: 1, 2 Review: 11.2
11.3	Relate the structure of cardiac muscle to its function.	Cardiac muscle tissue: p. 391 Figure 11.8		*Text:* Review: 11.3 *Workbook:* Multiple Select: 1
11.4	Explain why the heart does not fatigue.	Cardiac muscle tissue: pp. 391–392 Figure 11.8		*Text:* Spot Check: 3 Review: 11.4 *Workbook:* Completion: 1, 2
11.5	Trace blood flow through the heart.	Blood flow through the heart: pp. 392–395 Figures 11.9–11.10	*Lab exercises and activities:* Blood flow through the heart	*Text:* Spot Check: 4, 5 Review: 11.5
11.6	Describe the heart's electrical conduction system.	Cardiac conduction system: pp. 395–396 Figure 11.11	*Concept maps:* Cardiac conduction system Figure 11.6	*Text:* Spot Check: 6 Review: 11.6
11.7	Describe the events that produce the heart's cycle of contraction and relaxation.	Cardiac cycle: pp. 396–399 Figures 11.12–11.13	*Concept maps:* Cardiac cycle Figure 11.7	*Text:* Spot Check: 7 Review: 11.7 *Workbook:* Completion: 3, 4 Multiple Select: 8
11.8	Interpret a normal EKG, explaining what is happening electrically in the heart.	Electrocardiogram: pp. 399–401 Figures 11.14–11.15	*Concept maps:* Electrocardiogram Figure 11.8	*Text:* Spot Check: 8 Review: 11.8 *Workbook:* Multiple Select: 9
11.9	Calculate cardiac output given heart rate and stroke volume.	Cardiac output: p. 401		*Text:* Spot Check: 9 Review: 11.9 *Workbook:* Completion: 5
11.10	Explain the factors that govern cardiac output.	Cardiac output: pp. 401–402 Figure 11.16	*Concept maps:* Stroke volume Figure 11.9	*Text:* Spot Check: 11 Review: 11.10 *Workbook:* Multiple Select: 3
11.11	Summarize nervous and chemical factors that alter heart rate, stroke volume, and cardiac output.	Heart regulation: pp. 403–404 Figure 11.17		*Text:* Spot Check: 10 Review: 11.11 *Workbook:* Multiple Select: 2 Critical Thinking: 2

Learning Outcomes		Text Readings, Figures, and Tables	Workbook Activities	Text and Workbook Assessments
11.12	Locate and identify the major arteries and veins of the body.	Vessel anatomy: pp. 404–406 Figures 11.18–11.19	*Lab exercises and activities:* Labeling vessels Figures 11.4–11.5	*Text:* Review: 11.12
11.13	Compare the anatomy of the three types of blood vessels.	Vessel anatomy: pp. 407–409 Figures 11.20–11.21	*Concept maps:* Types of vessels; Vessel walls Figures 11.10–11.11	*Text:* Review: 11.13 *Workbook:* Matching: 1–5
11.14	Describe coronary and systemic circulatory routes.	Circulatory routes: pp. 409–413 Figures 11.22–11.25		*Text:* Review: 11.14 *Workbook:* Critical Thinking: 1
11.15	Explain how blood in veins is returned to the heart.	Venous return: p. 414 Figure 11.26		*Text:* Review: 11.15 *Workbook:* Multiple Select: 7
11.16	Explain the relationship between blood pressure, resistance, and flow.	Blood pressure, resistance, and flow: p. 415		*Text:* Spot Check: 12 Review: 11.16 *Workbook:* Critical Thinking: 2 Multiple Select: 6
11.17	Describe how blood pressure is expressed and how mean arterial pressure and pulse pressure are calculated.	Blood pressure, resistance, and flow: pp. 415–416	*Lab exercises and activities:* Blood pressures Tables 11.1–11.5	*Text:* Spot Check: 13 Review: 11.17 *Workbook:* Matching: 6–10
11.18	Explain how blood pressure and flow are regulated.	Regulation of blood pressure and flow: pp. 417–419	*Lab exercises and activities:* Blood pressures Tables 11.1–11.5	*Text:* Review: 11.18 *Workbook:* Multiple Select: 5
11.19	Explain the effect of exercise on cardiac output.	Effects of exercise on cardiac output: pp. 419–420 Figure 11.27	*Lab exercises and activities:* Blood pressures Tables 11.1–11.5	*Text:* Review: 11.19 *Workbook:* Multiple Select: 4
11.20	Summarize the effects of aging on the cardiovascular system.	Effects of aging on the cardiovascular system: p. 420		*Text:* Review: 11.20 *Workbook:* Multiple Select: 10
11.21	Describe common diagnostic tests used to diagnose heart and vessel disorders.	Diagnostic tests for heart and vessel disorders: pp. 420–422 Table 11.1		*Text:* Spot Check: 14 Review: 11.21 *Workbook:* Case Study: 3
11.22	Describe heart and vessel disorders and relate abnormal function to pathology.	Disorders of the heart and vessels: pp. 422–430 Figures 11.28–11.34 Table 11.2		*Text:* Spot Check: 15 Review: 11.22 *Workbook:* Critical Thinking: 3 Case Study: 1, 2

12 The Lymphatic System

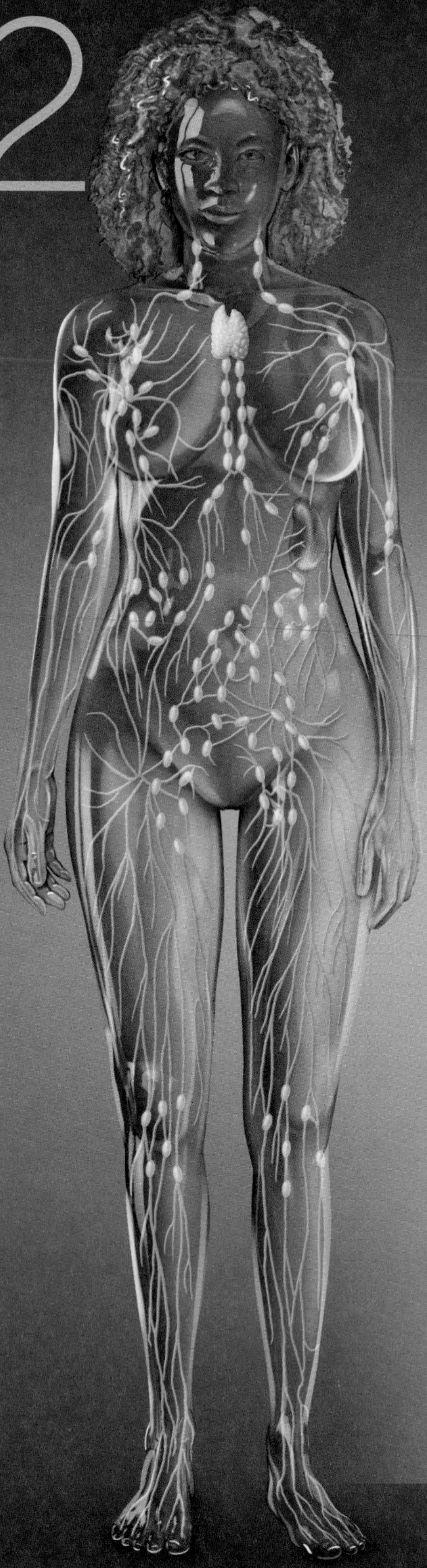

Major Organs and Structures:
thymus gland, spleen, tonsils

Accessory Structures:
thoracic duct, right lymphatic duct, lymph nodes, lymph vessels, MALT, Peyer's patches

Functions:
fluid balance, immunity, lipid absorption, defense against disease

learning outcomes

This chapter of the workbook is designed to help you learn the anatomy and physiology of the lymphatic system. After completing this chapter in the text and this workbook, you should be able to:

12.1 Use medical terminology related to the lymphatic system.

12.2 Explain the origin and composition of lymph.

12.3 Describe lymph vessels.

12.4 Explain the route of lymph from the blood and back again.

12.5 Describe cells of the lymphatic system and list their functions.

12.6 Identify lymphoid tissues and organs and explain their functions.

12.7 Summarize three lines of defense against pathogens.

12.8 Contrast nonspecific resistance and specific immunity.

12.9 Describe the body's nonspecific defenses.

12.10 Explain the role of an APC in specific immunity.

12.11 Explain the process of humoral immunity.

12.12 Explain the process of cellular immunity.

12.13 Compare the different forms of acquired immunity.

12.14 Explain the importance of T_{helper} cells to specific and nonspecific defense.

12.15 Explain the functions of the lymphatic system.

12.16 Summarize the effects of aging on the lymphatic system.

12.17 Describe common diagnostic tests used for lymphatic system disorders.

12.18 Describe lymphatic system disorders and relate abnormal function to pathology.

word roots & combining forms

immun/o: protection

lymph/o: lymph

lymphaden/o: lymph node

splen/o: spleen

thym/o: thymus gland

Lymphatic Drainage

Figure 12.1 shows part of the lymphatic drainage system. Color the box next to each term. Use the same color for the corresponding structure in the figure. Lightly color the part of the body drained by the right lymphatic duct.

- ☐ Lymph vessels(A)
- ☐ Axillary lymph nodes(B)
- ☐ Right lymphatic duct(C)
- ☐ Thoracic duct(D)
- ☐ Subclavian veins(E)
- ☐ Inguinal lymph nodes(F)

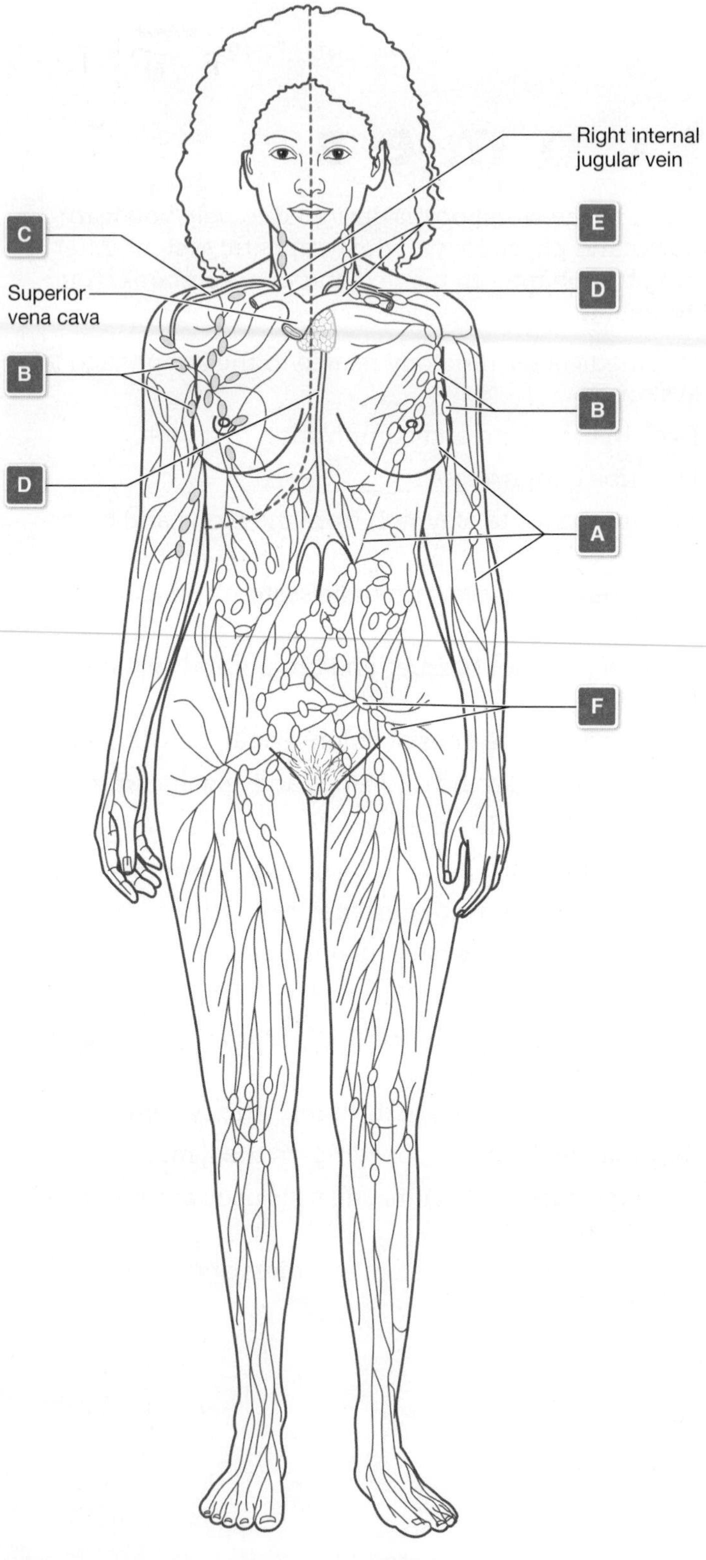

Right internal jugular vein
C
E
Superior vena cava
D
B
B
D
A
F

FIGURE 12.1 Lymphatic system drainage.

COLORING BOOK

Lymphoid Tissues

Figure 12.2 shows the lymphoid tissues of the lymphatic system. Color the box next to each term. Use the same color for the corresponding structure in the figure.

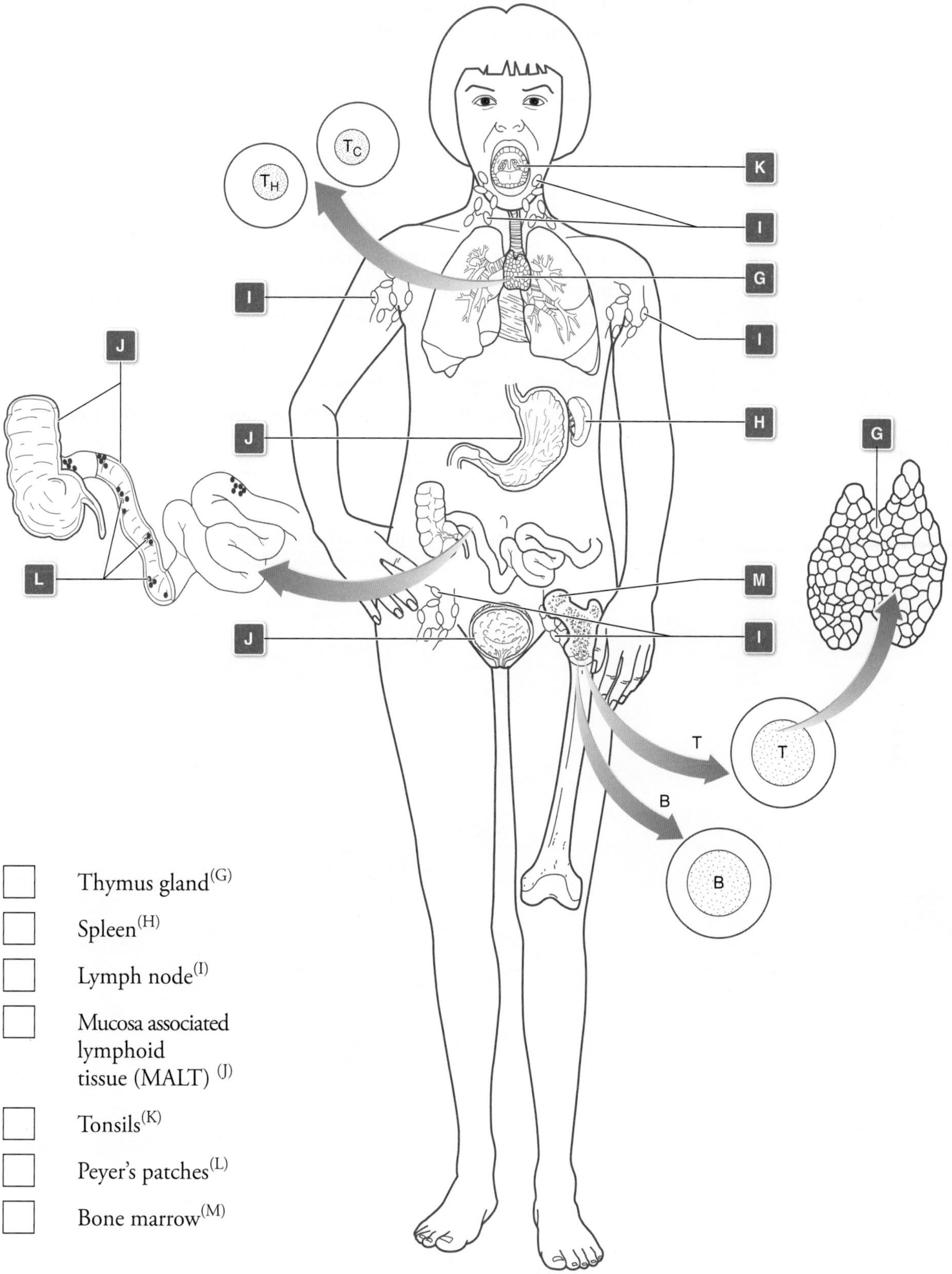

- [ ] Thymus gland(G)
- [ ] Spleen(H)
- [ ] Lymph node(I)
- [ ] Mucosa associated lymphoid tissue (MALT) (J)
- [ ] Tonsils(K)
- [ ] Peyer's patches(L)
- [ ] Bone marrow(M)

FIGURE 12.2 Lymphoid tissues of a child.

Specific Immunity

In this exercise you are to compare humoral and cellular immunity. Both of these forms of specific immunity involve three major steps—*recognize, react, and remember.* Complete Table 12.1 by writing each of the following bulleted steps in the appropriate box of the table. Some boxes of the table may contain more than one step.

- B_{memory} cells provide long-lasting protection against future exposures to the same pathogen.
- $T_{cytotoxic}$ cell binds to cells (other than APCs) displaying a foreign antigen and delivers a lethal hit, which destroys the cell.
- T_{helper} cell releases interleukin-2, which tells the B cell to clone itself.
- B_{plasma} cells make specific antibodies to travel in the bloodstream.
- If there is a subsequent exposure, B_{memory} cells will clone more B_{plasma} cells in hours and there will be a peak of antibody production in 2 days.
- T_{helper} cells recognize a foreign antigen and release interleukin-1, which tells $T_{cytotoxic}$ cells to clone themselves to produce $T_{cytotoxic}$, T_{helper}, and T_{memory} cells.
- If the foreign antigen returns, T_{memory} cells release more interleukin-1 to start more cloning of T cells.
- A virally infected cell presents an antigen from its internal environment on an MHC protein.
- An APC presents an epitope from its external environment on its MHC protein.

LABORATORY EXERCISES AND ACTIVITIES

TABLE 12.1 Specific immunity

	Humoral Immunity	Cellular Immunity
Recognize		
React		
Remember		

Inflammation

Inflammation is one of the body's nonspecific defenses. In the table below, put the steps of the inflammatory process in order. Refer to **Figure 12.3**.

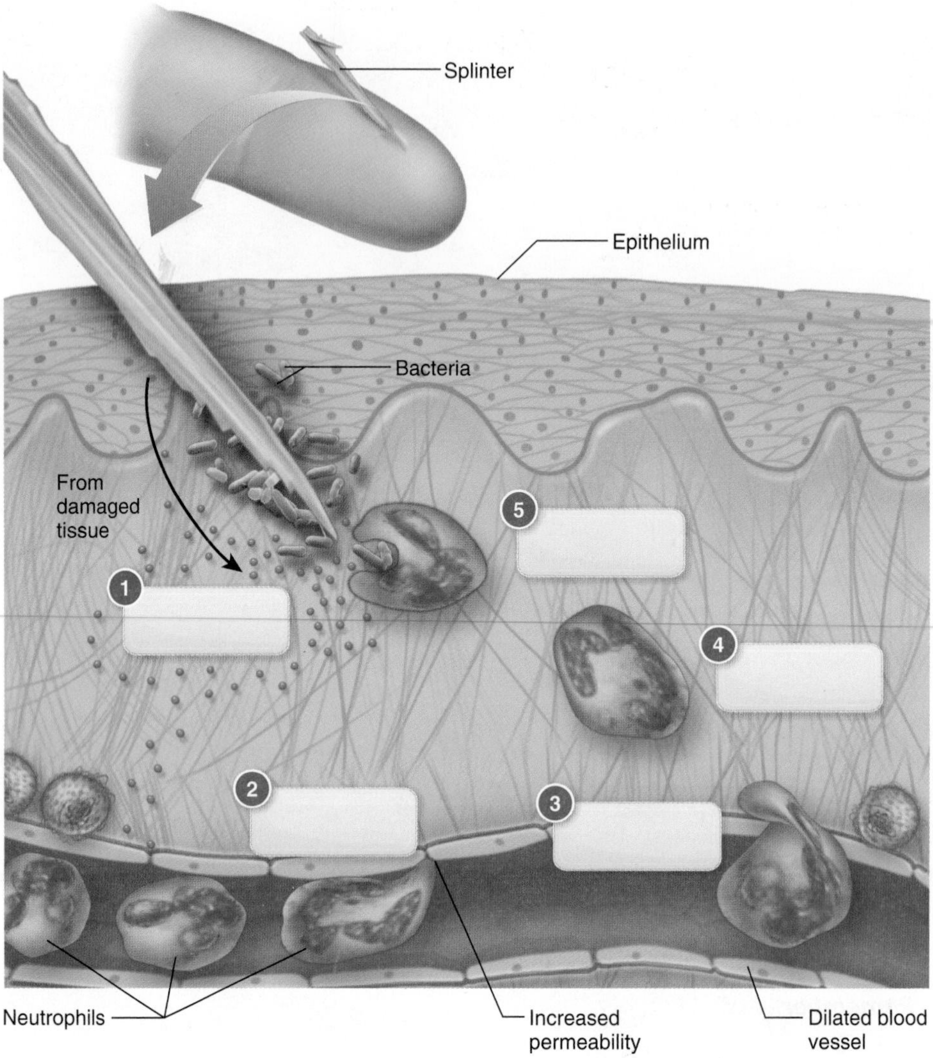

FIGURE 12.3 Inflammation.

1. ______________________	a. Margination
2. ______________________	b. Chemotaxis
3. ______________________	c. Release of vasodilators
4. ______________________	d. Diapedesis
5. ______________________	e. Phagocytosis

LABORATORY EXERCISES AND ACTIVITIES

Word Deconstruction: *In the textbook, you built words to fit a definition using the combining forms, prefixes, and suffixes. Here you are to break down the term into its parts (prefixes, roots, and suffixes) and give a definition. Prefixes and suffixes can be found inside the back cover of the textbook.*

FOR EXAMPLE Dermatitis: dermat/itis—inflammation of the skin

1. Thymic: ____________________
2. Lymphadenosis: ____________________
3. Lymphemia: ____________________
4. Immunology: ____________________
5. Splenomegaly: ____________________

Multiple Select: *Select the correct choices for each statement. The choices may be all correct, all incorrect, or any combination of correct or incorrect.*

1. What are the origin and composition of lymph?
   a. Lymph is plasma with the nutrients and wastes removed.
   b. Lymph is plasma with added proteins.
   c. Lymph is mostly water.
   d. Lymph contains glucose.
   e. Lymph contains carbon dioxide.
2. How would you describe lymph vessels?
   a. Lymph capillary walls have three tunics.
   b. Collecting ducts are open-ended to receive lymph from the tissues.
   c. Vessels carrying lymph get larger before lymph is returned to the bloodstream.
   d. Lymph vessels do not have valves.
   e. Lymph vessel walls contain smooth muscle that constricts to move lymph.
3. Which of the following statements pertain(s) to cells of the lymphatic system?
   a. $T_{cytotoxic}$ cells are important in humoral immunity and nonspecific defense.
   b. Neutrophils are important in nonspecific defense.
   c. Dendritic cells are found mostly in lymph nodes.
   d. B cells are in constant circulation.
   e. B cells are produced in the red bone marrow and are cloned in germinal centers of lymph nodes.
4. What are the three lines of defense?
   a. Mucous membranes are part of the first line of defense.
   b. Dendritic cells are important in the third line of defense.
   c. Fever is controlled by the thalamus.
   d. Fever helps speed cell division for tissue repair.
   e. The complement system is 20 inactive proteins important for blood clotting.
5. How would you contrast specific and nonspecific defenses?
   a. Specific immunity works only after nonspecific defenses have failed.
   b. Nonspecific defenses and specific immunity require a prior exposure to be effective.
   c. B lymphocytes work in both nonspecific defense and specific immunity.
   d. T lymphocytes work in both nonspecific defense and specific immunity.
   e. Macrophages work in both nonspecific defense and specific immunity.
6. What is the role of APC in specific immunity?
   a. An APC samples its external environment.
   b. An APC digests a sample from its external environment.

c. An APC posts an MHC protein on its epitope.
d. An APC releases interleukin-2 if the sample is foreign.
e. An APC releases interleukin-1 if the sample is foreign.

7. How would you describe humoral immunity?
   a. Humoral immunity involves B and T cells.
   b. Humoral immunity results in direct cell-to-cell attacks.
   c. Humoral immunity works against all pathogens equally.
   d. Humoral immunity results in the production of antibodies.
   e. Humoral immunity requires $T_{cytotoxic}$ cells.
8. How would you describe cellular immunity?
   a. Cellular immunity involves B and T cells.
   b. Cellular immunity results in direct cell-to-cell attacks.
   c. Cellular immunity works against all pathogens equally.
   d. Cellular immunity results in the production of antibodies.
   e. Cellular immunity requires $T_{cytotoxic}$ cells.
9. Which of the following statements describe(s) a form of acquired immunity?
   a. A Rhogam injection is an example of artificial active immunity.
   b. A Rhogam injection is an example of natural passive immunity.
   c. Nursing an infant to provide the antibodies in breast milk is an example of natural active immunity.
   d. A flu nasal spray of weakened flu viruses is meant to provide natural active immunity.
   e. Catching the flu from your significant other should result in natural active immunity.
10. What is the importance of T_{helper} cells?
   a. T_{helper} cells release interleukins.
   b. T_{helper} cells are necessary to recognize what is foreign.
   c. T_{helper} cells activate macrophages.
   d. T_{helper} cells release interferons.
   e. T_{helper} cells release pyrogens.

Matching: *Match the term to the correct definition.*

_______	1. Leukocytes stick to vessel walls.	a. Margination
_______	2. Leukocytes move toward chemicals.	b. Histamine
_______	3. Leukocytes engulf (eat) bacteria.	c. Chemotaxis
_______	4. Leukocytes move through vessel walls.	d. Phagocytosis
_______	5. This chemical causes vasodilation.	e. Diapedesis

Matching: *Match the following structures to their functions.*

_______	6. Collects lymph directly from tissues.	a. Thoracic duct
_______	7. Contains valves to prevent backflow.	b. Lymph vessel
_______	8. Filters lymph.	c. Right lymphatic duct
_______	9. Drains lymph from the left arm into a subclavian vein.	d. Lymph capillary
_______	10. Drains lymph from the right arm into a subclavian vein.	e. Lymph node

Completion: *Fill in the blanks to complete the following statements.*

1. Autoimmune diseases may result from ________________, in which the immune system attacks self-antigens because they resemble foreign antigens previously fought.
2. Macrophages release ________________ that cause the hypothalamus to raise the temperature set point.
3. Virally infected cells may release ______________ so that other healthy cells make antiviral proteins to protect themselves.
4. B cells, macrophages, and dendritic cells function as _____________ in the lymphatic system.
5. A(n) ______________ is a piece of antigen from the external environment of an APC that is displayed on an MHC protein.

Critical Thinking

1. What would be the effect on the functions of the lymphatic system if the thymus was removed from an infant? Explain.

____

____

____

____

____

____

2. As stated in the introduction to Chapter 12 in the text, Deborah is a 58-year-old woman who received a smallpox vaccination when she was 5 years old. What type of acquired immunity does she have for smallpox? Would it be effective today? Explain.

____

____

____

____

3. Dorothy went to the prom with Daniel in early May. It was the perfect date, and everything seemed magical. Even though they had not planned to have sex, they did. Because they had not planned for it, the sex was unprotected. Their human biology teacher began class the following week with a discussion of HIV. The discussion made Dorothy and Daniel anxious about their prom night, so they decided to be tested for HIV infection. Where could they be tested if

they lived in your area? What is the test? When should they be tested? Do they need follow-up testing? Why or why not? What can they do in the future to reduce their risk of HIV infection? Explain.

CHAPTER 12 REVIEW QUESTIONS

Case Study

A mantoux test is administered to healthcare workers to check for exposure to the pathogen that causes tuberculosis. A small amount of weakened tuberculin antigens is injected into the skin of the anterior forearm, and 48 hours later, the test is "read" by observing the location of the injection.

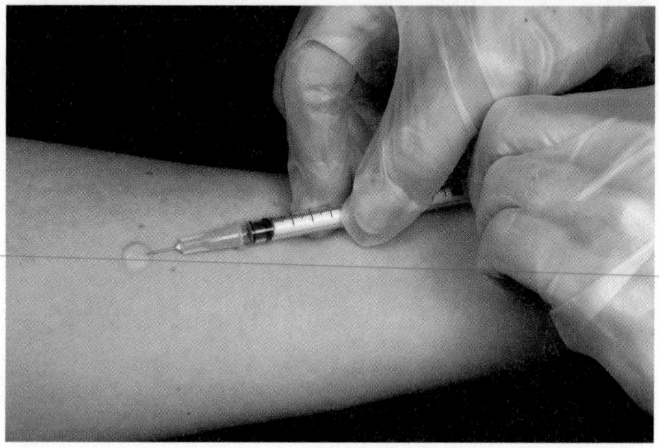

1. Explain why it takes 48 hours for the test to produce results. Why is the timing so important?

2. What might inflammation at the injection site indicate?

3. Explain why there would be inflammation at the site 48 hours after the test was administered if the individual had been previously exposed to the pathogen.

This section of the chapter is designed to help you find where each outcome is covered in the textbook and workbook.

Learning Outcomes		Text Readings, Figures, and Tables	Workbook Activities	Text and Workbook Assessments
12.1	Use medical terminology related to the lymphatic system.	Word roots & combining forms, pronunciation key: p. 437	Word roots & combining forms	*Text:* Review: 12.1 *Workbook:* Word Deconstruction: 1–5
12.2	Explain the origin and composition of lymph.	Lymph and lymph vessels: p. 438		*Text:* Review: 12.2 *Workbook:* Multiple Select: 1
12.3	Describe lymph vessels.	Lymph and lymph vessels: p. 438 Figures 12.2–12.3		*Text:* Review: 12.3 *Workbook:* Multiple Select: 2
12.4	Explain the route of lymph from the blood and back again.	Lymph and lymph vessels: pp. 438–442 Figures 12.4–12.5	*Coloring book:* Lymphatic drainage Figure 12.1	*Text:* Spot Check: 1, 2 Review: 12.4 *Workbook:* Matching: 6–10
12.5	Describe cells of the lymphatic system and list their functions.	Cells of the lymphatic system: pp. 442–443	*Concept maps:* Cells of the lymphatic system Figure 12.4	*Text:* Review: 12.5 *Workbook:* Multiple Select: 3 Completion: 4
12.6	Identify lymphoid tissues and organs and explain their functions.	Lymphoid tissues and organs: p. 443–449 Figures 12.6–12.12	*Coloring book:* Lymphoid tissues Figure 12.2	*Text:* Spot Check: 3, 4 Review: 12.6 *Workbook:* Critical Thinking: 1
12.7	Summarize three lines of defense against pathogens.	Three lines of defense: p. 449	*Concept maps:* Three lines of defense Figure 12.5	*Text:* Review: 12.7 *Workbook:* Multiple Select: 4
12.8	Contrast nonspecific resistance and specific immunity.	Nonspecific resistance versus specific immunity: p. 450		*Text:* Review: 12.8 *Workbook:* Multiple Select: 5

Learning Outcomes		Text Readings, Figures, and Tables	Workbook Activities	Text and Workbook Assessments
12.9	Describe the body's nonspecific defenses.	Nonspecific defenses: pp. 450–453 Figures 12.13–12.14	*Lab exercises and activities:* Inflammation Figure 12.3 *Concept maps:* Inflammation Figure 12.6	*Text:* Spot Check: 5–7 Review: 12.9 *Workbook:* Matching: 1–5 Completion: 2, 3
12.10	Explain the role of an APC in specific immunity.	Specific immunity: pp. 453–454 Figure 12.15	*Lab exercises and activities:* Specific immunity Table 12.1	*Text:* Review: 12.10 *Workbook:* Multiple Select: 6 Completion: 5
12.11	Explain the process of humoral immunity.	Humoral immunity: pp. 454–455 Figures 12.16–12.17	*Lab exercises and activities:* Specific immunity Table 12.1 *Concept maps:* Humoral immunity Figure 12.7	*Text:* Spot Check: 8, 9 Review: 12.11 *Workbook:* Multiple Select: 7
12.12	Explain the process of cellular immunity.	Cellular immunity: pp. 455–457 Figures 12.18–12.19	*Lab exercises and activities:* Specific immunity Table 12.1 *Concept maps:* Cellular immunity Figure 12.8	*Text:* Spot Check: 9 Review: 12.12 *Workbook:* Multiple Select: 8
12.13	Compare the different forms of acquired immunity.	Forms of acquired immunity: pp. 457–459		*Text:* Spot Check: 10 Review: 12.13 *Workbook:* Multiple Select: 9 Critical Thinking: 2
12.14	Explain the importance of T_{helper} cells to specific and nonspecific defense.	Importance of T_{helper} cells in nonspecific resistance and specific immunity: pp. 459–460 Figure 12.20		*Text:* Review: 12.14 *Workbook:* Multiple Select: 10
12.15	Explain the functions of the lymphatic system.	Functions of the lymphatic system: pp. 460–461 Figure 12.21		*Text:* Review: 12.15 *Workbook:* Critical Thinking: 1

Learning Outcomes		Text Readings, Figures, and Tables	Workbook Activities	Text and Workbook Assessments
12.16	Summarize the effects of aging on the lymphatic system.	Effects of aging on the lymphatic system: p. 461		*Text:* Review: 12.16 *Workbook:* Critical Thinking: 2
12.17	Describe common diagnostic tests used for lymphatic system disorders.	Diagnostic tests used for lymphatic system disorders: pp. 461–462 Table 12.1		*Text:* Spot Check: 11 Review: 12.17 *Workbook:* Case Study: 1–3
12.18	Describe lymphatic system disorders and relate abnormal function to pathology.	Lymphatic system disorders: p. 462 Table 12.2 Figures 12.22–12.24		*Text:* Spot Check: 12 Review: 12.18 *Workbook:* Completion: 1 Critical Thinking: 3

13 The Respiratory System

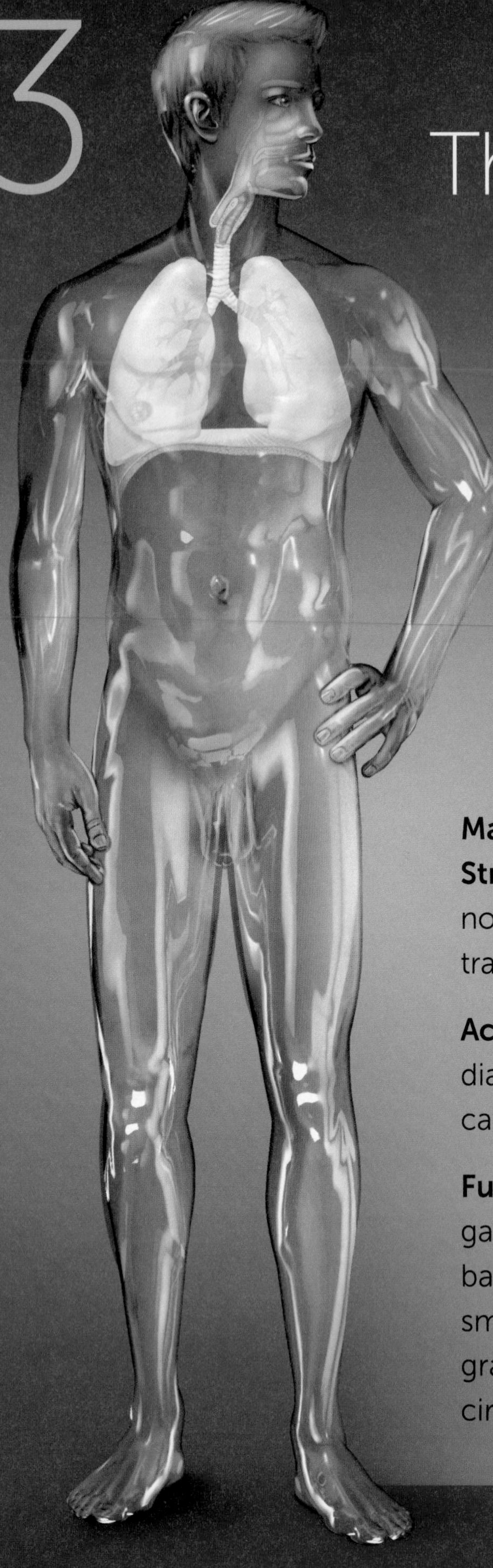

Major Organs and Structures:
nose, pharynx, larynx, trachea, bronchi, lungs

Accessory Structures:
diaphragm, sinuses, nasal cavity

Functions:
gas exchange, acid-base balance, speech, sense of smell, creation of pressure gradients necessary to circulate blood and lymph

learning outcomes

This chapter of the workbook is designed to help you learn the anatomy and physiology of the respiratory system. After completing this chapter in the text and this workbook, you should be able to:

13.1 Use medical terminology related to the respiratory system.

13.2 Trace the flow of air from the nose to the pulmonary alveoli and relate the function of each part of the respiratory tract to its gross and microscopic anatomy.

13.3 Explain the role of surfactant.

13.4 Describe the respiratory membrane.

13.5 Explain the mechanics of breathing in terms of anatomy and pressure gradients.

13.6 Define the measurements of pulmonary function.

13.7 Define *partial pressure* and explain its relationship to a gas mixture such as air.

13.8 Explain gas exchange in terms of the partial pressures of gases at the capillaries and the alveoli and at the capillaries and the tissues.

13.9 Compare the composition of inspired and expired air.

13.10 Explain the factors that influence the efficiency of alveolar gas exchange.

13.11 Describe the mechanisms for transporting O_2 and CO_2 in the blood.

13.12 Explain how respiration is regulated to homeostatically control blood gases and pH.

13.13 Explain the functions of the respiratory system.

13.14 Summarize the effects of aging on the respiratory system.

13.15 Describe common diagnostic tests used for respiratory system disorders.

13.16 Describe respiratory system disorders and relate abnormal function to pathology.

word roots & combining forms

alveol/o: alveolus, air sac

bronch/o: bronchial tube

bronchi/o: bronchus

bronchiol/o: bronchiole

capn/o: carbon dioxide

cyan/o: blue

laryng/o: larynx

lob/o: lobe

nas/o: nose

pharyng/o: pharynx

phren/o: diaphragm

pneum/o, pneumon/o: air

pulmon/o: lung

rhin/o: nose

sinus/o: sinus

spir/o: breathing

thorac/o: chest

trache/o: trachea

Upper Respiratory Tract

Figure 13.1 shows the anatomy of the upper respiratory tract. Color the box next to each term. Use the same color for the corresponding structure in the figure.

- ☐ Frontal sinus(A)
- ☐ Sphenoid sinus(B)
- ☐ Nasal cavity(C)
- ☐ Nasopharynx(D)
- ☐ Oropharynx(E)
- ☐ Laryngo-pharynx(F)
- ☐ Epiglottis(G)
- ☐ Larynx(H)
- ☐ Vocal cord(I)
- ☐ Trachea(J)

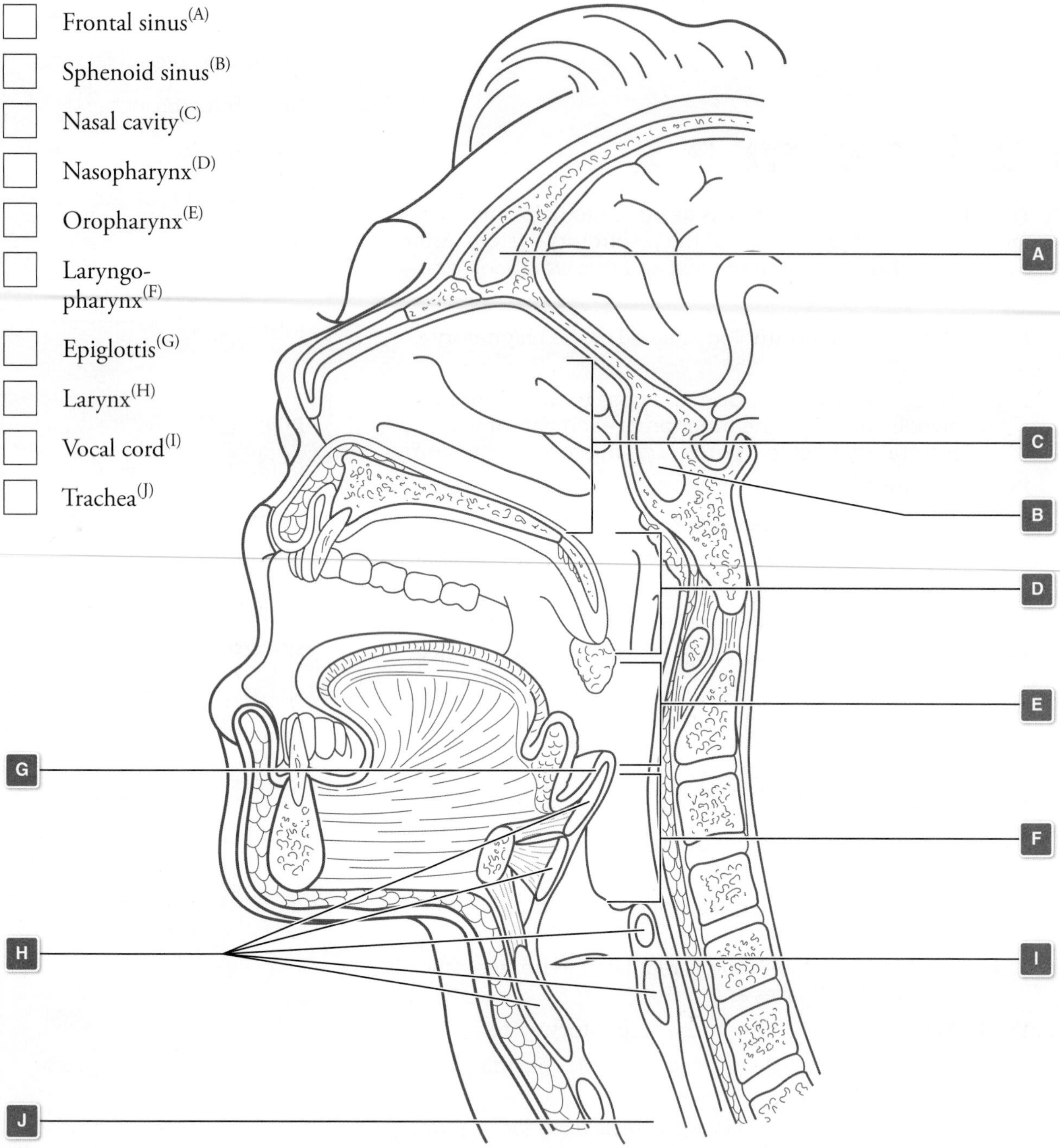

FIGURE 13.1 The upper respiratory tract.

COLORING BOOK

Lower Respiratory Tract

Figure 13.2 shows the anatomy of the lower respiratory tract. Color the box next to each term. Use the same color for the corresponding structure in the figure.

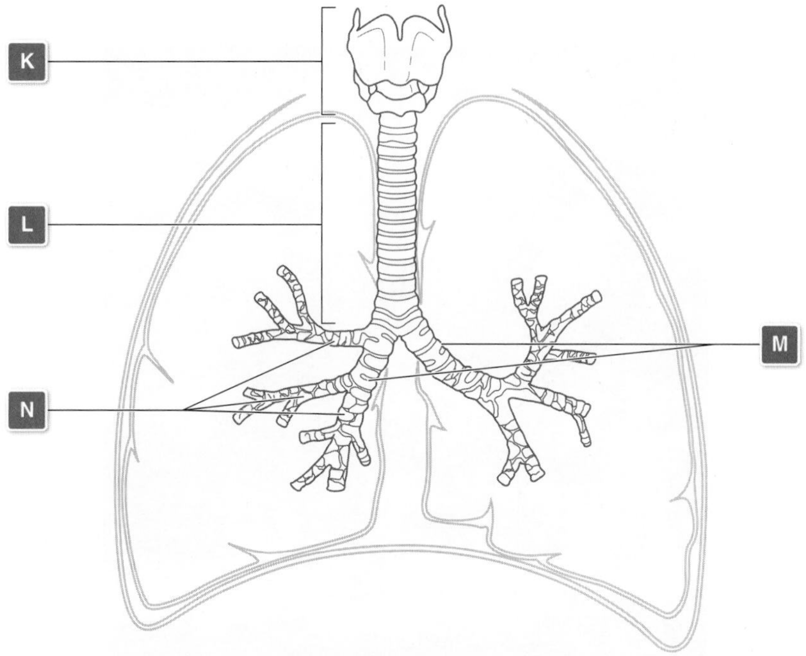

- ☐ Larynx(K)
- ☐ Trachea(L)
- ☐ Main bronchi(M)
- ☐ Lobar bronchi(N)

FIGURE 13.2 The lower respiratory tract.

Respiratory Membrane

Figure 13.3 shows the anatomy of the respiratory membrane. Color the box next to each term. Use the same color for the corresponding structure in the figure.

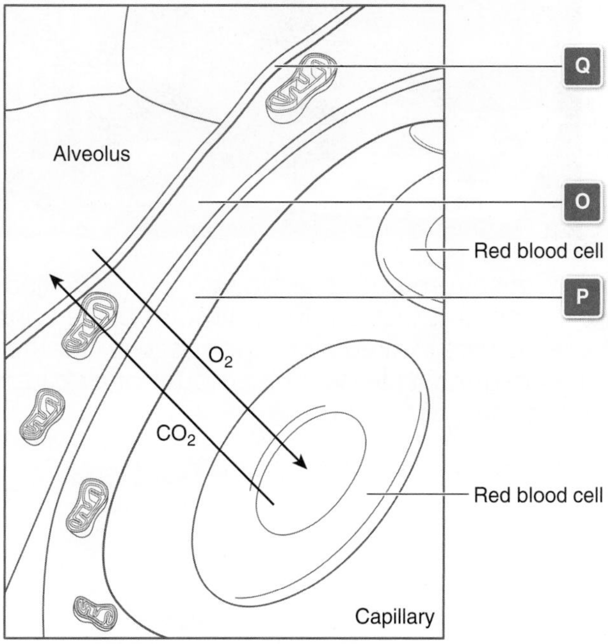

- ☐ Alveolar wall(O)
- ☐ Capillary wall(P)
- ☐ Layer of water with surfactant(Q)

FIGURE 13.3 The respiratory membrane.

COLORING BOOK

Spirometry

Figure 13.4 shows a wet spirometer and explains how it is used. Students in the anatomy and physiology lab were given the instructions below for their spirometry lab. You are to interpret a student's graph of her lab results.

FIGURE 13.4 Wet spirometer. The blue container is filled with water to the fill line. A one-time-use cardboard mouthpiece is inserted in the hose. The hose delivers expired air to the space under the inverted white container, causing it to rise. The amount of air causing the rise is measured on the scale at the top.

INSTRUCTIONS GIVEN TO STUDENTS IN A&P LAB

Procedure

Each student will use a spirometer in the following manner:

1. *Breathe in normally and expire normally into the disposable cardboard mouthpiece of the spirometer provided for just your own use. Record the result. Do this three times and graph the average three times (see "Graphing," below).*
2. *Breathe in normally and expire as much as possible into the mouthpiece of the spirometer. Record the result. Do this three times and graph the average three times.*
3. *Breathe in as much as possible and expire as much as possible into the mouthpiece of the spirometer. Record the result. Do this three times and graph the average three times.*

Graphing

The graph of similar tests in your textbook shows how much air is left in the lung after a complete exhalation (residual volume). You cannot determine residual volume in the A&P lab; therefore, you have to graph a little differently. The graph cannot have an X or Y axis with a 0 volume from which to start. So, instead, establish a scale: 1 square = 0.1 L. Then graph anywhere on your sheet of graph paper as long as you are consistent with using the scale. If the value can be known, a bar is shown where the volume or capacity starts and ends. If the complete value cannot be determined because of residual volume, a bar is shown where it begins and an arrow is shown at the end to indicate that the end is unknown.

Define the following volumes and capacities in your own words and indicate where they are shown on your graph. Record the value for each volume and capacity.

- *Tidal volume*
- *Residual volume*
- *Vital capacity*
- *Total lung capacity*
- *Inspiratory capacity*
- *Inspiratory reserve volume*
- *Functional residual capacity*
- *Expiratory reserve volume*

Figure 13.5 shows a student graph for the spirometry lab described above. For this exercise, you are to answer the following questions to interpret her graph.

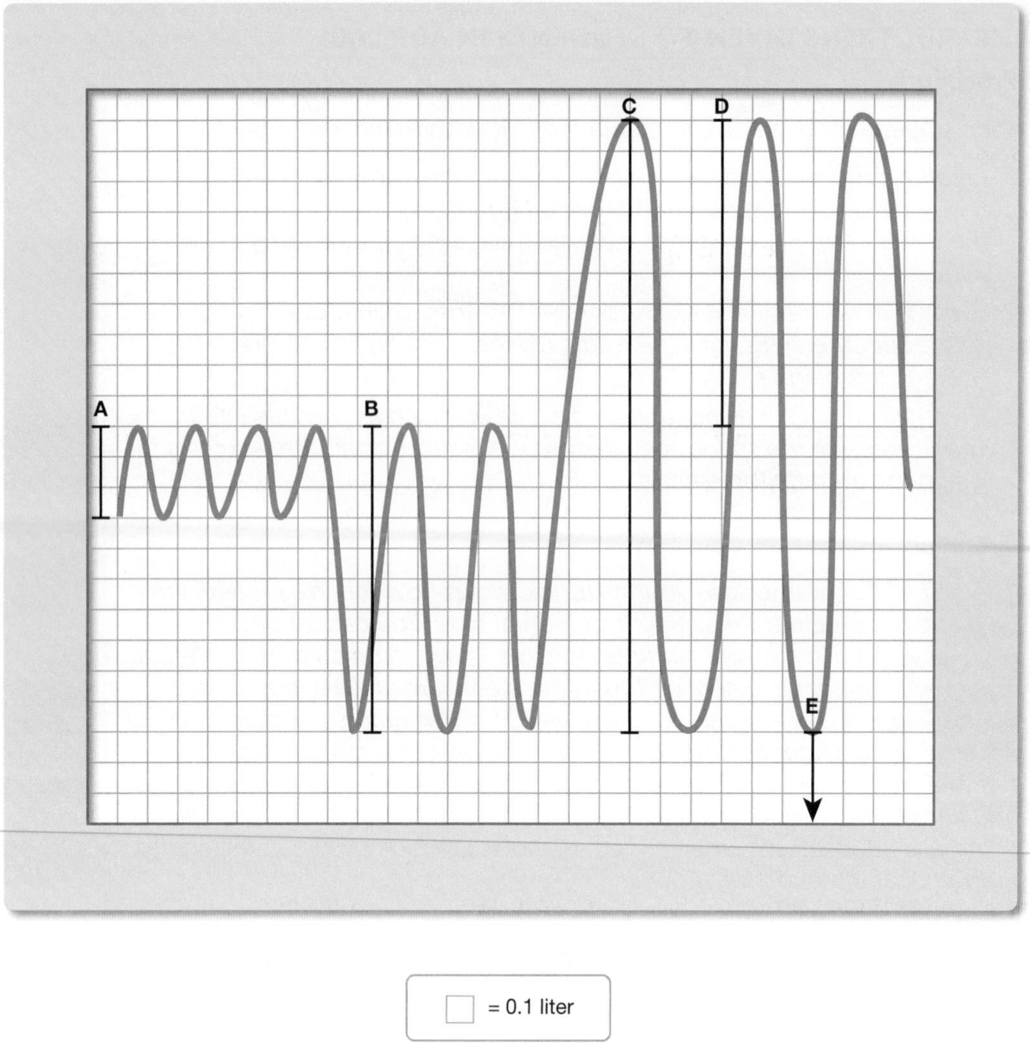

FIGURE 13.5 Student spirometry graph.

1. What volume or capacity is represented by A? ______
2. What is the definition for A? ______

______

3. What is the value for A (give number; make sure to include the units)? ______
4. The student made a mistake with B. She intended B to represent functional residual capacity. Mark on the graph where the functional residual capacity can be found. What is the definition of functional residual capacity? ______

______

5. What is the value of the functional residual capacity for this student? ______
6. What volume or capacity is represented by C? ______
7. What is the definition for C? ______

______

8. What is the value for C (give number; make sure to include the units)? ____________

9. What volume or capacity is represented by D? ____________

10. What is the definition for D? ____________

____

11. What is the value for D (give number; make sure to include the units)? ____________

12. What volume or capacity is represented by E? ____________

13. What is the definition for E? ____________

____

14. What is the value for E (give number; make sure to include the units)? ____________

15. Mark on the graph where the total lung capacity can be found (F). What is the definition of total lung capacity? ____________

____

16. What is the value of the total lung capacity for this student? ____________

17. Mark on the graph where the inspiratory capacity can be found (G). What is the definition of inspiratory capacity? ____________

____

18. What is the value of the inspiratory capacity for this student? ____________

19. Mark on the graph where the expiratory reserve volume can be found (H). What is the definition of expiratory reserve volume? ____________

____

20. What is the value of the expiratory reserve volume for this student? ____________

____

21. Why is it impossible to determine the residual volume using spirometry in an A&P lab?

____

____

____

Gas Exchange

Gas exchange happens across the respiratory membrane between the alveoli and the capillaries of the lung and out in the body between the tissues and the capillaries located there. See **Figure 13.6**. The amount of gas that is exchanged is expressed in partial pressures. Complete the statements on the next page. Use greater than (>), less than (<), or equals (=) symbols for the partial pressures.

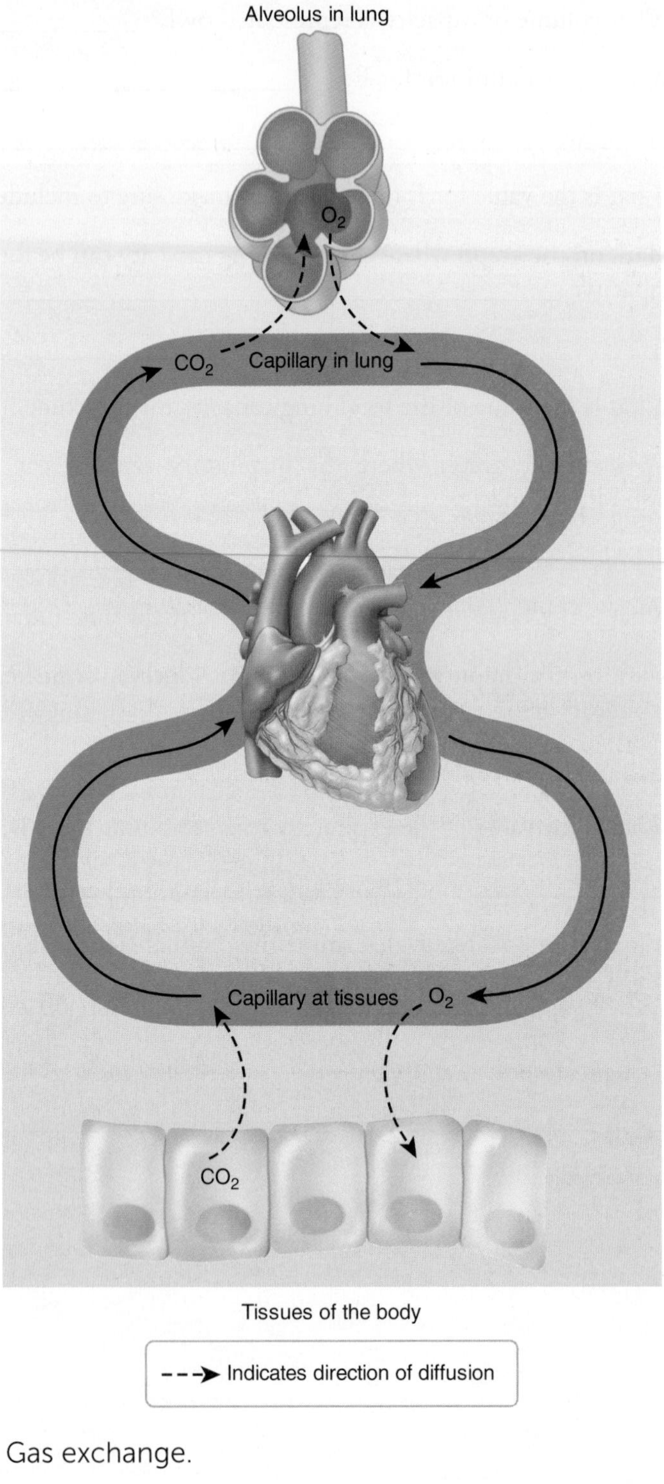
Alveolus in lung
O_2
CO_2
Capillary in lung
Capillary at tissues
O_2
CO_2
Tissues of the body
Indicates direction of diffusion

FIGURE 13.6 Gas exchange.

Word Deconstruction *In the textbook, you built words to fit a definition using the combining forms, prefixes, and suffixes. Here you are to break down the term into its parts (prefixes, roots, and suffixes) and give a definition. Prefixes and suffixes can be found inside the back cover of the textbook.*

FOR EXAMPLE Dermatitis: dermat/itis—inflammation of the skin

1. Capnia: ______________________________
2. Bronchoscopy: ______________________________
3. Lobectomy: ______________________________
4. Spirometry: ______________________________
5. Tracheostomy: ______________________________

Multiple Select *Select the correct choices for each statement. The choices may be all correct, all incorrect, or any combination of correct and incorrect.*

1. What is the respiratory membrane?
   - **a.** It is composed of two layers of simple squamous cells with water and surfactant between the layers.
   - **b.** It is where gas exchange takes place at the tissues.
   - **c.** It is ciliated mucous membranes.
   - **d.** It is composed of pleural membranes.
   - **e.** It is located in the bronchial tree.
2. What are pulmonary volumes and capacities?
   - **a.** Inspiratory capacity is the maximum amount of air that can be inspired including a normal breath.
   - **b.** Expiratory reserve volume is the amount of air that can be maximally exhaled after expiration of a normal breath.
   - **c.** Vital capacity is the maximum amount of air that can be moved in the lungs.
   - **d.** Residual volume is the amount of air that still remains in the lungs after maximum expiration.
   - **e.** Total lung volume is the maximum amount of air the lungs can hold and includes air that can and cannot be moved.
3. What happens at the respiratory membrane?
   - **a.** CO_2 diffuses across the capillary wall to the alveolar wall to a layer of water with surfactant.
   - **b.** O_2 diffuses across the capillary wall to the alveolar wall to a layer of water with surfactant.
   - **c.** CO_2 diffuses across the thin layer of water with surfactant to the alveolar wall to the capillary wall to the blood.
   - **d.** O_2 diffuses across the thin layer of water with surfactant to the alveolar wall to the capillary wall to the blood.
   - **e.** N_2 does not move across the respiratory membrane at normal atmospheric pressure.
4. What happens when you take a deep breath and blow out the candles of your birthday cake?
   - **a.** Intercostal muscles expand the chest, and the diaphragm flattens during inspiration.
   - **b.** Intercostal muscles and the diaphragm simply relax as you blow out the candles.
   - **c.** Intercostal muscles and the diaphragm contract as you blow out the candles.
   - **d.** The parietal pleura pulls on the visceral pleura because of surfactant.
   - **e.** The phrenic nerve is involved in inspiration and expiration.
5. How does inspired air compare to expired air?
   - **a.** Inspired air has more nitrogen than expired air.
   - **b.** Inspired air has more oxygen than expired air.

c. Inspired air has more carbon dioxide than expired air.
d. Inspired air has the same amount of nitrogen as expired air.
e. Both inspired air and expired air contain water vapor.

6. Which of the following will have a long-term effect on Peggy's respiratory system?
   a. Taking calcium supplements and vitamin D.
   b. Regular exercise.
   c. Surgery to improve her scoliosis.
   d. Keeping her weight within normal limits.
   e. Scuba diving.

7. Where does air travel as it is expired?
   a. Air travels from a bronchus to a bronchiole as it is expired.
   b. Air travels from the laryngopharynx to the oropharynx as it is expired.
   c. Air travels from the trachea to the larynx as it is expired.
   d. Air travels through the glottis to the larynx as it is expired.
   e. Air travels past the vocal cords before going through the glottis as it is expired.

8. Which of the following would *not* affect alveolar gas exchange?
   a. A lobectomy to remove a cancerous tumor.
   b. Pulmonary edema.
   c. Mountain climbing.
   d. An asthma attack.
   e. Snorkeling (swimming with a breathing tube in shallow water to observe fish).

9. If the atmospheric pressure of air is 700 mmHg, what else is true?
   a. The partial pressure of oxygen is approximately 147 mmHg.
   b. The partial pressure of nitrogen is approximately 553 mmHg.
   c. The partial pressure of carbon dioxide is approximately 0.04 percent.
   d. The partial pressure of nitrogen is approximately 79 percent.
   e. The partial pressure of oxygen is approximately 21 percent.

10. Peter is a five-year-old who is stubbornly giving his baby-sitter trouble. He is threatening to "hold his breath so he turns blue and dies" if he does not get his way. His wise baby-sitter (an A&P student) allows him to hold his breath. What does the baby-sitter know?
    a. Voluntary control of breathing is limited.
    b. Breathing is controlled by respiratory centers in the medulla oblongata.
    c. As Peter holds his breath, carbon dioxide levels will rise in his blood.
    d. Peter's central and peripheral chemoreceptors will detect a rise in pH.
    e. If Peter is stubborn enough to pass out from holding his breath, his medulla oblongata will start his breathing again as soon as Peter is unconscious.

Matching *Match the disorder to the description. Some of the choices may be used more than once. Some of the descriptions fit more than one choice.*

______	**1.** Bronchioles are hyperreactive to a stimulus.	**a.** Tuberculosis
______	**2.** Protection is achieved through a DPT shot.	**b.** Asthma
______	**3.** Affected cells metastasize easily.	**c.** Influenza
______	**4.** This infection is caused by bacteria.	**d.** Oat cell carcinoma
______	**5.** This infection is caused by a virus.	**e.** Pertussis

Matching *Match the gas to the description. Some of the choices may be used more than once. Some of the descriptions fit more than one choice.*

________	**6.** Binds to hemoglobin in the lungs	**a.**	Oxygen
________	**7.** Mixes with water at the tissues to form an acid	**b.**	Carbon dioxide
________	**8.** Becomes soluble at pressures higher than normal	**c.**	Carbon monoxide
________	**9.** Is transported through the blood as an ion	**d.**	Nitrogen
________	**10.** Separates from hemoglobin at the tissues		

Completion *Fill in the blanks to complete the following statements.*

1. ________________ is secreted by great alveolar cells, and it reduces the ________________ of water.

2. The definition of partial pressure is ____

____

____.

3. ________________ is matching airflow to blood flow in the lung.

4. A(n) ________________ occurs if air is introduced between the pleural membranes.

5. The mucous membranes of the upper respiratory tract function to ________________________,

____________________, and ____________________.

Critical Thinking

1. An uncontrolled diabetic, who is using fat and protein as an energy source because he cannot use glucose, may develop a condition called *ketoacidosis.* This condition causes excess hydrogen ions to accumulate in the blood. What would be the effect of ketoacidosis on his respiratory rate? Explain which receptors will detect this condition and how his respiratory rate is regulated in this case.

____

____

____

____

____

2. A drowning victim may be administered cardiopulmonary resuscitation (CPR). During the process of CPR, the person administering the CPR breathes in and then forcefully expires into the mouth of the recipient. Compare the amount of alveolar gas exchange that occurs in the recipient to that of the person administering CPR. Explain in terms of partial pressures of inspired and expired air.

3. If the respiratory system fails, a patient may be kept alive with a respirator. Will a respirator be able to fulfill all the functions of the respiratory system? Explain.

Case Study

Jimmy is a 46-year-old male who has COPD. Although Jimmy no longer smokes cigarettes, he started smoking when he was 14 and continued well into adulthood. Currently, in conjunction with medication to treat his condition, he also receives supplemental oxygen.

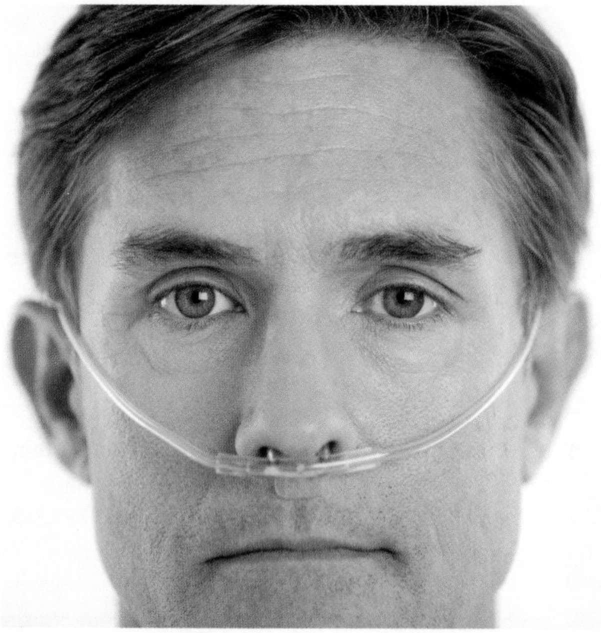

1. What effect has years of smoking had on Jimmy's lungs?

2. With Jimmy's diagnosis of COPD, what respiratory disorders is he most likely suffering from?

3. Why would Jimmy need supplemental oxygen to help with his condition? Explain your answer.

4. Can Jimmy's COPD be treated and cured? Explain your answer.

This section of the chapter is designed to help you find where each outcome is covered in the textbook and workbook.

Learning Outcomes	Text Readings, Figures, and Tables	Workbook Activities	Text and Workbook Assessments
13.1 Use medical terminology related to the respiratory system.	Word roots & combining forms; pronunciation key: pp. 473–474	Word roots & combining forms	*Text:* Review: 13.1 *Workbook:* Word Deconstruction: 1–5
13.2 Trace the flow of air from the nose to the pulmonary alveoli and relate the function of each part of the respiratory tract to its gross and microscopic anatomy.	Anatomy of the respiratory system: pp. 474–485 Figures 13.2–13.13	*Coloring book:* Upper respiratory tract; Lower respiratory tract Figures 13.1–13.2	*Text:* Spot Check: 1–6 Review: 13.2 *Workbook:* Multiple Select: 7 Completion: 5
13.3 Explain the role of surfactant.	Surfactant: pp. 485–486		*Text:* Review: 13.3 *Workbook:* Completion: 1
13.4 Describe the respiratory membrane.	Respiratory membrane: p. 486 Figure 13.12	*Coloring book:* Respiratory membrane Figure 13.3	*Text:* Review: 13.4 *Workbook:* Multiple Select: 1, 3
13.5 Explain the mechanics of breathing in terms of anatomy and pressure gradients.	Mechanics of taking a breath: pp. 486–489 Figures 13.14–13.15		*Text:* Review: 13.5 *Workbook:* Multiple Select: 4 Completion: 4
13.6 Define the measurements of pulmonary function.	Measurements of pulmonary function: pp. 490–491 Figures 13.16–13.17 Table 13.1	*Lab exercises and activities:* Spirometry Figures 13.4–13.5	*Text:* Review: 13.6 *Workbook:* Multiple Select: 2 Case Study: 1
13.7 Define *partial pressure* and explain its relationship to a gas mixture such as air.	Composition of air: p. 492	*Concept maps:* Partial pressure Figure 13.7	*Text:* Spot Check: 7 Review: 13.7 *Workbook:* Multiple Select: 9 Completion: 2
13.8 Explain gas exchange in terms of the partial pressures of gases at the capillaries and the alveoli and at the capillaries and the tissues.	Gas exchange: pp. 492–494 Figures 13.18–13.19	*Lab exercises and activities:* Gas exchange Figure 13.6 *Concept maps:* Gas exchange Figure 13.8	*Text:* Review: 13.8 *Workbook:* Multiple Select: 3 Critical Thinking: 2 Case Study: 3, 4

Learning Outcomes		Text Readings, Figures, and Tables	Workbook Activities	Text and Workbook Assessments
13.9	Compare the composition of inspired and expired air.	Comparison of inspired and expired air: p. 494 Figure 13.19	*Concept maps:* Partial pressure Figure 13.7	*Text:* Review: 13.9 *Workbook:* Multiple Select: 5 Critical Thinking: 2 Case Study: 3, 4
13.10	Explain the factors that influence the efficiency of alveolar gas exchange.	Factors that influence gas exchange: pp. 494–497 Figures 13.20–13.22		*Text:* Spot Check: 8 Review: 13.10 *Workbook:* Multiple Select: 3, 8 Matching: 8 Completion: 3
13.11	Describe the mechanisms for transporting O_2 and CO_2 in the blood.	Gas transport: pp. 498–500 Figures 13.23–13.24	*Concept maps:* Oxygen transport; Carbon dioxide transport Figures 13.9–13.10	*Text:* Review: 13.11 *Workbook:* Matching: 6, 7, 9, 10
13.12	Explain how respiration is regulated to homeostatically control blood gases and pH.	Regulation of respiration: pp. 500–502 Figures 13.25–13.27	*Concept maps:* Regulation of respiration Figure 13.11	*Text:* Review: 13.12 *Workbook:* Multiple Select: 10 Critical Thinking: 1
13.13	Explain the functions of the respiratory system.	Functions of the respiratory system: pp. 502–503 Figure 13.27		*Text:* Review: 13.13 *Workbook:* Critical Thinking: 3
13.14	Summarize the effects of aging on the respiratory system.	Effects of aging on the respiratory system: pp. 503–504		*Text:* Review: 13.14 *Workbook:* Multiple Select: 6
13.15	Describe common diagnostic tests used for respiratory system disorders.	Diagnostic tests for respiratory system disorders: pp. 504–505 Table 13.2		*Text:* Spot Check: 9 Review: 13.15 *Workbook:* Matching: 1–5
13.16	Describe respiratory system disorders and relate abnormal function to pathology.	Respiratory system disorders: pp. 505–511 Figures 13.28–13.31 Table 13.3		*Text:* Spot Check: 10 Review: 13.16 *Workbook:* Case Study: 1–4

14 The Digestive System

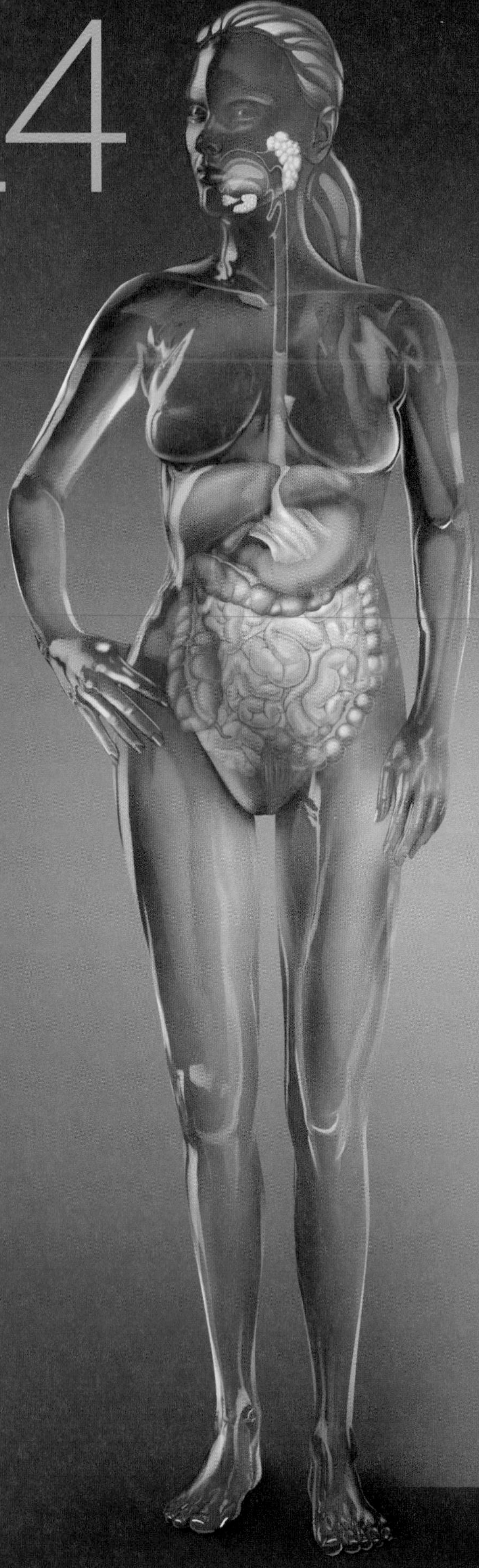

Major Organs and Structures:
esophagus, stomach, small intestine, large intestine

Accessory Structures:
liver, pancreas, gall-bladder, cecum, teeth, salivary glands

Functions:
ingestion, digestion, absorption, defecation

learning outcomes

This chapter of the workbook is designed to help you learn the anatomy and physiology of the digestive system. After completing this chapter in the text and this workbook, you should be able to:

14.1 Use medical terminology related to the digestive system.

14.2 Differentiate between mechanical digestion and chemical digestion.

14.3 Describe the digestive anatomy of the oral cavity.

14.4 Explain the physiology of mechanical and chemical digestion in the mouth.

14.5 Describe the digestive anatomy from the mouth to the stomach.

14.6 Explain how materials move from the mouth to the stomach.

14.7 Describe the digestive anatomy of the stomach.

14.8 Explain the physiology of mechanical and chemical digestion in the stomach.

14.9 Explain the feedback mechanism of how food moves from the stomach to the small intestine.

14.10 Describe the anatomy of the digestive accessory organs connected to the duodenum by ducts.

14.11 Describe the digestive anatomy of the small intestine.

14.12 Explain the physiology of chemical digestion in the duodenum, including the hormones and digestive secretions involved.

14.13 Explain how nutrients are absorbed in the small intestine.

14.14 Describe the anatomy of the large intestine.

14.15 Explain the physiology of the large intestine in terms of absorption, preparation of feces, and defecation.

14.16 Summarize the types of nutrients absorbed by the digestive system from the diet.

14.17 Trace the circulation of the nutrients once they have been absorbed.

14.18 Explain the control of digestion.

14.19 Summarize the functions of digestion.

14.20 Summarize the effects of aging on the digestive system.

14.21 Describe common diagnostic tests used for digestive system disorders.

14.22 Describe digestive system disorders and relate abnormal function to pathology.

word **roots** & combining **forms**

chol/e: gall, bile

col/o: colon

cyst/o: bladder, sac

duoden/o: duodenum

emet/o: vomit

enter/o: intestine

esophag/o: esophagus

gastr/o: stomach

gingiv/o: gums

gloss/o: tongue

hepat/o: liver

peps/o: digestion

rect/o: rectum

sigmoid/o: sigmoid colon

Digestive Tract

Figure 14.1 shows the anatomy of the digestive tract. Color the box next to each term. Use the same color for the corresponding structure in the figure.

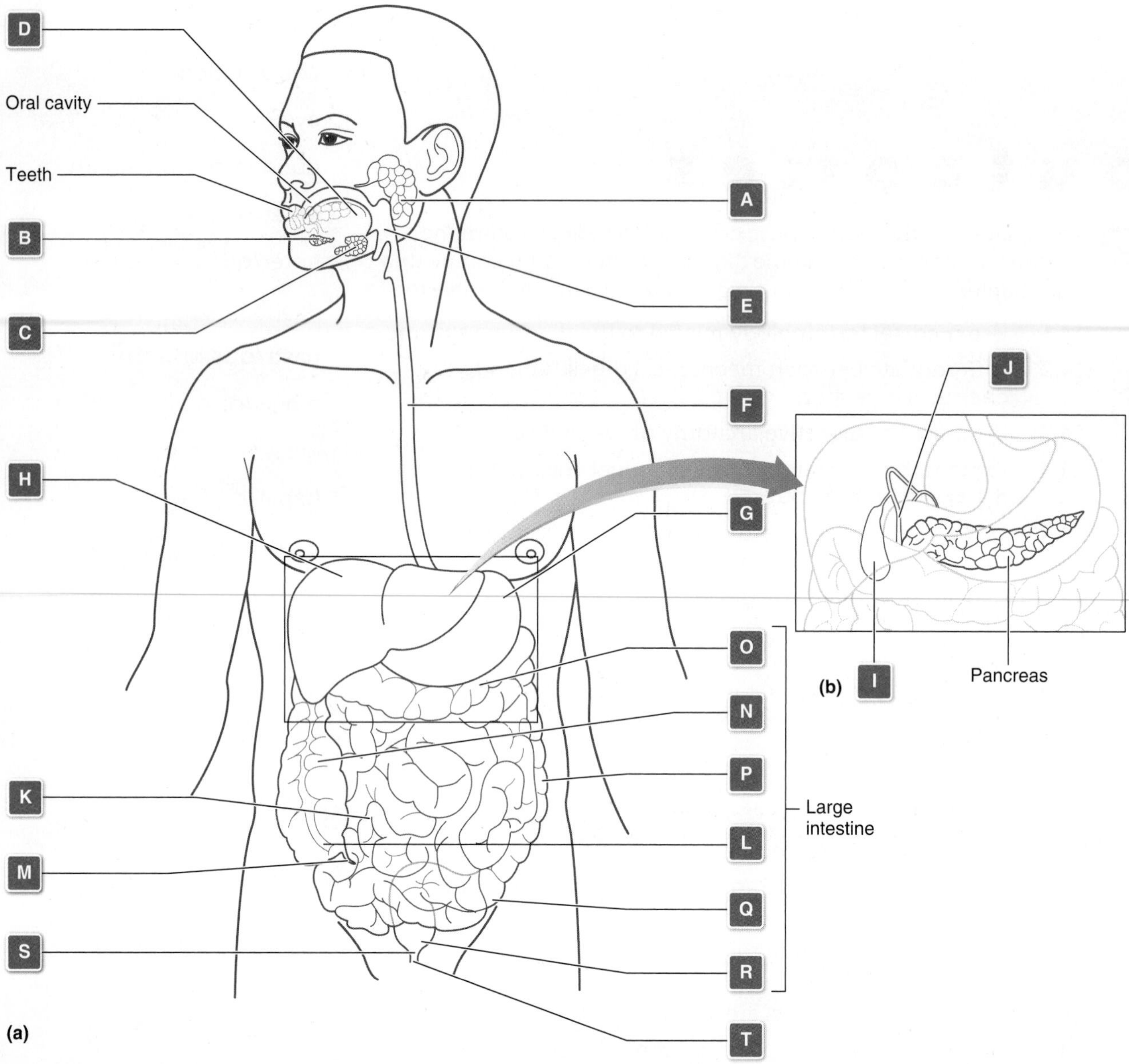

FIGURE 14.1 The digestive tract.

- ☐ Parotid gland(A)
- ☐ Sublingual gland(B)
- ☐ Submandibular gland(C)
- ☐ Tongue(D)
- ☐ Pharynx(E)
- ☐ Esophagus(F)
- ☐ Stomach(G)
- ☐ Liver(H)
- ☐ Gallbladder(I)
- ☐ Common bile duct(J)
- ☐ Small intestine(K)
- ☐ Cecum(L)
- ☐ Appendix(M)
- ☐ Ascending colon(N)
- ☐ Transverse colon(O)
- ☐ Descending colon(P)
- ☐ Sigmoid colon(Q)
- ☐ Rectum(R)
- ☐ Anal canal(S)
- ☐ Anus(T)

COLORING BOOK

Anatomy of a Tooth

Figure 14.2 shows the anatomy of a tooth. Color the box next to each term. Use the same color for the corresponding structure in the figure.

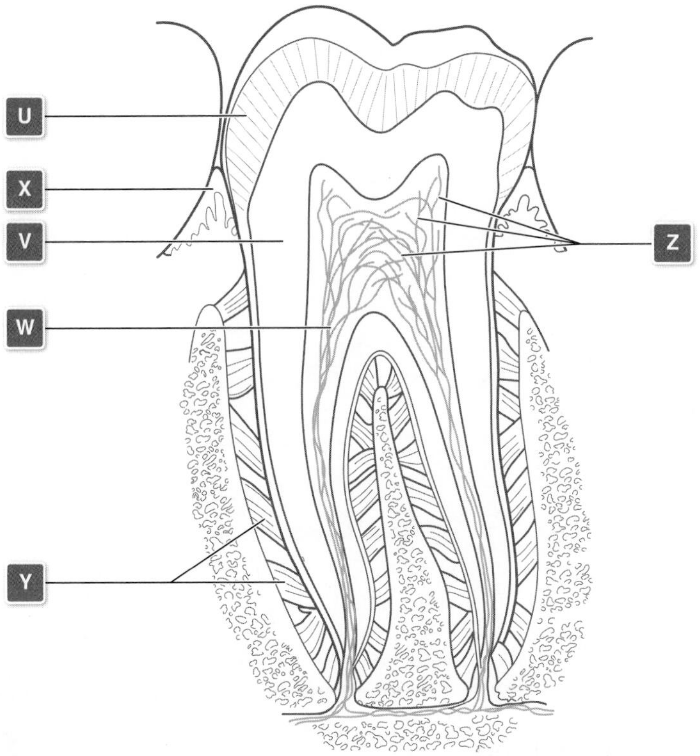

- ☐ Enamel(U)
- ☐ Dentin(V)
- ☐ Pulp in pulp cavity(W)
- ☐ Gingiva(X)
- ☐ Periodontal ligament(Y)
- ☐ Artery, nerve, vein(Z)

FIGURE 14.2 The anatomy of a tooth.

Salivary Glands

Figure 14.3 shows the salivary glands. Color the box next to each term. Use the same color for the corresponding structure in the figure.

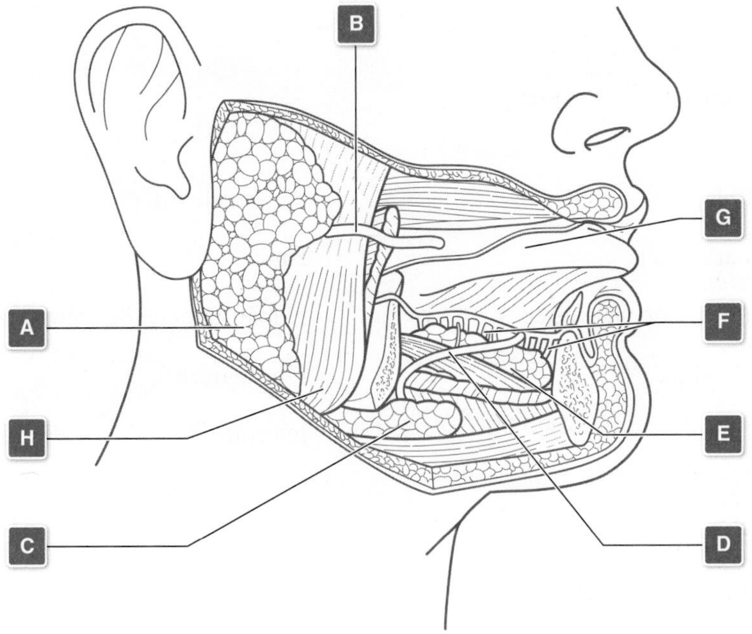

- ☐ Parotid gland(A)
- ☐ Parotid duct(B)
- ☐ Submandibular gland(C)
- ☐ Submandibular duct(D)
- ☐ Sublingual gland(E)
- ☐ Sublingual ducts(F)
- ☐ Tongue(G)
- ☐ Masseter muscle(H)

FIGURE 14.3 The salivary glands.

COLORING BOOK

Gallbladder, Pancreas, and Ducts

Figure 14.4 shows the anatomy of the gallbladder, pancreas, and their related ducts. Color the box next to each term. Use the same color for the corresponding structure in the figure.

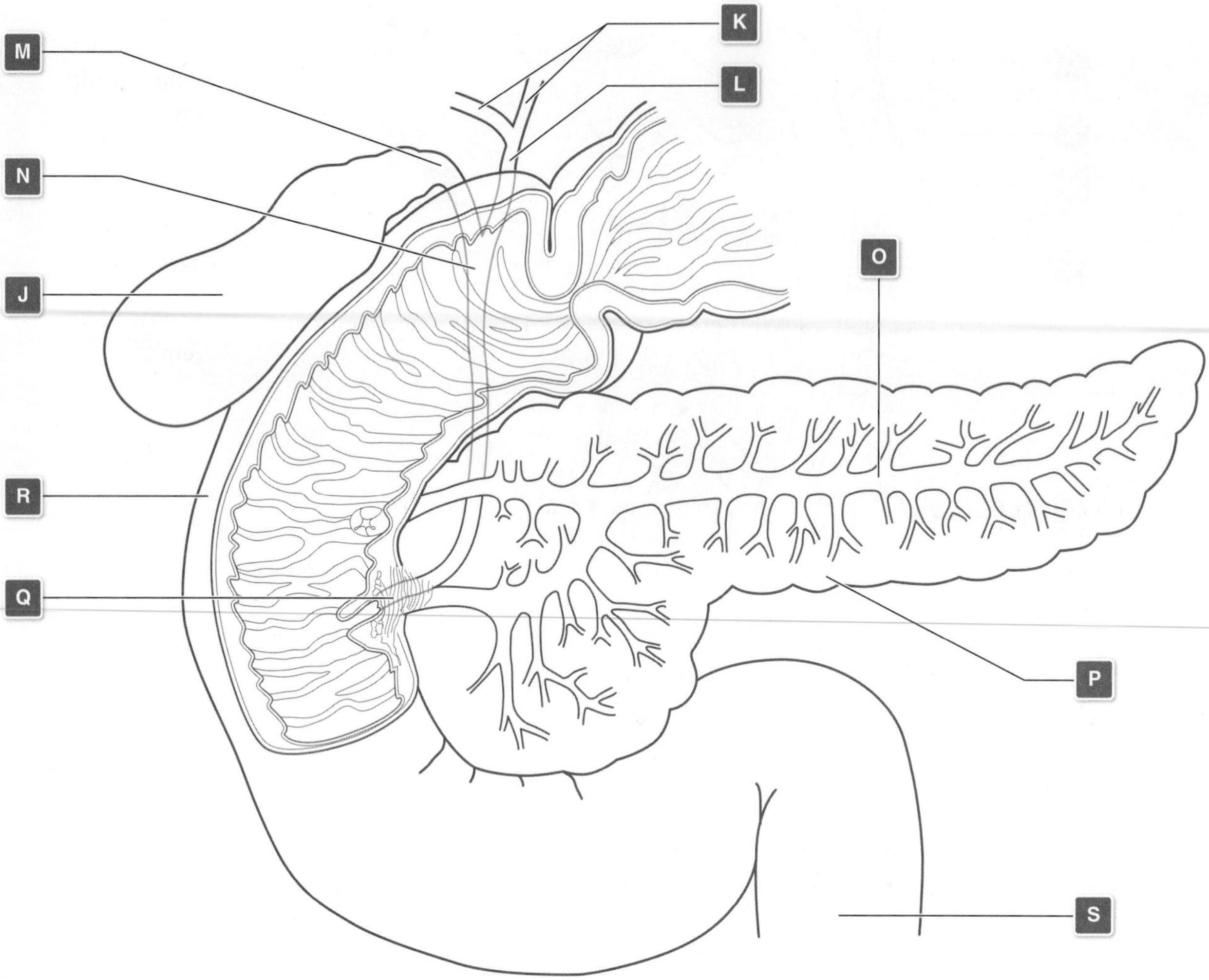

FIGURE 14.4 The gallbladder, pancreas, and ducts.

☐ Gallbladder(J)	☐ Pancreatic duct(O)
☐ Hepatic ducts(K)	☐ Pancreas(P)
☐ Common hepatic duct(L)	☐ Hepatopancreatic sphincter(Q)
☐ Cystic duct(M)	☐ Duodenum(R)
☐ Common bile duct(N)	☐ Jejunum(S)

Soda Cracker

A soda cracker is composed of starch, a complex carbohydrate. For this exercise, you are to take a bite of a soda cracker and chew it for two minutes without swallowing. Answer the following questions.

1. What function of digestion involves taking a bite of the soda cracker?

____

2. What is the term for chewing? ______________________________

3. What type of digestion involves chewing the cracker? ______________

____

4. What structures secrete digestive juices into the oral cavity? Be specific.

____

____

5. What part of these secretions is involved in chemical digestion of the soda cracker in the mouth? ______________________________

____

6. What organic molecule is chemically digested in the mouth? __________

____

7. What is the product of chemical digestion of this type of organic molecule?

____

8. Record your observation of the taste of the soda cracker over time, while you were chewing. ______________________________

____

____

9. What accounts for the change (if any) in the taste? ______________

____

____

10. What is the term for the bite of cracker once you are ready to swallow it?

____

LABORATORY EXERCISES AND ACTIVITIES

LABORATORY EXERCISES AND ACTIVITIES

Digestion of a Cheeseburger

A cheeseburger moves through the alimentary canal as it is digested. See **Figure 14.5**. Complete Table 14.1 by first putting the structures of the alimentary canal in order (refer to the list provided). Then indicate what (if any) type of digestion takes place in each structure (some parts of the table will be blank). Finally, indicate (where appropriate) what is chemically digested in each structure.

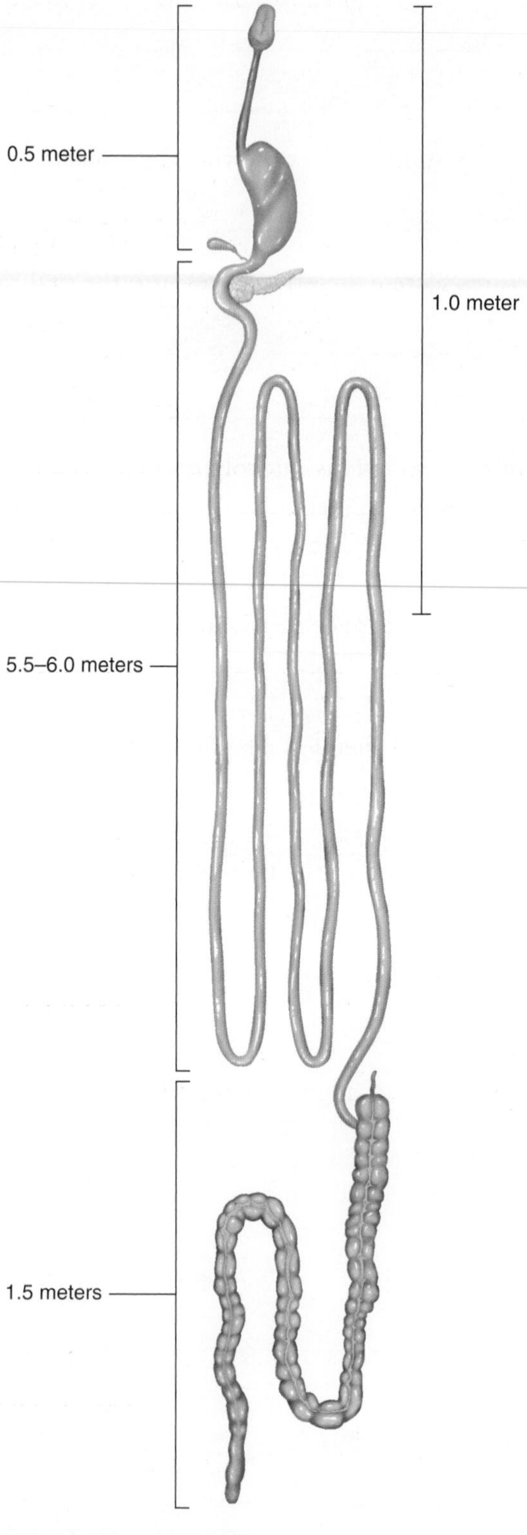

0.5 meter
1.0 meter
5.5–6.0 meters
1.5 meters

FIGURE 14.5 The alimentary canal.

Word Deconstruction *In the textbook, you built words to fit a definition using the combining forms, prefixes, and suffixes. Here you are to break down the term into its parts (prefixes, roots, and suffixes) and give a definition. Prefixes and suffixes can be found inside the back cover of the textbook.*

FOR EXAMPLE Dermatitis: dermat/itis—inflammation of the skin

1. Emetic: ______________________________
2. Hepatopancreatic: ______________________________
3. Gastroenterologist: ______________________________
4. Dyspepsia: ______________________________
5. Colitis: ______________________________

Multiple Select *Select the correct choices for each statement. The choices may be all correct, all incorrect, or any combination of correct and incorrect.*

1. What is the digestive anatomy of the oral cavity?
   - **a.** There are 28 teeth in a full set of permanent teeth.
   - **b.** A tooth fits into an alveolus.
   - **c.** There is enamel in the root of a tooth.
   - **d.** Salivary ducts empty into the oral cavity.
   - **e.** The uvula is an extension of the hard palate.
2. What happens in the mouth?
   - **a.** Mastication is the start of chemical digestion.
   - **b.** The chemical digestion of proteins is started.
   - **c.** Lysosomes begin chemical digestion of lipids.
   - **d.** Water is absorbed.
   - **e.** The chemical digestion of carbohydrates is started.
3. Which of the following types of dietary nutrients is (are) absorbed by the digestive system?
   - **a.** Some electrolytes are absorbed throughout the small and large intestines.
   - **b.** Most minerals are absorbed at a constant rate.
   - **c.** Fat-soluble vitamins are absorbed with the products of lipid digestion.
   - **d.** Water-soluble vitamins are absorbed through facilitated diffusion.
   - **e.** The amount of calcium and iron absorbed by the body is regulated to meet a need.
4. What is the anatomy of the stomach?
   - **a.** There are four layers of smooth muscle in the walls of the stomach.
   - **b.** Gastric pits extend to form gastric glands.
   - **c.** The opening to the stomach from the esophagus is controlled by the pyloric sphincter.
   - **d.** Rugae give extra surface area for villi in the stomach.
   - **e.** Regenerative cells in the gastric pits and gastric glands may develop into parietal or chief cells.
5. What happens in the stomach?
   - **a.** Lingual lipase is activated to begin lipid digestion.
   - **b.** Both mechanical digestion and chemical digestion take place.
   - **c.** Pepsin is changed to pepsinogen to start the digestion of proteins.
   - **d.** Carbohydrates are completely digested.
   - **e.** The chemical digestion of carbohydrates continues because of gastric lipase.
6. What causes chyme to move from the stomach to the small intestine?
   - **a.** An increase in pH.
   - **b.** Distension of the stomach.
   - **c.** Cholecystokinin.

d. Secretin.
e. A decrease in pH and peristaltic contractions of the stomach.

7. Where does bile go?
   a. From hepatic ductules to the hepatic ducts.
   b. From the common bile duct to the cystic duct.
   c. From the cystic duct to the common bile duct.
   d. From the pancreatic duct to the common bile duct.
   e. Through the pyloric sphincter.
8. What is produced in the stomach?
   a. Chief cells produce pepsinogen.
   b. Parietal cells produce vitamin B_{12}.
   c. Hydrochloric acid is produced by parietal cells.
   d. Intrinsic factor is produced so that vitamin B_{12} can be absorbed.
   e. Endocrine cells produce cholecystokinin and secretin.
9. How are nutrients absorbed in the intestines?
   a. Fatty acids and glycerol are absorbed by the epithelial cells of villi through active transport.
   b. Monosaccharides are absorbed through facilitated diffusion.
   c. All nutrients are absorbed into capillaries of the villi.
   d. All nutrients are absorbed by the epithelium of the villi by facilitated diffusion.
   e. Only amino acids are exocytosed to lacteals.
10. What happens in the large intestine?
   a. Peyer's patches destroy bacteria.
   b. Vitamin B_{12} is produced.
   c. Water is absorbed.
   d. Na^+ and Cl^- are excreted.
   e. Mass movements are initiated in the rectum.

Matching *Match the function of the digestive system to the location where it takes place. Some answers may be used more than once.*

_____	**1.** Takes place in the mouth	**a.** Ingestion
_____	**2.** Takes place in the stomach	**b.** Mechanical digestion
_____	**3.** Takes place in the small intestine	**c.** Absorption
_____	**4.** Takes place in the large intestine	**d.** Defecation
_____	**5.** Initiated in the large intestine	**e.** Chemical digestion

Matching: *Match the disorder to its description.*

_____	**6.** May result from constipation	**a.** *Staphylococcus*
_____	**7.** Food poisoning resulting from a toxin	**b.** Botulism
_____	**8.** Food poisoning from ingestion of bacteria	**c.** *Salmonella*
_____	**9.** May result from decreased mucus production	**d.** Hemorrhoids
_____	**10.** May result from a weak cardiac sphincter	**e.** Peptic ulcer
		f. GERD

Completion *Fill in the blanks to complete the following statements.*

1. The ___________________ division of the autonomic nervous system stimulates digestion, and the ___________________ division inhibits digestion.
2. ___________________ breaks large pieces of complex molecules into smaller pieces of complex molecules, but ___________________ produces simpler molecules.
3. The ___________________ begins with skeletal muscle, but it ends with smooth muscle in its walls.
4. A bolus moves through the esophagus because of wavelike contractions called ___________________.
5. The products of ___________________ (type of organic molecule) digestion travel through the blood to the heart before traveling to the liver.

Critical Thinking

1. What can you do to minimize the effects of aging on the digestive system?

____

____

____

____

2. Alcoholism can lead to cirrhosis of the liver and liver failure. What would be the effects on the digestive system if the liver failed to function? Which of the systems of the body (that you have studied so far) do you think would be affected the most by liver failure? Explain.

____

____

____

____

____

____

3. The anatomy of the large intestine is quite different from the anatomy of the small intestine. How do the anatomical differences fit with the physiology of each of these two structures?

____

____

____

____

____

____

Case Study

Curtis is a 35-year-old male who suffers from Crohn's disease. While in college, Curtis had a very successful football career but was unable to continue playing the sport due to severe symptoms associated with Crohn's disease. Now that Curtis has been diagnosed and treated, he is able to manage his symptoms and lead a normal life.

1. What is Crohn's disease?

____

____

2. What symptoms does the disease cause?

____

____

3. What diagnostic tests are used to determine whether a patient has Crohn's disease?

____

____

4. How is Crohn's disease treated?

____

____

____

____

This section of the chapter is designed to help you find where each outcome is covered in the textbook and workbook.

Learning Outcomes		Text Readings, Figures, and Tables	Workbook Activities	Text and Workbook Assessments
14.1	Use medical terminology related to the digestive system.	Word roots & combining forms; pronunciation key: pp. 517–519 Overview Figures 14.2–14.3	Word roots & combining forms	*Text:* Review: 14.1 *Workbook:* Word Deconstruction: 1–5
14.2	Differentiate between mechanical and chemical digestion.	Overview: pp. 519–520	*Laboratory exercises and activities:* Digestion of a cheeseburger Figure 14.5 Table 14.1 *Concept maps:* Digestion begins in the mouth Figure 14.6	*Text:* Spot Check: 1 Review: 14.2 *Workbook:* Completion: 2 Matching: 1–3
14.3	Describe the digestive anatomy of the oral cavity.	Anatomy in the mouth: pp. 520–523 Figures 14.2–14.8	*Coloring book:* Digestive tract; Anatomy of a tooth; Salivary glands Figures 14.1–14.3	*Text:* Spot Check: 2 Review: 14.3 *Workbook:* Multiple Select: 1
14.4	Explain the physiology of mechanical and chemical digestion in the mouth.	Physiology of digestion in the mouth: p. 523	*Laboratory exercises and activities:* Digestion of a cheeseburger Figure 14.5 Table 14.1 *Concept maps:* Digestion begins in the mouth Figure 14.6	*Text:* Spot Check: 3 Review: 14.4 *Workbook:* Multiple Select: 2
14.5	Describe the digestive anatomy from the mouth to the stomach.	Anatomy from the mouth to the stomach: pp. 523–524 Figure 14.9	*Coloring book:* Digestive tract Figure 14.1	*Text:* Spot Check: 4 Review: 14.5 *Workbook:* Completion: 3
14.6	Explain how materials are moved from the mouth to the stomach.	Physiology of digestion from the mouth to the stomach: pp. 524–526 Figure 14.10	*Concept maps:* From the mouth to the stomach Figure 14.7	*Text:* Review: 14.6 *Workbook:* Completion: 4
14.7	Describe the digestive anatomy of the stomach.	Anatomy of the stomach: pp. 526–529 Figures 14.11–14.12 Table 14.1		*Text:* Review: 14.7 *Workbook:* Multiple Select: 4

CHAPTER 14 MAPPING

Learning Outcomes	Text Readings, Figures, and Tables	Workbook Activities	Text and Workbook Assessments
14.8 Explain the physiology of mechanical and chemical digestion in the stomach.	Physiology of digestion in the stomach: p. 529	*Laboratory exercises and activities:* Digestion of a cheeseburger Figure 14.5 Table 14.1 *Concept maps:* Digestion in the stomach Figure 14.8	*Text:* Spot Check: 5 Review: 14.8 *Workbook:* Multiple Select: 5, 8
14.9 Explain the feedback mechanism of how food moves from the stomach to the small intestine.	Physiology of digestion in the stomach: p. 530	*Concept maps:* Digestion in the stomach Figure 14.8	*Text:* Spot Check: 7 Review: 14.9 *Workbook:* Multiple Select: 6
14.10 Describe the anatomy of the digestive accessory organs connected to the duodenum by ducts.	Anatomy of digestive accessory structures: pp. 530–534 Figures 14.13–14.15 Table 14.2	*Coloring book:* Digestive tract; Gallbladder, pancreas, and ducts Figures 14.1–14.4	*Text:* Spot Check: 6 Review: 14.10 *Workbook:* Multiple Select: 7
14.11 Describe the digestive anatomy of the small intestine.	Anatomy of the small intestine: pp. 534–536 Figures 14.16–14.17	*Coloring book:* Digestive tract Figure 14.1	*Text:* Review: 14.11 *Workbook:* Critical Thinking: 3
14.12 Explain the physiology of chemical digestion in the duodenum including the hormones and digestive secretions involved.	Physiology of digestion in the small intestine: pp. 536–537 Figure 14.18	*Laboratory exercises and activities:* Digestion of a cheeseburger Figure 14.5 Table 14.1 *Concept maps:* Digestion in the small intestine Figure 14.9	*Text:* Spot Check: 8 Review: 14.12 *Workbook:* Critical Thinking: 2
14.13 Explain how nutrients are absorbed in the small intestine.	Absorption of nutrients in the small intestine: p. 538	*Concept maps:* Absorption of nutrients in the small intestine Figure 14.10	*Text:* Review: 14.13 *Workbook:* Multiple Select: 9
14.14 Describe the anatomy of the large intestine.	Anatomy of the large intestine: pp. 538–540 Figure 14.19	*Coloring book:* Digestive tract Figure 14.1	*Text:* Spot Check: 9 Review: 14.14 *Workbook:* Critical Thinking: 3
14.15 Explain the physiology of the large intestine in terms of absorption, preparation of feces, and defecation.	Physiology of digestion in the large intestine: pp. 540–541 Figure 14.20	*Concept maps:* Digestion in the large intestine Figure 14.11	*Text:* Review: 14.15 *Workbook:* Multiple Select: 10

Learning Outcomes	Text Readings, Figures, and Tables	Workbook Activities	Text and Workbook Assessments
14.16 Summarize the types of nutrients absorbed by the digestive system from the diet.	Types of absorbed nutrients: pp. 541–542		*Text:* Review: 14.16 *Workbook:* Multiple Select: 3
14.17 Trace the circulation of the nutrients once they have been absorbed.	Circulation of absorbed nutrients: p. 542 Figure 14.21	*Concept maps:* Absorption of nutrients in the small intestine Figure 14.10	*Text:* Review: 14.17 *Workbook:* Completion: 5
14.18 Explain the control of digestion.	Control of digestion: p. 543		*Text:* Review: 14.18 *Workbook:* Completion: 1
14.19 Summarize the functions of digestion.	Functions of the digestive system: p. 543 Figure 14.22		*Text:* Review: 14.19 *Workbook:* Matching: 1–5
14.20 Summarize the effects of aging on the digestive system.	Effects of aging on the digestive system: pp. 543–544		*Text:* Review: 14.20 *Workbook:* Critical Thinking: 1
14.21 Describe common diagnostic tests used for digestive system disorders.	Diagnostic tests for digestive system disorders: p. 544 Table 14.3		*Text:* Review: 14.21 *Workbook:* Case Study: 3
14.22 Describe digestive system disorders and relate abnormal function to pathology.	Digestive system disorders: pp. 545–552 Figures 14.23–14.29 Table 14.4		*Text:* Review: 14.22 *Workbook:* Matching: 6–10 Case Study: 1, 2, 4

15 The Excretory/Urinary System

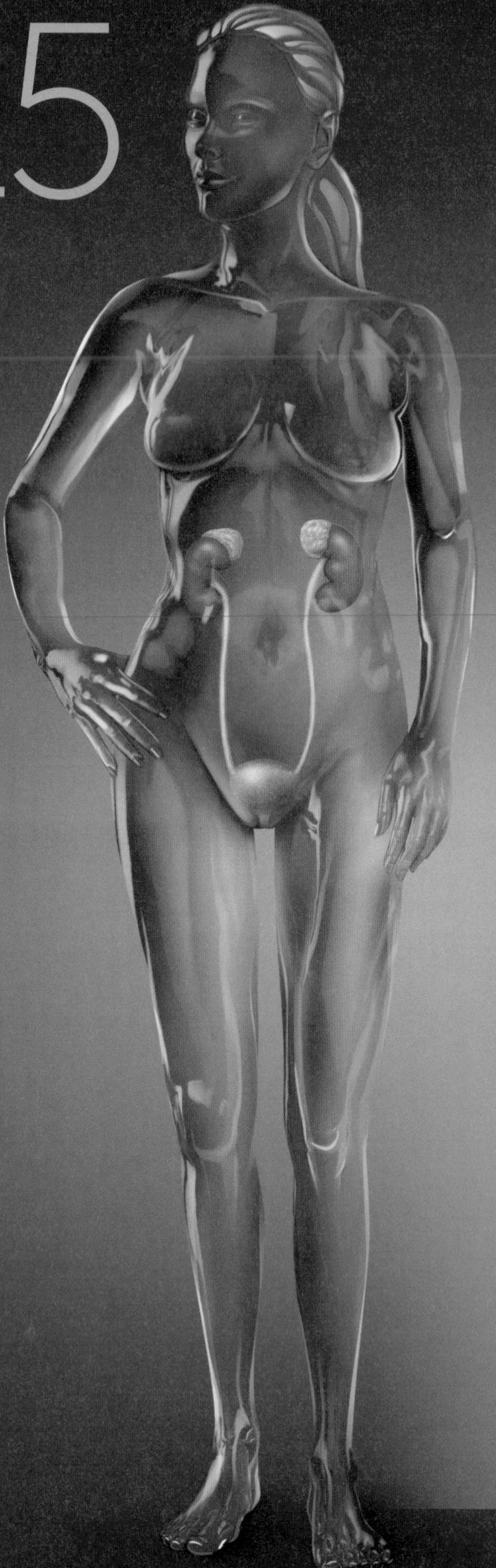

Major Organs and Structures:
kidneys, ureters, urinary bladder, urethra

Accessory Structures:
lungs, skin, liver

Functions:
removal of metabolic wastes, fluid and electrolyte balance, acid-base balance, blood pressure regulation

learning outcomes

This chapter of the workbook is designed to help you learn the anatomy and physiology of the excretory/urinary system. After completing this chapter in the text and this workbook, you should be able to:

15.1 Use medical terminology related to the excretory system.

15.2 Define *excretion* and identify the organs that excrete waste.

15.3 List the body's major nitrogenous wastes and their sources.

15.4 List the functions of the kidneys in addition to urine production.

15.5 Describe the external and internal anatomy of the kidneys.

15.6 Describe the anatomy of a nephron.

15.7 Trace the components of urine through a nephron.

15.8 Trace the flow of blood through a nephron.

15.9 Describe filtration, reabsorption, and secretion in the kidneys with regard to the products moving in each process, the direction of movement, and the method of movement.

15.10 Describe the fluid compartments of the body and how water moves between them.

15.11 Explain how urine volume and concentration are regulated.

15.12 Explain how diuretics, such as medications, caffeine, and alcohol, affect urine production.

15.13 Describe the anatomy of the ureters, urinary bladder, and male and female urethras.

15.14 Describe the micturition reflex and explain how the nervous system and urinary sphincters control the voiding of urine.

15.15 Summarize the functions of the excretory system.

15.16 Summarize the effects of aging on the excretory system.

15.17 Describe common diagnostic tests used for excretory system disorders.

15.18 Describe excretory system disorders and relate abnormal function to pathology.

word **roots** & combining **forms**

azot/o: nitrogen

cyst/o: urinary bladder

glomerul/o: glomerulus

nephr/o: kidney

pyel/o: renal pelvis

ren/o: kidney

ur/o: urinary tract, urine

ureter/o: ureter

urethr/o: urethra

Urinary System

Figure 15.1 shows the anatomy of the urinary system. Color the box next to each term. Use the same color for the corresponding structure in the figure.

- ☐ Kidneys(A)
- ☐ Ureters(B)
- ☐ Urinary bladder(C)
- ☐ Urethra(D)

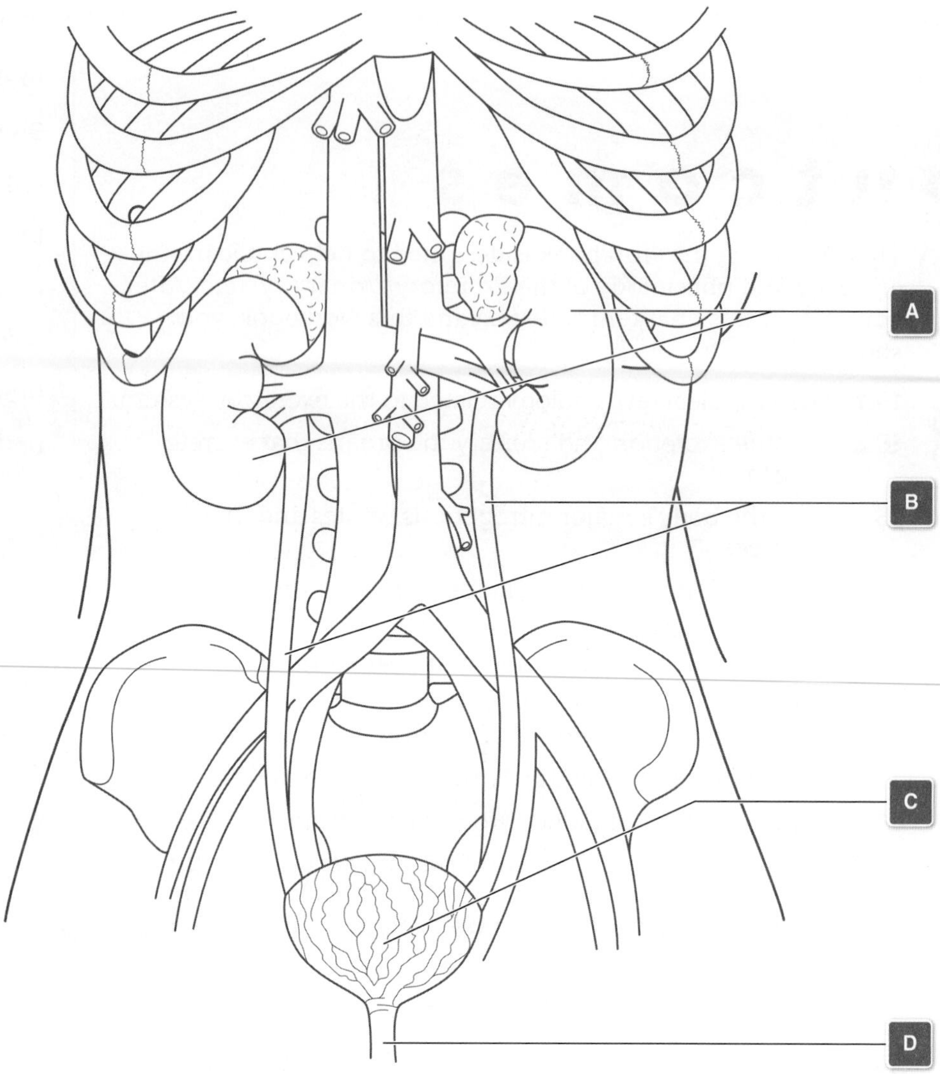

FIGURE 15.1 Anatomy of the urinary system.

COLORING BOOK

Kidneys

Figure 15.2 shows the anatomy of the kidneys. Color the box next to each term. Use the same color for the corresponding structure in the figure.

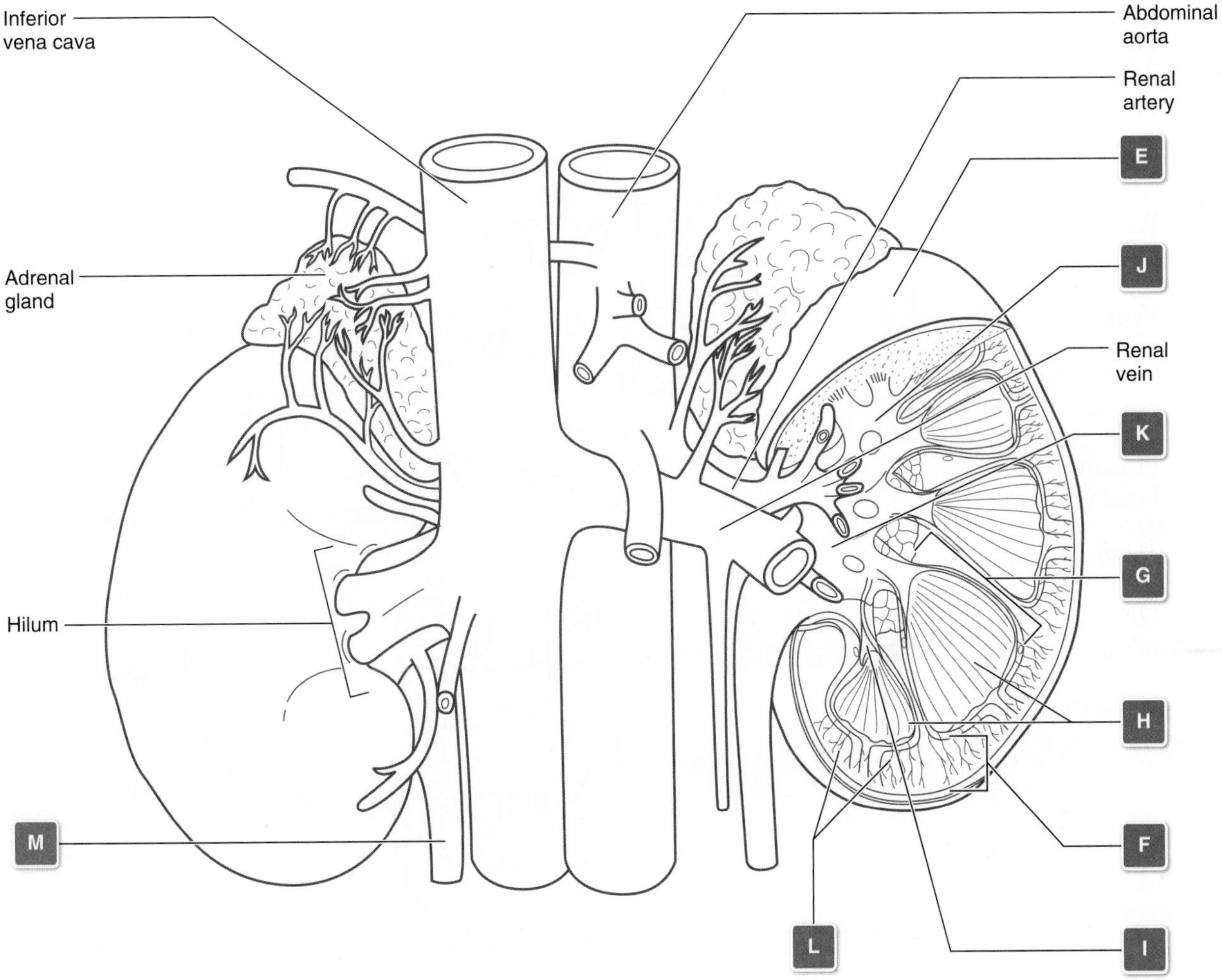

FIGURE 15.2 Anatomy of the kidneys.

- ☐ Fibrous capsule(E)
- ☐ Renal cortex(F)
- ☐ Renal medulla(G)
- ☐ Renal pyramids(H)
- ☐ Minor calyx(I)
- ☐ Major calyx(J)
- ☐ Renal pelvis(K)
- ☐ Renal blood vessels(L)
- ☐ Ureter(M)

Nephron

Figure 15.3 shows the anatomy of a nephron. Color the box next to each term. Use the same color for the corresponding structure in the figure.

- ☐ Afferent arteriole (N)
- ☐ Glomerulus (O)
- ☐ Efferent arteriole (P)
- ☐ Peritubular capillaries (Q)
- ☐ Venule (R)
- ☐ Glomerular capsule (S)
- ☐ Proximal convoluted tubule (T)
- ☐ Nephron loop (U)
- ☐ Distal convoluted tubule (V)
- ☐ Collecting duct (W)

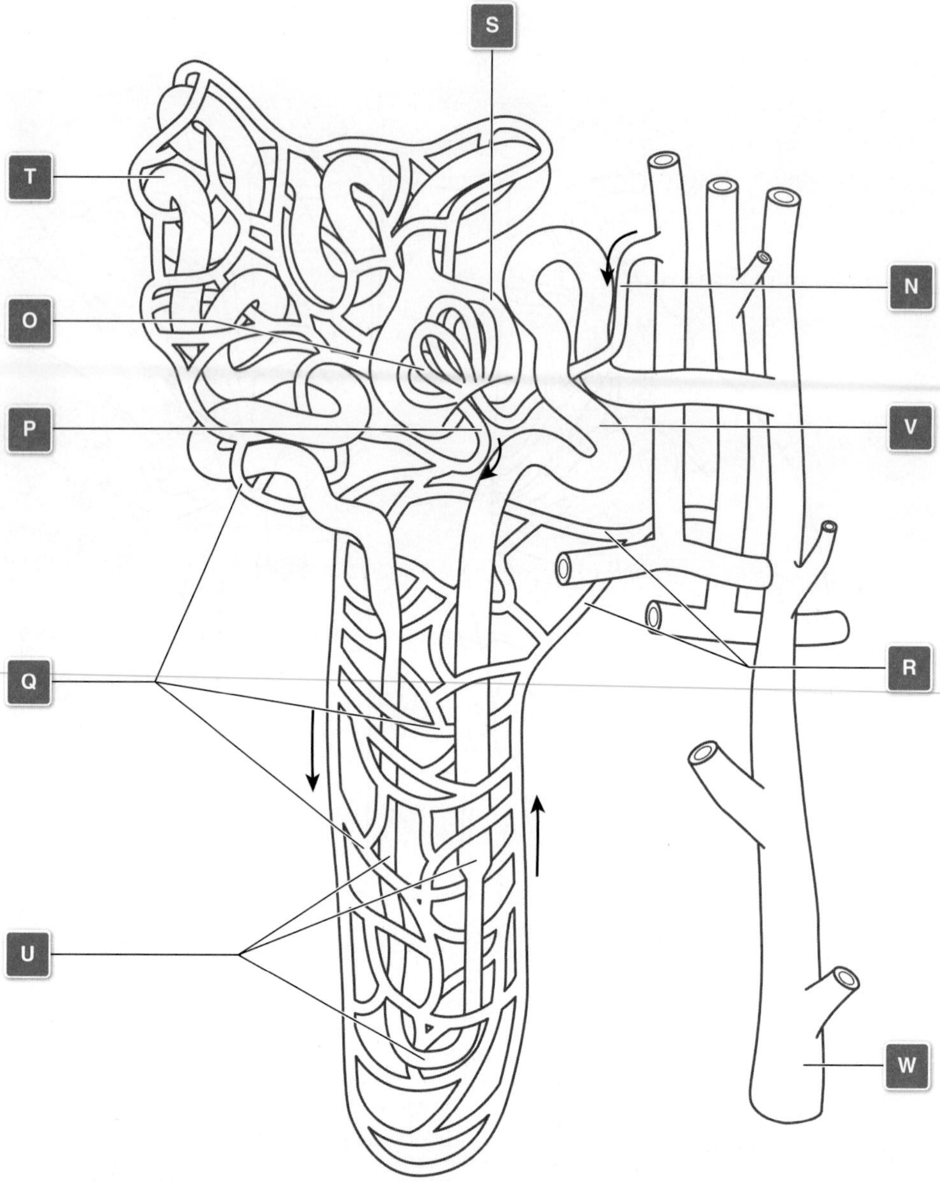

FIGURE 15.3 Anatomy of a nephron.

Ureters, Urinary Bladder, and Urethra

Figure 15.4 shows the anatomy of the ureters, urinary bladder, and urethra in a female and a male. Color the box next to each term. Use the same color for the corresponding structure in the figure.

- ☐ Urinary bladder (A)
- ☐ Ureter (B)
- ☐ Detrusor muscle (C)
- ☐ Ureteral openings (D)
- ☐ Prostatic urethra (E)
- ☐ Internal urethral sphincter (F)
- ☐ Pelvic floor (G)
- ☐ External urethral sphincter (H)
- ☐ Membranous urethra (I)
- ☐ Penile urethra (J)
- ☐ External urethral orifice (K)

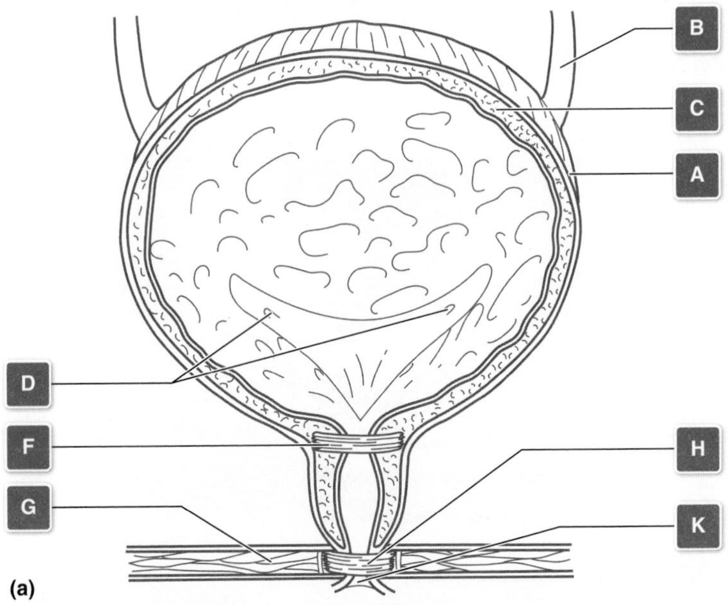

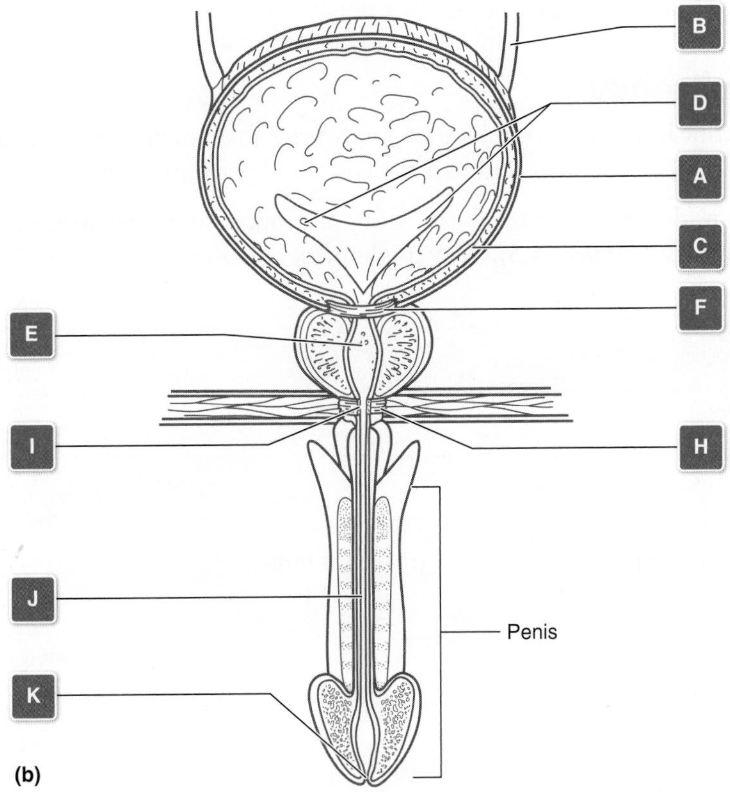

FIGURE 15.4 The ureters, urinary bladder, and urethra: (a) female urinary anatomy, (b) male urinary anatomy.

COLORING BOOK

Urinalysis

A urine sample (shown in **Figure** 15.5) is often requested in a routine physical because it can provide basic information on the overall status of the kidneys and urinary tract. The urine is analyzed using several tests that are explained below. In this activity, you are to answer questions concerning the anatomy and physiology on which the tests are based. If you have trouble with any of the areas, you may want to review urine production in the text.

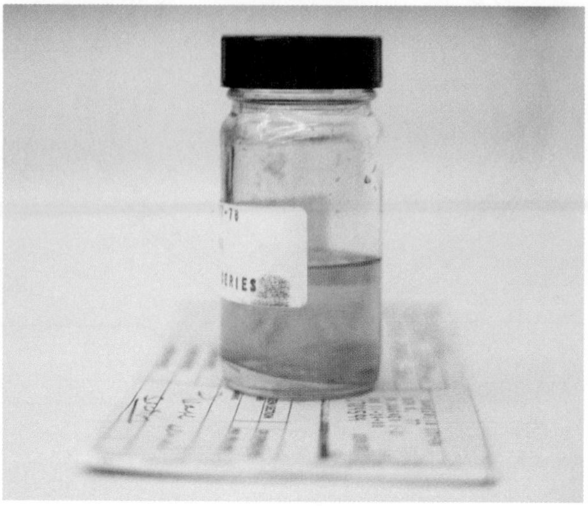

FIGURE 15.5 Urine sample for urinalysis.

Transparency

Normal urine is clear to partly cloudy. Cloudy urine may contain fat globules, mucus, epithelial cells, or microbes. This is a test you could perform on your own urine.

1. Consider the pathway of urine from its production to the urine sample container (see **Figure** 15.5). Why might urine contain mucus and epithelial cells?

____

____

____

____

2. What would the presence of microbes indicate?

____

____

Color

Normal urine is transparent yellow, while yellow-brown to greenish urine may occur when high concentrations of bile pigments are present. Red to dark brown indicates the presence of blood in the urine. The color of the first urine passed in the morning is usually the most intense. This is another test you could conduct on your own urine.

3. What causes the yellow color of normal urine?

4. What disorder may be indicated by reddish urine? Explain.

5. Why do you think urine collected right away after waking has the most intense color?

pH of urine

The pH of urine normally falls in the range of 4.8 to 7.5, depending greatly upon the diet. Foods such as cranberry juice result in more acidic urine. Values decrease below this range in acidosis, emphysema, and starvation. Values increase above the normal range in urinary tract infections and severe alkalosis.

6. Which of the three processes of urine production determines the pH of urine?

7. How does acidic urine contribute to homeostasis of the body?

8. How might acidic urine contribute to the health of the urinary tract?

Specific Gravity

In urinalysis, specific gravity is the ratio of the density of urine in relation to water. It is used as a measurement of the concentration of urine (amount of solutes). A hydrometer is used to conduct the test. See **Figure 15.6**. The urine is placed in a cylinder, and a float is placed in the urine. A scale is read at the point where the top of the urine meets the scale on the float. Normal values fall in the range

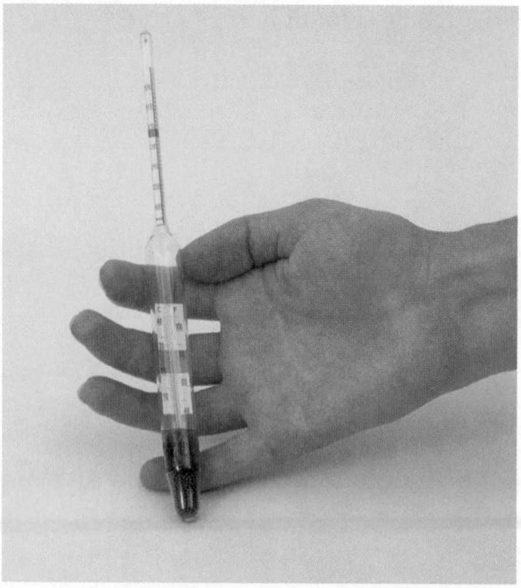

FIGURE 15.6 A hydrometer used to test the specific gravity of urine. The hydrometer shown is placed in a cylinder of urine deep enough for the hydrometer to float. The scale on the neck of the hydrometer is read at the point where the surface of the urine sample intersects with the floating hydrometer.

of 1.020 to 1.028 (normal values may differ slightly among laboratories). Values below normal may indicate excessive fluid intake or diabetes insipidus. Values above normal may indicate dehydration or diabetes mellitus.

9. What hormone is likely to contribute to increased values for specific gravity? Explain.

____

____

____

10. What hormone is likely to contribute to decreased values for specific gravity? Explain.

____

____

____

11. If specific gravity is increased due to diabetes mellitus, what solute is most likely to be in high concentration?

____

____

____

12. If specific gravity is decreased due to diabetes insipidus, what hormone is likely to be involved? Explain.

____

____

____

Glucose Test

A glucose test is performed by dipping a reagent strip into the urine sample and removing it immediately. The color of the reagent strip is then compared to a color chart on the container. See **Figure 15.7**. The presence of glucose in the urine may indicate diabetes mellitus.

13. Why might the presence of glucose in the urine indicate diabetes mellitus? Explain.

____

____

____

____

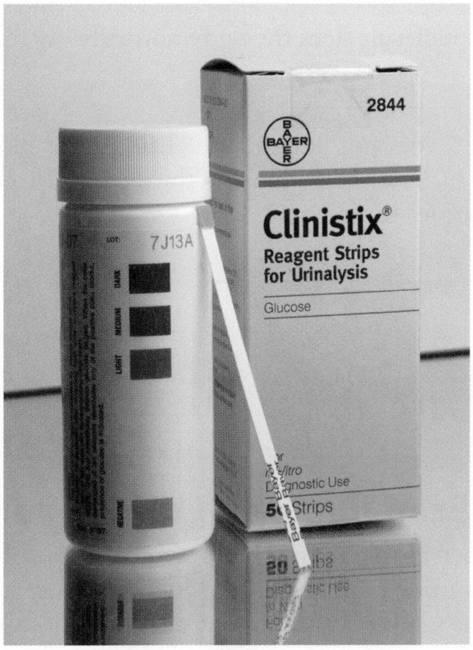

FIGURE 15.7 Glucose test strips used for urinalysis.

Ketones

Ketones are chemical by-products made by the body when it uses fat for energy. Small amounts of ketones in the blood and urine are not harmful. However, large amounts of ketones may lead to acidosis because ketones are acidic. This test is performed in the same way as the glucose test. A reagent strip that detects ketones is dipped into the urine and immediately removed. The color of the reagent strip is then compared to a color chart on the strip container. See **Figure 15.8**.

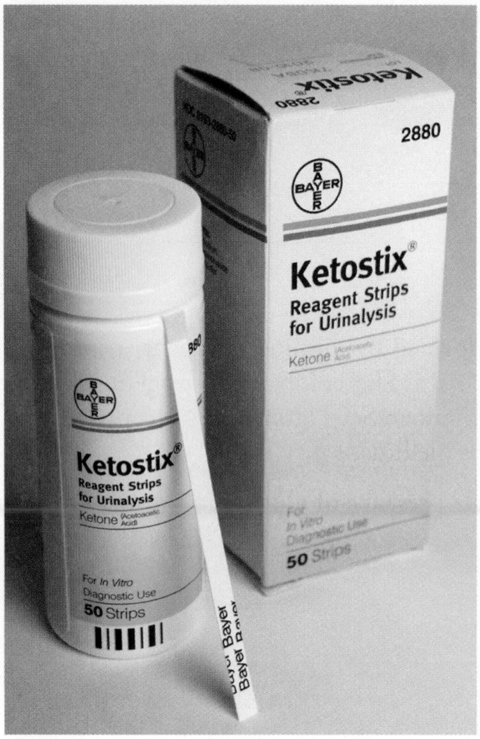

FIGURE 15.8 Ketone test strips used for urinalysis.

14. What organic molecule does the body normally use for energy?

____

15. A diabetic who does not have good control of his blood sugar levels may have ketones in his urine. If this is the case, what else will probably be found in his urine? Explain.

____

____

____

____

Microscopic Examination of Urine Sediments

The final test described in this activity is a microscopic examination of urine sediments. In this test, a urine sample is placed in a centrifuge and spun so that any solid sediments in the sample accumulate at the bottom of the centrifuge tube. A dropper is used to obtain a few drops of the urine at the bottom of the tube, and these drops are placed on a clean microscope slide. The urine drops are viewed through the microscope at low and high power.

Sediments may include tubular accumulations (called *casts*) held together by proteins. These casts may be composed of mucous proteins (hyaline casts), epithelial cells (epithelial casts), red blood cells (erythrocyte casts), white blood cells (leukocyte casts), or tubular cells (granular or waxy casts). An occasional cast in urinary sediments is considered normal, but increased numbers of casts indicate damage to the renal tubule. See **Figure 15.9**.

Other sediments found in urine include mucous threads (see **Figure 15.9d**) and crystals (see **Figure 15.10**). Calcium oxalate crystals look like tiny envelopes

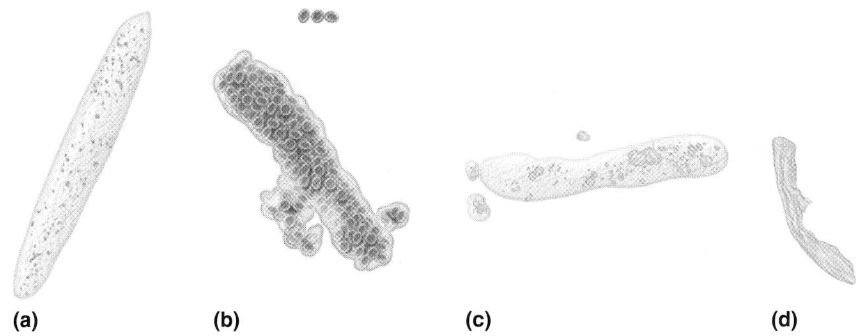

FIGURE 15.9 Urine sediments: (a) hyaline cast, (b) erythrocyte cast, (c) leukocyte cast, (d) mucous thread.

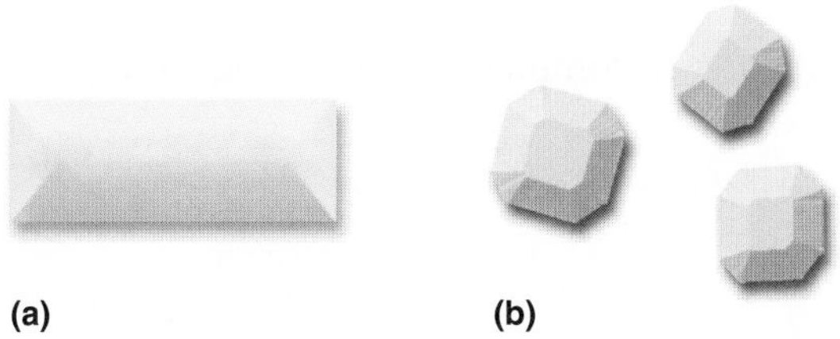

FIGURE 15.10 Crystalline urine sediments: (a) calcium oxalate crystals, (b) cystine crystals.

and are commonly seen even in healthy urine; cystine (an amino acid) crystals are shaped like stop signs and are rarely seen.

16. Cystine is an amino acid. Should it be found in urine? Explain.

____

____

____

17. Why is it normal to have mucus threads in urine?

____

____

____

18. An increase in urinary sediments often indicates damage to nephrons. A kidney stone occurs when crystals form stones in the renal pelvis, well past the nephrons. Why might a kidney stone also cause an increase in casts and blood in the urine?

____

____

____

____

Key Words

The following terms are defined in the glossary of the textbook.

cystectomy
cystitis
cystoscopy
dialysis
diuretic
excretion
fluid compartments
glomerulonephritis
hyperkalemia
hyponatremia
lithotripsy
metabolic acidosis
metabolic alkalosis
metabolic waste
metabolic water
micturition
nephron
nitrogenous wastes
renal corpuscle
renal tubule
respiratory acidosis
respiratory alkalosis
secretion

Concept Maps

Use key words and other bold words from the chapter to complete the following concept maps (**Figures 15.11** to 15.15).

b. ADH causes sodium to be reabsorbed.
c. Aldosterone does not cause sodium to be reabsorbed.
d. ANH results in an increased volume of dilute urine.
e. ADH and aldosterone result in reduced urine volume.

7. How do diuretics work?
   a. Alcohol inhibits aldosterone production.
   b. Caffeine decreases blood flow to the kidneys.
   c. Diuretic drugs cause increased potassium reabsorption.
   d. Diuretic drugs increase reabsorption of sodium.
   e. Caffeine increases the glomerular filtration rate.

8. How would you describe the anatomy of the urinary system?
   a. Ureters are retroperitoneal.
   b. A ureter is much longer in a male than in a female.
   c. The detrusor muscle forms the external urethral sphincter.
   d. The detrusor muscle is an involuntary smooth muscle.
   e. The renal pelvis leads into the urethra.

9. How does micturition work?
   a. The medulla oblongata regulates micturition.
   b. The pons regulates micturition.
   c. Micturition is a reflex.
   d. Micturition is a reflex and a voluntary activity.
   e. Sympathetic neurons stimulate the passing of urine.

10. What can be expected of the excretory system as it ages?
   a. Benign prostatic hyperplasia leads to incontinence in males.
   b. Weakened pelvic-floor muscles make the passing of urine difficult for women.
   c. The glomerular filtration rate decreases with age.
   d. Elderly people are less responsive to ADH.
   e. Drugs are less likely to be effective because they are cleared by the excretory system too quickly.

Matching: *Match the process of urine production to the description. Some choices may be used more than once. Some questions may have more than one answer.*

_______ 1. Happens in the renal corpuscle
_______ 2. Happens in the renal tubules
_______ 3. Moves water by osmosis
_______ 4. Moves materials from the peritubular capillaries to the tubules
_______ 5. Moves materials from the tubules to the peritubular capillaries

a. Filtration
b. Reabsorption
c. Secretion

Matching: *Match the structure to its description. Some choices may be used more than once. Some questions may have more than one answer.*

_______ 6. Collects the products of filtration
_______ 7. Receives the products of secretion
_______ 8. Contains the glomerulus and the glomerular capsule
_______ 9. Takes blood away from the glomerulus
_______ 10. Contains part of the juxtaglomerular apparatus

a. Afferent arteriole
b. Efferent arteriole
c. Glomerulus
d. Glomerular capsule
e. Renal corpuscle
f. Distal convoluted tubule

Completion: *Fill in the blanks to complete the following statements.*

1. ______________ is the removal of metabolic wastes from the body.

2. ______________ is a waste from the breakdown of creatine phosphate.

3. ______________ is a highly toxic waste from the breakdown of amino acids.

4. ______________ is a common waste from the ultimate breakdown of proteins.

5. ______________ is a waste from the breakdown of nucleic acids.

Critical Thinking

1. Peter is an A&P student who is studying the excretory system. He decided to conduct an experiment on his own. He collected his morning urine and then drank 2 liters of water in the next hour and collected his urine again. What would you predict would be his next urine output and his urine concentration (based on a color comparison with his first urine sample)? What hormone is likely at work regulating his urine production? Explain.

2. In terms of function, why is a kidney transplant preferable to dialysis?

3. Select one of the disorders covered in the chapter. Predict how that disorder would affect urinalysis results.

Case Study

Shelly is 30 weeks pregnant with a baby girl and learned during a routine ultrasound that her baby's right kidney was enlarged. Once the baby was born, the pediatric urologist discovered the baby had hydronephrosis.

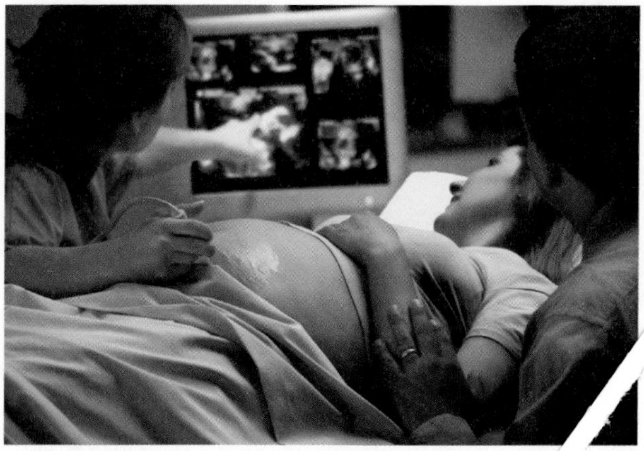

1. What is hydronephrosis?

____

____

2. How is the condition diagnosed?

____

3. Considering the fact that the physician discovered this condition during an ultrasound before the baby was born, what is the likely cause of hydronephrosis in this infant?

____

4. Can the baby's condition be treated? Please explain your answer.

____

____

____

CHAPTER 15 MAPPING

This section of the chapter is designed to help you find where each outcome is covered in the textbook and workbook.

Learning Outcomes		Text Readings, Figures, and Tables	Workbook Activities	Text and Workbook Assessments
15.1	Use medical terminology related to the excretory system.	Word roots & combining forms; pronunciation key: p. 559	Word roots & combining forms	*Text:* Review: 15.1 *Workbook:* Word Deconstruction: 1–5
15.2	Define *excretion* and identify the organs that excrete waste.	Overview: pp. 560–561 Figure 15.2		*Text:* Review: 15.2 *Workbook:* Completion: 1
15.3	List the body's major nitrogenous wastes and their sources.	Overview: p. 561 Figure 15.3		*Text:* Review: 15.3 *Workbook:* Completion: 2–5
15.4	List the functions of the kidney in addition to urine production.	Overview: p. 561		*Text:* Review: 15.4 *Workbook:* Multiple Select: 1 Critical Thinking: 2
15.5	Describe the external and internal anatomy of the kidney.	Anatomy of the kidney: pp. 562–564 Figures 15.4–15.5	*Coloring book:* Kidneys Figure 15.2	*Text:* Spot Check: 1 Review: 15.5 *Workbook:* Multiple Select: 2
15.6	Describe the anatomy of a nephron.	Anatomy of a nephron: pp. 564–566 Figures 15.6–15.7	*Coloring book:* Nephron Figure 15.3	*Text* Spot Check: 2 Review: 15.6 *Workbook:* Matching: 8, 10
15.7	Trace the components of urine through a nephron.	Flow of urine components through a nephron: pp. 566–567 Figure 15.6–15.7		*Text:* Review: 15.7 *Workbook:* Multiple Select: 3
15.8	Trace the flow of blood through a nephron.	Blood flow to a nephron: p. 567 Figure 15.8		*Text:* Spot Check: 3 Review: 15.8 *Workbook:* Multiple Select: 4 Matching: 9
15.9	Describe filtration, reabsorption, and secretion in the kidney with regard to the products moving in each process, the direction of movement, and the method of movement.	Physiology of urine production: pp. 567–570 Figures 15.9–15.10	*Lab exercises and activities:* Urinalysis Figures 15.5–15.10 *Concept maps:* Filtration; Reabsorption; Secretion Figures 15.11–15.13	*Text:* Spot Check: 4 Review: 15.9 *Workbook:* Matching: 1–7

Learning Outcomes	Text Readings, Figures, and Tables	Workbook Activities	Text and Workbook Assessments
15.10 Describe the fluid compartments of the body and how water moves between them.	Water in the body: pp. 570–572 Figures 15.11–15.12		*Text:* Review: 15.10 *Workbook:* Multiple Select: 5
15.11 Explain how urine volume and concentration are regulated.	Regulation of urine volume and concentration: pp. 573–577 Figures 15.13–15.16	*Lab exercises and activities:* Urinalysis Figures 15.5–15.10 *Concept maps:* Hormonal control Figure 15.14	*Text:* Spot Check: 5, 6, 7 Review: 15.11 *Workbook:* Multiple Select: 6 Critical Thinking: 1
15.12 Explain how diuretics, such as medications, caffeine, and alcohol, affect urine production.	Diuretics: p. 577		*Text:* Review: 15.12 *Workbook:* Multiple Select: 7
15.13 Describe the anatomy of the ureters, urinary bladder, and male and female urethras.	Anatomy of ureters, urinary bladder, and male and female urethras: pp. 577–579 Figures 15.17–15.19	*Coloring book:* Urinary system; Ureters, urinary bladder, and urethra Figures 15.1, 15.4	*Text:* Review: 15.13 *Workbook:* Multiple Select: 8
15.14 Describe the micturition reflex and explain how the nervous system and urinary sphincters control the voiding of urine.	Physiology of the passing of urine: pp. 580–581 Figure 15.20	*Concept maps:* Micturition Figure 15.15	*Text:* Review: 15.14 *Workbook:* Multiple Select: 9
15.15 Summarize the functions of the excretory system.	Function of the excretory system: pp. 581–582 Figure 15.21	*Lab exercises and activities:* Urinalysis Figures 15.5–15.10	*Text:* Review: 15.15 *Workbook:* Multiple Select: 1 Critical Thinking: 2
15.16 Summarize the effects of aging on the excretory system.	Effects of aging on the excretory system: p. 582		*Text:* Review: 15.16 *Workbook:* Multiple Select: 10
15.17 Describe common diagnostic tests used for excretory system disorders.	Diagnostic tests for excretory system disorders: pp. 582–583 Table 15.1	*Lab exercises and activities:* Urinalysis Figures 15.5–15.10	*Text:* Spot Check: 8 Review: 15.17 *Workbook:* Case Study: 2
15.18 Describe excretory system disorders and relate abnormal function to pathology.	Disorders of the excretory system: pp. 583–587 Figures 15.22–15.24 Table 15.2		*Text:* Spot Check: 9 Review: 15.18 *Workbook:* Critical Thinking: 3 Case Study: 1, 3, 4

CHAPTER 15 MAPPING

16 The Male Reproductive System

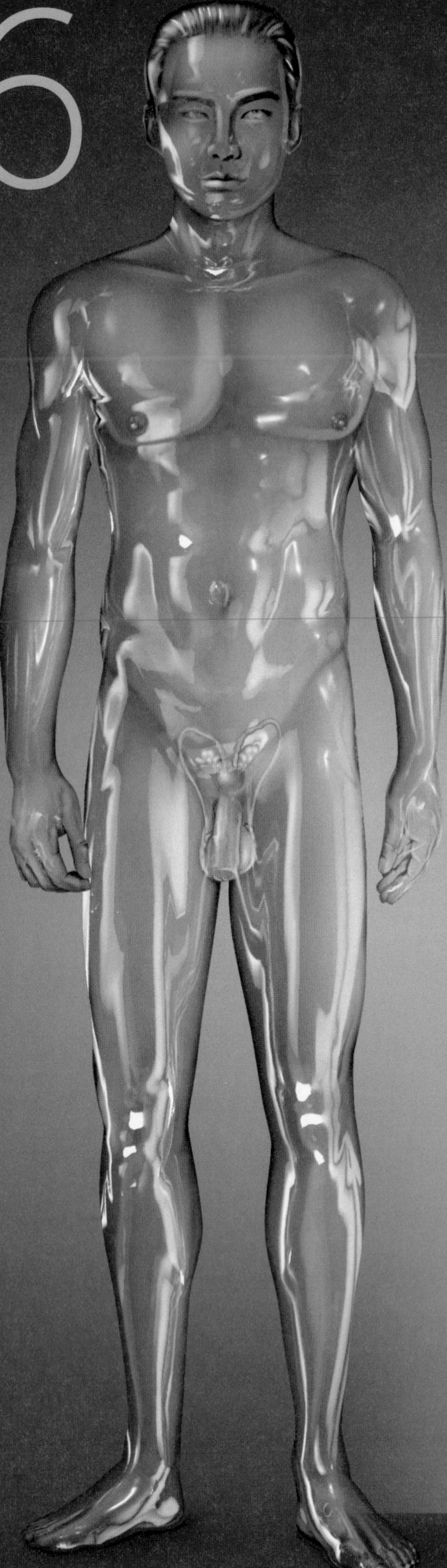

Major Organs and Structures:
testes

Accessory Structures:
scrotum, spermatic ducts (epididymis, ductus deferens), accessory glands (seminal vesicles, prostate gland, bulbourethral glands), penis

Functions:
production and delivery of sperm, secretion of sex hormones

learning outcomes

This chapter of the workbook is designed to help you learn the anatomy and physiology of the male reproductive system. After completing this chapter in the text and this workbook, you should be able to:

16.1 Use medical terminology related to the male reproductive system.

16.2 Explain what is needed for male anatomy to develop.

16.3 Describe the anatomy of the testes.

16.4 Describe the male secondary sex organs and structures and their respective functions.

16.5 Describe the anatomy of a sperm.

16.6 Explain the hormonal control of puberty and the resulting changes in the male.

16.7 Explain the stages of meiosis and contrast meiosis to mitosis.

16.8 Explain the processes of sperm production and differentiate between spermatogenesis and spermiogenesis.

16.9 Explain the hormonal control of the adult male reproductive system.

16.10 Trace the path a sperm takes from its formation to its ejaculation.

16.11 Describe the stages of the male sexual response.

16.12 Explain the effects of aging on the male reproductive system.

16.13 Describe common diagnostic tests used for male reproductive system disorders.

16.14 Describe male reproductive system disorders and relate abnormal function to pathology.

word **roots** & combining **forms**

andr/o: male

crypt/o: hidden

epididym/o: epididymis

orch/o, orchi/o, orchid/o: testis, testicle

pen/o: penis

prostat/o: prostate

semin/i: semen

sperm/o, spermat/o: sperm

test/o: testis, testicle

vas/o: duct, vas deferens

Male Reproductive System

Figure 16.1 shows the anatomy of the male reproductive system. Color the box next to each term. Use the same color for the corresponding structure in the figure.

- ☐ Ureter(A)
- ☐ Ampulla(B)
- ☐ Seminal vesicles(C)
- ☐ Ejaculatory duct(D)
- ☐ Prostate gland(E)
- ☐ Prostatic urethra(F)
- ☐ Bulbourethral gland(G)
- ☐ Ductus deferens(H)
- ☐ Epididymis(I)
- ☐ Testis(J)
- ☐ Penile urethra(K)
- ☐ Urinary bladder(L)
- ☐ Bulb of penis(M)
- ☐ Crus of penis(N)
- ☐ Corpus cavernosum(O)
- ☐ Lobule of testis(P)
- ☐ Corpus spongiosum(Q)
- ☐ Glans(R)

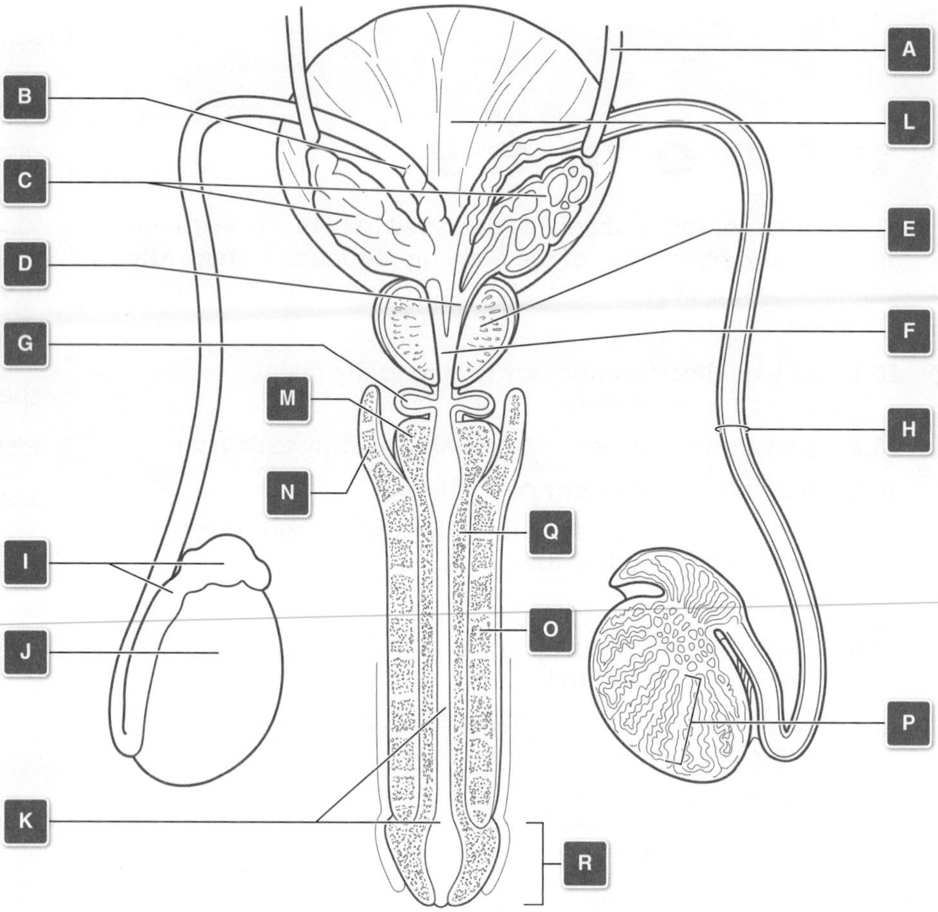

FIGURE 16.1 Male reproductive system, posterior view.

Testis and Spermatic Cord

Figure 16.2 shows the anatomy of a testis and spermatic cord. Color the box next to each term. Use the same color for the corresponding structure in the figure.

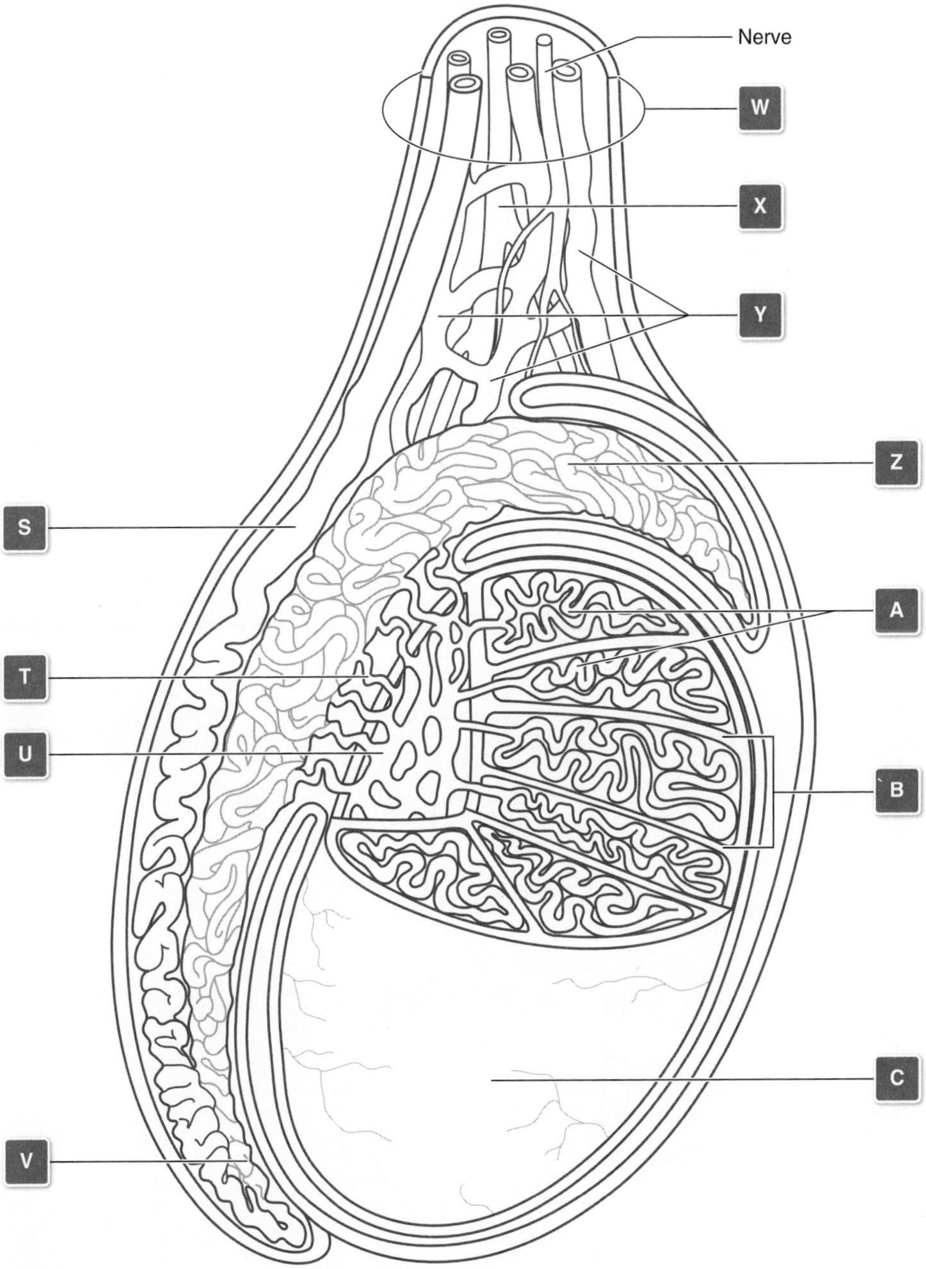

- ☐ Ductus deferens(S)
- ☐ Efferent ductules(T)
- ☐ Rete testis(U)
- ☐ Tail of epididymis(V)
- ☐ Spermatic cord(W) (color circle)
- ☐ Testicular artery(X)
- ☐ Pampiniform plexus(Y)
- ☐ Head of epididymis(Z)
- ☐ Seminiferous tubules(A)
- ☐ Lobule of testis(B)
- ☐ Tunica albuginea(C)

FIGURE 16.2 Anatomy of a testis and spermatic cord. The cremaster muscle is missing from this drawing so other structures may be shown.

Where in this drawing would the cremaster muscle be located?

____

Sperm

Figure 16.3 shows the anatomy of a spermatozoon. Color the box next to each term. Use the same color for the corresponding structure in the figure.

- ☐ Acrosome(D)
- ☐ Head(E)
- ☐ Midpiece of tail(F)
- ☐ Mitochondria(G)
- ☐ Tail(H)
- ☐ Flagellum(I)

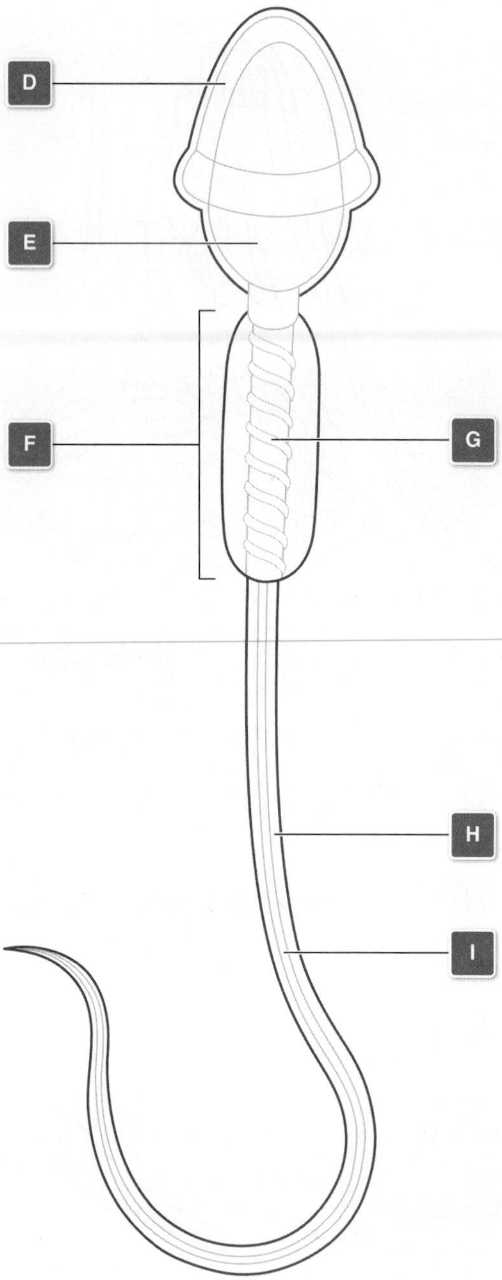

FIGURE 16.3 Anatomy of a spermatozoon.

Sperm Production

Sperm production involves two processes—spermatogenesis and spermiogenesis. Place the following statements in order in Table 16.1 to describe each process.

- Sperm is fully formed.
- The primary spermatocyte goes through meiosis I to form secondary spermatocytes.
- A spermatogonium near the basement membrane of the seminiferous tubule first divides by *mitosis.*
- Spermatids shed cytoplasm.
- Spermatogonium B enlarges to become a primary spermatocyte with 46 chromosomes.
- The spermatogonium (type B) migrates slightly away from the wall toward the lumen (hollow center) of the seminiferous tubule.
- Each secondary spermatocyte undergoes meiosis II, which produces in total four spermatids from the one original spermatogonium type B.
- Acrosome and flagellum form.

TABLE 16.1 Sperm production

SPERM PRODUCTION	
Spermatogenesis	**1.**
	2.
	3.
	4.
	5.
Spermiogenesis	**6.**
	7.
	8.

Male Sexual Response

The male sexual response involves four stages—arousal, emission, ejaculation, and resolution. Place the following statements in Table 16.2 to describe each stage.

- Penile arteries constrict.
- Arteries of penis dilate.
- Bulbourethral gland secretes a lubricating fluid.
- Bulbocavernosus muscle contracts.
- Erectile tissues engorge with blood.
- Prostate secretes an alkaline fluid.
- Penile veins are compressed.
- Penis becomes flaccid.
- Seminal vesicles release a nourishing fluid.
- Ductus deferens moves sperm to the ampulla, which then contracts to move sperm through the ejaculatory duct to the urethra.
- Prostate releases additional fluid.
- Excess blood is squeezed out of the penis.
- Internal urethral sphincter contracts.
- Penis becomes erect.
- Semen is ejaculated.
- Seminal vesicles release additional fluid.
- Trabecular muscles contract.

TABLE 16.2 Stages of the male sexual response

MALE SEXUAL RESPONSE	
Arousal	•
	•
	•
	•
	•

c. The internal urethral sphincter contracts during emission.
d. Trabecular muscles of erectile tissues relax during resolution.
e. The ductus deferens contracts in peristaltic waves during expulsion.

7. Herb is a 55-year-old farmer who gets frustrated because he does not seem to have the strength that he used to have. He also is not sleeping well because he needs to go to the bathroom several times a night. His wife reports that he is moodier than usual and is not as interested in having sex as he used to be. Although he is reluctant, Herb decides to tell all of this to his doctor at his annual checkup. Which of the following statements is (are) consistent with this description?
   a. The doctor will probably perform a digital rectal exam to check for BPH.
   b. Herb may be experiencing andropause.
   c. Herb's testosterone levels have fallen due to aging.
   d. Herb's FSH and LH levels are likely elevated.
   e. Herb's testosterone level may affect his libido.
8. What must happen for male reproductive anatomy to develop?
   a. The zygote must have an X and a Y chromosome.
   b. Testosterone must be produced by the developing gonads in the fetus.
   c. The X chromosome must have an SRY gene.
   d. Testosterone must be produced in childhood.
   e. Both sex chromosomes must have an SRY gene.
9. What is the anatomy of the testes?
   a. Each testis is composed of 4 lobules that contain 250 to 300 sem[illegible]iferous tubules.
   b. Each testis has sustentacular cells between seminiferous tubules.
   c. Interstitial cells have receptors for inhibin.
   d. Sustentacular cells have receptors for LH.
   e. Sustentacular cells produce inhibin to inhibit production of FSH.
10. What is the anatomy of a spermatozoon?
   a. The midpiece contains ribosomes.
   b. The acrosome cap contains a lubricant.
   c. The head contains mitochondria for energy production.
   d. The tail has cilia for movement
   e. There are 23 chromosomes in the acrosome cap.

Matching: *Match the cell to the correct number of chromosomes. The choi[illegible] may be used more than once.*

_______	1. Spermatid	a. [illegible]omosomes
_______	2. Type B spermatogonium	b. [illegible]mosomes
_______	3. Primary spermatocyte	
_______	4. Secondary spermatocyte	
_______	5. Spermatozoon	

Matching: *Match the male reproductive disorder to its description. Some of the choi[illegible]ay be used more than once.*

_______	6. May be detected by a PSA blood test	a. Hypospa[illegible]
_______	7. May be detected by a self-exam	b. Testicular [illegible]ncer
_______	8. May be detected by a digital rectal exam	c. Prostate ca[illegible]r

_______ 9. A congenital defect of the urethra

_______ 10. Testes in the abdominal cavity

d. BPH

e. Cryptorchidism

f. Erectile dysfunction

Completion: *Fill in the blanks to complete the following statements.*

1. The _______________ gene on the _______________ chromosome codes for a protein that interacts with other genes so that _______________ receptors are formed in a fetus.
2. The _______________ pulls the testes to their ultimate position in the scrotum.
3. Primary sex organs produce _______________ (sex cells).
4. Each testis descends through an opening in the abdominal wall called the _______________.
5. The accessory glands in the male reproductive system include two _______________, two _______________, and one _______________.

Critical Thinking

1. James has a long commute each morning. He uses public transportation so that he can work on his laptop while on the road. He and his wife are trying to have a baby, but so far have been unsuccessful. Why might his laptop be the problem?

____

____

____

____

2. What happens to the testes if a male does not regularly ejaculate sperm? Explain.

____

____

____

____

3. Marco's sperm fertilized his wife's egg a week after his vasectomy. How could this happen?

____

____

____

____

Case Study

Aaron is a newborn baby boy. When his pediatrician examined him for the first time, they noticed that Aaron's right testicle had not descended into the scrotum.

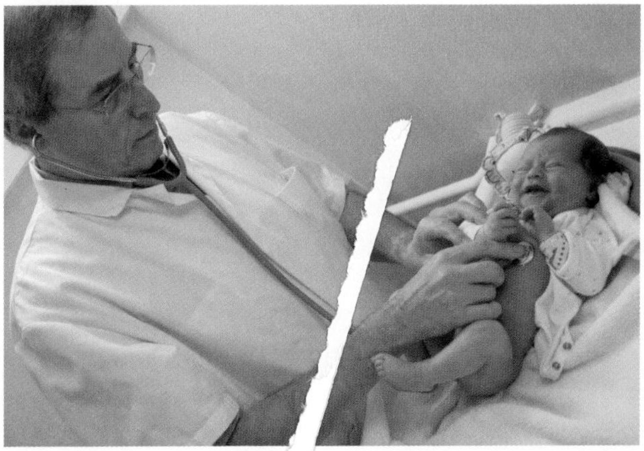

1. What is this condition called?

____

2. What diagnostic test(s) might physicians perform to diagnose Aaron's condition?

____

3. How can Aaron's condition be treated?

____

____

4. What might happen if the condition is left untreated and the testicle remains outside of the scrotum? Explain.

____

____

____

CHAPTER 16 MAPPING

This section of the chapter is designed to help you find where each outcome is covered in the textbook and workbook.

Learning Outcomes		Text Readings, Figures, and Tables	Workbook Activities	Text and Workbook Assessments
16.1	Use medical terminology related to the male reproductive system.	Word roots & combining forms; pronunciation key: p. 593	Word roots & combining forms	*Text:* Review: 16.1 *Workbook:* Word Deconstruction: 1–5
16.2	Explain what is needed for male anatomy to develop.	Overview: pp. 594–595 Figures 16.2–16.3		*Text:* Review: 16.2 *Workbook:* Multiple Select: 8 Completion: 1
16.3	Describe the anatomy of the testes.	Male reproductive anatomy; Testes: pp. 595–597 Figures 16.4–16.5	*Coloring book:* Testis and spermatic cord Figure 16.2	*Text:* Spot Check: 1, 2 Review: 16.3 *Workbook:* Multiple Select: 9 Completion: 2–4 Critical Thinking: 1
16.4	Describe the male secondary sex organs and structures and their respective functions.	Secondary sex organs and structures: pp. 597–602 Figures 16.6–16.10	*Coloring book:* Male reproductive system Figure 16.1	*Text:* Spot Check: 3, 4, 8 Review: 16.4 *Workbook:* Multiple Select: 1, 2 Completion: 5
16.5	Describe the anatomy of a sperm.	Anatomy of a sperm: p. 603 Figure 16.11	*Coloring book:* Sperm Figure 16.3	*Text:* Spot Check: 5 Review: 16.5 *Workbook:* Multiple Select: 10
16.6	Explain the hormonal control of puberty and the resulting changes in the male.	Physiology of the male reproductive system; Hormonal control at puberty: pp. 603–604	*Concept maps:* Hormonal control from development to puberty Figure 16.4	*Text:* Spot Check: 6 Review: 16.6 *Workbook:* Multiple Select: 3
16.7	Explain the stages of meiosis and contrast meiosis to mitosis.	Sperm production; Mitosis vs. meiosis: pp. 604–606 Figure 16.12	*Concept maps:* Meiosis Figure 16.5	*Text:* Review: 16.7 *Workbook:* Multiple Select: 4

Learning Outcomes		Text Readings, Figures, and Tables	Workbook Activities	Text and Workbook Assessments
16.8	Explain the processes of sperm production and differentiate between spermatogenesis and spermiogenesis.	Spermatogenesis: pp. 606–609 Figures 16.13–16.15	*Lab exercises and activities:* Sperm production Table 16.1 *Concept maps:* Sperm production Figure 16.6	*Text:* Spot Check: 7 Review: 16.8 *Workbook:* Matching: 1–5
16.9	Explain the hormonal control of the adult male reproductive system.	Hormonal control in the adult male: pp. 609–610 Figure 16.16	*Concept maps:* Hormonal control in the adult male Figure 16.7	*Text:* Spot Check: 9 Review: 16.9 *Workbook:* Multiple Select: 5
16.10	Trace the path a sperm takes from its formation to its ejaculation.	Pathway for sperm: p. 610		*Text:* Review: 16.10 *Workbook:* Critical Thinking: 3
16.11	Describe the stages of the male sexual response.	Sexual response in the male: pp. 611–612 Figure 16.17	*Lab exercises and activities:* Male sexual response Table 16.2	*Text:* Spot Check: 10, 11 Review: 16.11 *Workbook:* Multiple Select: 6
16.12	Explain the effects of aging on the male reproductive system.	Effects of aging on the male reproductive system: pp. 612–613		*Text:* Review: 16.12 *Workbook:* Multiple Select: 7
16.13	Describe common diagnostic tests used for male reproductive system disorders.	Diagnostic tests for male reproductive system disorders: p. 613 Table 16.1		*Text:* Review: 16.13 *Workbook:* Case Study: 3
16.14	Describe male reproductive system disorders and relate abnormal function to pathology.	Male reproductive system disorders: pp. 614–617 Figures 16.18–16.22 Table 16.2		*Text:* Review: 16.14 *Workbook:* Matching: 6–10 Case Study: 1, 2, 4

17 The Female Reproductive System

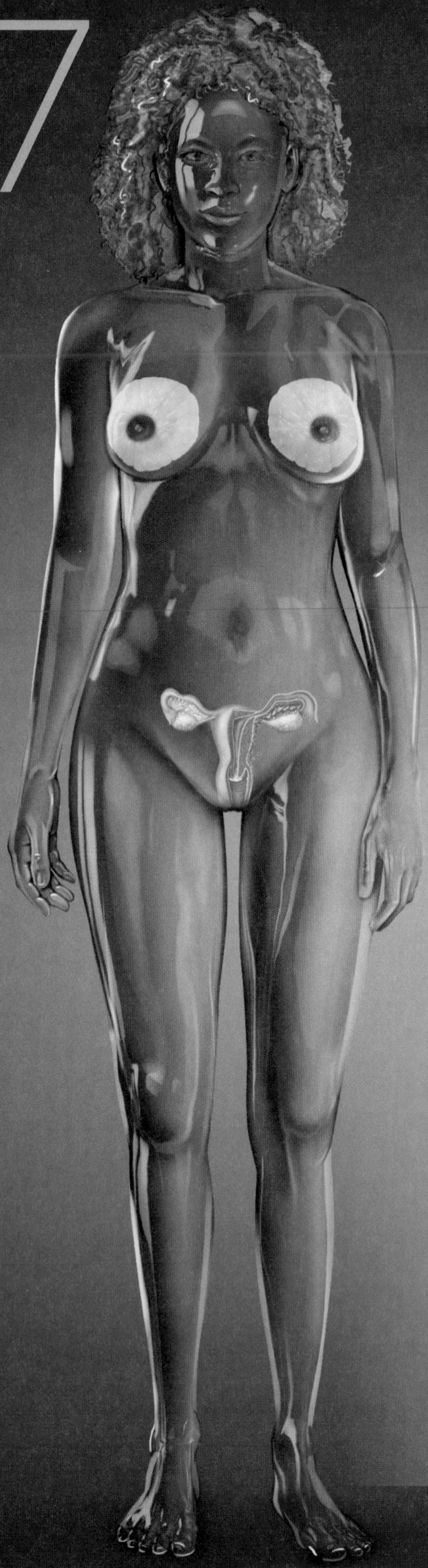

Major Organs and Structures:
ovaries

Accessory Structures:
uterus, uterine tubes, vagina, vulva, breasts

Functions:
production of an egg, housing of fetus, birth, lactation, secretion of sex hormones

learning outcomes

This chapter of the workbook is designed to help you learn the anatomy and physiology of the female reproductive system. After completing this chapter in the text and this workbook, you should be able to:

17.1 Use medical terminology related to the female reproductive system.

17.2 Explain what is needed for female anatomy to develop.

17.3 Describe the anatomy of the ovary and its functions.

17.4 Describe the female secondary reproductive organs and structures and their respective functions.

17.5 Explain the hormonal control of puberty and the resulting changes in the female.

17.6 Explain oogenesis in relation to meiosis.

17.7 Explain the hormonal control of the adult female reproductive system and its effect on follicles in the ovary and the uterine lining.

17.8 Describe the stages of the female sexual response.

17.9 Explain the effects of aging on the female reproductive system.

17.10 Describe common diagnostic tests for female reproductive system disorders.

17.11 Describe female reproductive disorders and relate abnormal function to pathology.

17.12 List the four requirements of pregnancy.

17.13 Trace the pathway for a sperm to fertilize an egg.

17.14 Describe the events necessary for fertilization and implantation.

17.15 Explain the hormonal control of pregnancy.

17.16 Explain the adjustments a woman's body makes to accommodate a pregnancy.

17.17 Explain the nutritional requirements for a healthy pregnancy.

17.18 Explain what initiates the birth process.

17.19 Describe the birth process.

17.20 Explain the process of lactation.

17.21 Describe disorders of pregnancy and relate abnormal function to pathology.

word **roots** & combining **forms**

amni/o: amnion

cervic/o: cervix, neck

chorion/o: chorion

episi/o: vulva

gynec/o: female

hyster/o: uterus

lact/o: milk

mamm/o: breast

mast/o: breast

men/o: menses, menstruation

metr/o, metri/o: uterus

o/o: egg

oophor/o: ovary

ov/o: egg

ovari/o: ovary

ovul/o: egg

salping/o: uterine tube

uter/o: uterus

vagin/o: vagina

vulv/o: vulva

An Ovary

Figure 17.1 shows the anatomy of an ovary. Color the box next to each term. Use the same color for the corresponding structure in the figure.

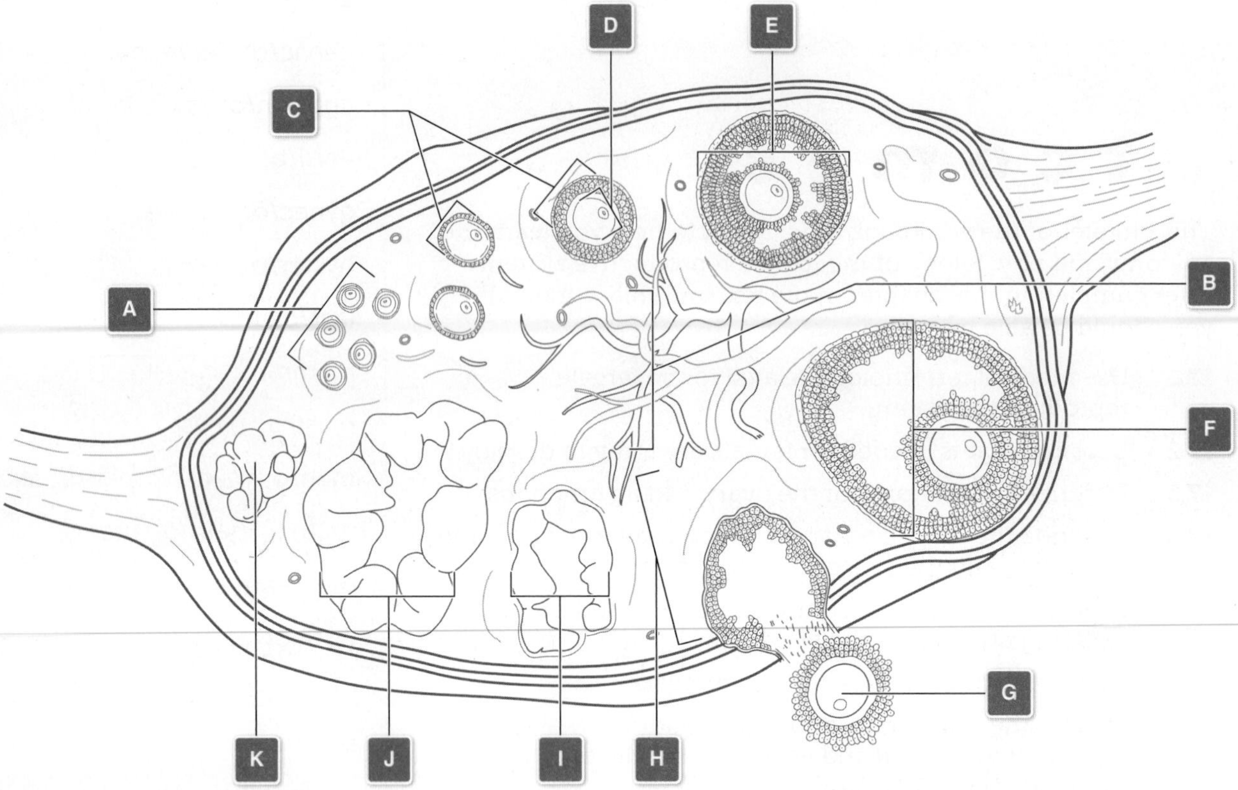

FIGURE 17.1 Anatomy of an ovary.

- ☐ Primordial follicle(A)
- ☐ Medulla of ovary(B)
- ☐ Primary follicles(C)
- ☐ Primary oocyte(D)
- ☐ Secondary follicle(E)
- ☐ Mature follicle(F)
- ☐ Ovulated secondary oocyte(G)
- ☐ Cortex of ovary(H)
- ☐ Developing corpus luteum(I) (color yellow)
- ☐ Corpus luteum(J) (color yellow)
- ☐ Corpus albicans(K) (color white)

COLORING BOOK

Internal Female Reproductive Anatomy

Figure 17.2 shows the anatomy of the internal female reproductive system. Color the box next to each term. Use the same color for the corresponding structure in the figure.

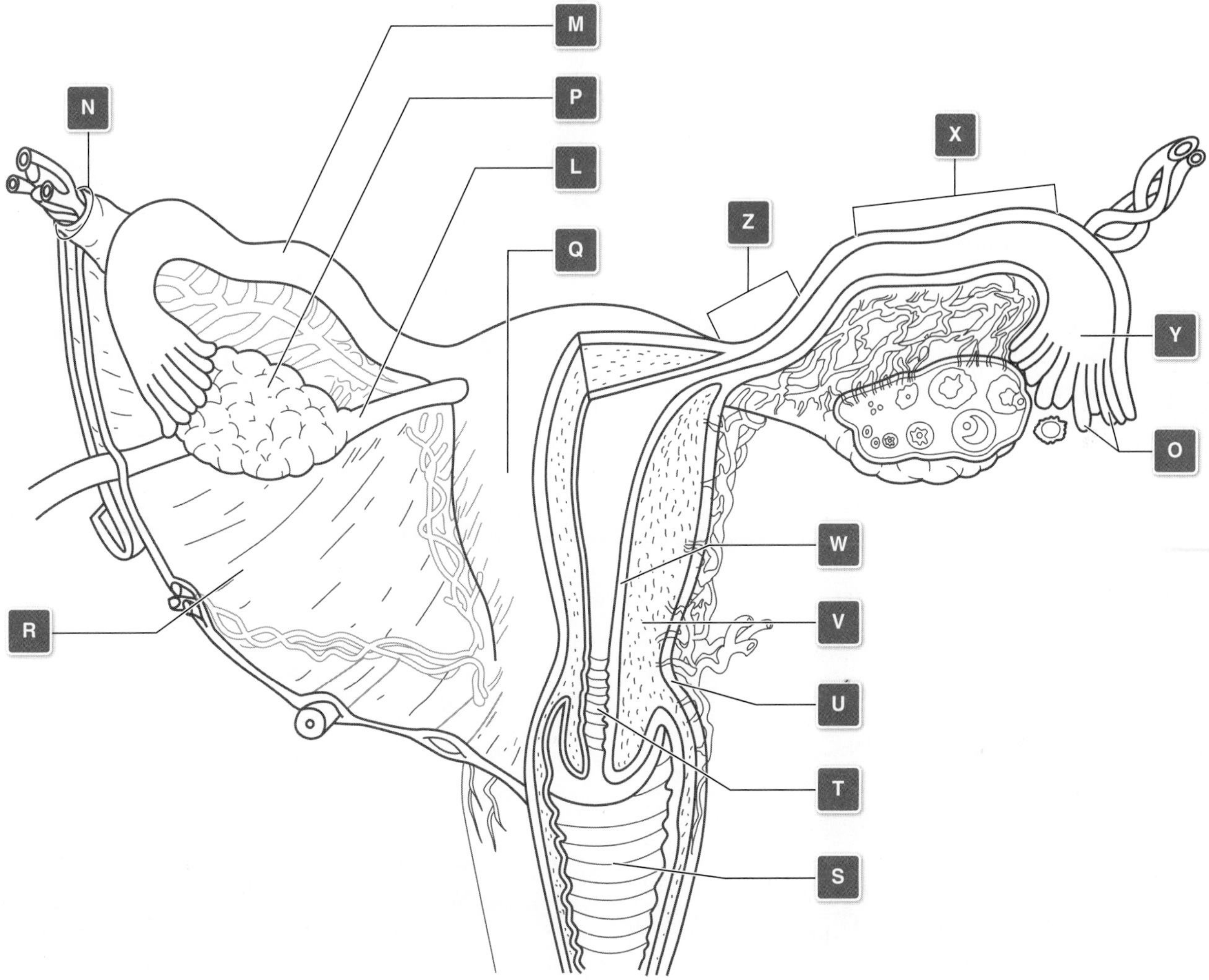

FIGURE 17.2 Internal female reproductive anatomy.

- ☐ Ovarian ligament(L)
- ☐ Uterine tube(M)
- ☐ Suspensory ligament(N)
- ☐ Fimbriae(O)
- ☐ Ovary(P)
- ☐ Body of uterus(Q)
- ☐ Broad ligament(R)
- ☐ Vagina(S)
- ☐ Cervical canal(T)
- ☐ Perimetrium(U)
- ☐ Myometrium(V)
- ☐ Endometrium(W)
- ☐ Ampulla(X)
- ☐ Infundibulum(Y)
- ☐ Isthmus(Z)

A Breast

Figure 17.3 shows the anatomy of a breast. Color the box next to each term. Use the same color for the corresponding structure in the figure.

- ☐ Suspensory ligaments(A)
- ☐ Lobe(B)
- ☐ Areolar gland(C)
- ☐ Nipple(D)
- ☐ Areola(E)
- ☐ Lactiferous sinus(F)
- ☐ Lactiferous ducts(G)
- ☐ Adipose tissue(H)

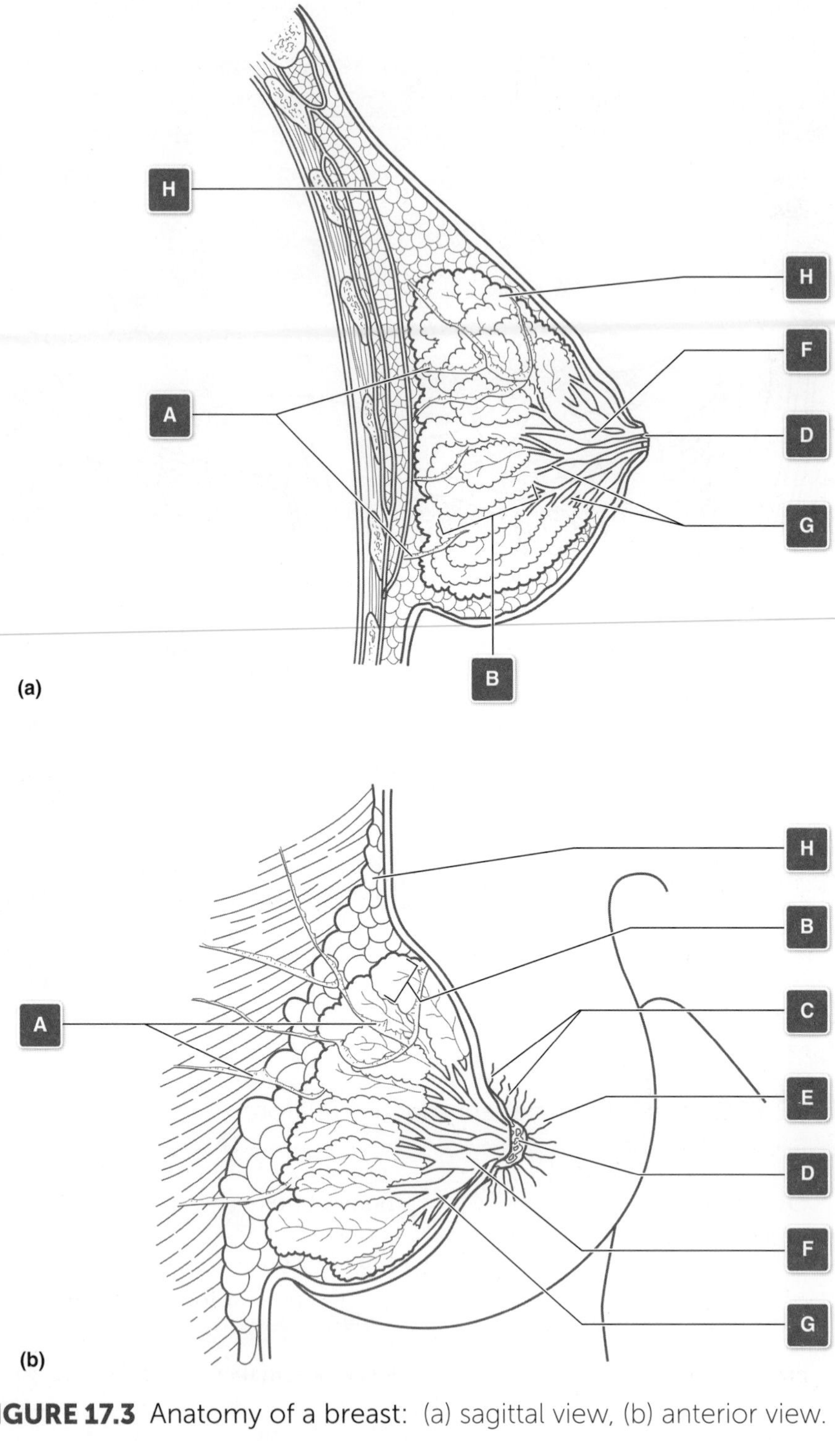

FIGURE 17.3 Anatomy of a breast: (a) sagittal view, (b) anterior view.

COLORING BOOK

Oogenesis Timeline

Oogenesis produces gametes in the female. Like spermatogenesis, oogenesis produces gametes with 23 chromosomes, but oogenesis differs greatly from spermatogenesis in the timing of the gamete production. In Table 17.1, complete the timeline for oogenesis using the information provided in the textbook.

TABLE 17.1 Oogenesis timeline

Stage of life	Age	Description of gametes (term for gamete, stage of division, number of gametes)
Conception to birth ↓	Fetus to 6 months old	
Puberty ↓		
Adolescence up to menopause ↓		
Menopause to end of life		

Contraceptives

Another way to study the anatomy and physiology of the reproductive system is to study how to prevent the system from working. Pregnancy may not be the desired outcome each time a couple engages in sexual intercourse. Therefore, many couples use methods of contraception to avoid a pregnancy. As you may recall, the four requirements for a pregnancy to occur are:

1. There must be a sperm and an egg.
2. The sperm must meet the egg.
3. The sperm must fertilize the egg.
4. The fertilized egg must implant in the uterus.

Contraceptive practices, medications, or devices may be used by a man or a woman to prevent a pregnancy. Complete Table 17.2 on contraceptives. As you can see, a few of the boxes in this table have been filled in for you. Here is additional information about each of the columns in the table:

Contraceptive: There are many contraceptive measures that can be used to avoid a pregnancy. Nine are listed in Table 17.2. You need to add another five methods to the table.

Description: In this column, describe how each method uses the anatomy and physiology of the reproductive system to avoid a pregnancy. Also include who uses it—the man or the woman.

Pregnancy requirement blocked: Select which of the four pregnancy requirements above is directly blocked by each method.

Effectiveness percentage: Notice in this column that two types of values are listed—theoretical effectiveness and actual effectiveness. *Theoretical* means how well the method should work if it is carried out perfectly. *Actual* means how well the method works given human error and product failure. There is often a big difference in the two values. Credible sources will have this information.

Source: It is always important to have credible sources. Document your sources in this column. You may want to consult with your instructor on what is considered a credible source.

TABLE 17.2 Contraceptives

Contraceptive	Description (How does it work? Who uses it: man or woman?)	Pregnancy requirement blocked	Effectiveness percentage	Source
Abstinence	The man and woman do not engage in sexual intercourse.	2. Sperm does not meet the egg.	Theoretical: 100% Actual: 100%	The author of this workbook
Withdrawal			Theoretical: Actual:	
Male condom			Theoretical: Actual:	
Vasectomy			Theoretical: Actual:	
Diaphragm with spermicide			Theoretical: Actual:	

TABLE 17.2 concluded

Contraceptive	Description (How does it work? Who uses it: man or woman?)	Pregnancy requirement blocked	Effectiveness percentage	Source
IUD			Theoretical: Actual:	
Birth control pill			Theoretical: Actual:	
Natural family planning			Theoretical: Actual:	
Rhythm			Theoretical: Actual:	
			Theoretical: Actual:	
			Theoretical: Actual:	
			Theoretical: Actual:	
			Theoretical: Actual:	
			Theoretical: Actual:	

Key Words

The following terms are defined in the glossary of the textbook.

afterbirth
atresia
capacitation
colostrum
crowning
effacement
episiotomy
folliculogenesis
gestation
labor
lactation
mammography
menopause
menstrual cycle
milk ejection reflex
oogenesis
ovarian cycle
ovulation
parturition
pelvic inflammatory disease
placenta previa
placental abruption
preeclampsia
prolapse

Concept Maps

Use key words and other bold words from the chapter to complete the following concept maps (**Figures** 17.4 to 17.8).

c. In the cervical canal.
d. In the lumen of the uterus.
e. In the isthmus of the vagina.

7. What needs to happen for fertilization to occur?
   a. The sperm must meet the egg within 24 hours of ovulation.
   b. The acrosome caps must release their enzymes so that sperm can make their way through the cells surrounding the egg.
   c. A single sperm must penetrate the egg.
   d. The nuclei of the sperm and the egg must rupture so that the chromosomes from each may mix.
   e. A new nucleus having 46 chromosomes is formed.
8. How does a female body adjust to a pregnancy?
   a. Her bone mass may decrease due to PTH.
   b. Her cardiac output decreases.
   c. She becomes more sensitive to carbon dioxide because of increased estrogen.
   d. Her skin may darken.
   e. Her breasts may double in size due to increased FSH and LH.
9. What initiates a birth?
   a. The fetus produces glucocorticoids that target the placenta.
   b. The fetal glucocorticoids cause the placenta to level off its production of progesterone and start secreting prostaglandins.
   c. Uterine stretch receptors target the fetal hypothalamus so that more hormones are released.
   d. Oxytocin is released from the mother's posterior pituitary.
   e. Together, oxytocin and progesterone stimulate uterine contractions.
10. What happens during a normal birth process?
    a. The afterbirth goes through involution.
    b. The cervix widens, a process called *effacement.*
    c. The head is seen first because the fetus usually assumes a head-down position in the seventh month.
    d. The cervix dilates to 10 cm during stage I.
    e. About 3.5 L of blood is usually lost when the placenta detaches.

Matching: *Match the follicle to its description. Some choices may be used more than once.*

_______	**1.** Type of follicle found in a two-year-old	**a.** Corpus luteum
_______	**2.** Follicle that has completed involution	**b.** Corpus albicans
_______	**3.** Type of follicle found in an 80-year-old	**c.** Primordial follicle
_______	**4.** Follicle that ruptures to release an egg	**d.** Primary follicle
_______	**5.** Follicle containing a secondary oocyte	**e.** Mature follicle

Matching: *Match the pregnancy hormone to its description. Some choices may be used more than once.*

_______	**6.** Targets the corpus luteum to keep it functioning	**a.** Estrogen
_______	**7.** Increases fat metabolism for the mother	**b.** Progesterone
_______	**8.** Suppresses uterine contractions	**c.** Oxytocin
_______	**9.** Causes the uterus to contract	**d.** HCS
_______	**10.** Makes the uterus more likely to contract	**e.** HCG

Completion: *Fill in the blanks to complete the following statements.*

1. The sex chromosomes (23rd pair) in a female zygote are _______________.
2. Surges in _______________ each time an infant nurses ensure that ample milk will be produced.
3. A healthy pregnancy requires a balanced diet with emphasis on four nutrients: _______________, _______________, _______________, and _______________.
4. _______________ is pregnancy-induced hypertension accompanied by protein in the urine.
5. _______________ is an overgrowth of the uterine lining in places other than in the uterus.

Critical Thinking

1. What can a woman do to reduce the effects of aging on her reproductive system?

____

____

____

____

2. Which is the only requirement of pregnancy that cannot be directly blocked by a method of contraception?

____

____

3. What would a working mother need to do to make sure her milk production provided enough milk for her infant? Explain.

____

____

____

____

____

Case Study

Karen is in her 50s and experiencing major changes related to her reproductive system.

1. What is the name for this stage in Karen's life?

___

2. What are the changes related to aging Karen is likely experiencing?

___

___

___

___

3. Specifically, what will result in Karen's inability to have more children? Explain.

___

___

This section of the chapter is designed to help you find where each outcome is covered in the textbook and workbook.

Learning Outcomes		Text Readings, Figures, and Tables	Workbook Activities	Text and Workbook Assessments
17.1	Use medical terminology related to the female reproductive system.	Word roots & combining forms; pronunciation key: p. 623	Word roots & combining forms	*Text:* Review: 17.1 *Workbook:* Word Deconstruction: 1–5
17.2	Explain what is needed for female anatomy to develop.	Overview: p. 624		*Text:* Review: 17.2 *Workbook:* Completion: 1
17.3	Describe the anatomy of the ovary and its functions.	Ovaries: pp. 624–626 Figures 17.2–17.3	*Coloring book:* An ovary Figure 17.1	*Text:* Spot Check: 1 Review: 17.3 *Workbook:* Matching: 1–5
17.4	Describe the female secondary reproductive organs and structures and their respective functions.	Secondary female reproductive organs and structures: pp. 626–631 Figures 17.3–17.7	*Coloring book:* Internal female reproductive anatomy; A breast Figures 17.2–17.3	*Text:* Spot Check: 2 Review: 17.4 *Workbook:* Multiple Select: 1
17.5	Explain the hormonal control of puberty and the resulting changes in the female.	Hormonal control at puberty: pp. 631–632	*Concept maps:* Hormonal control at puberty Figure 17.5	*Text:* Spot Check: 3 Review: 17.5 *Workbook:* Multiple Select: 2
17.6	Explain oogenesis in relation to meiosis.	Oogenesis: pp. 632–635 Figures 17.8–17.9	*Lab exercises and activities:* Oogenesis timeline Table 17.1 *Concept maps:* Oogenesis Figure 17.4	*Text:* Spot Check: 4, 5 Review: 17.6 *Workbook:* Multiple Select: 3
17.7	Explain the hormonal control of the adult female reproductive system and its effect on follicles in the ovary and the uterine lining.	Hormonal control in the adult female: pp. 635–639 Figures 17.10–17.12	*Concept maps:* Hormonal control in an adult female Figure 17.6	*Text:* Spot Check: 6 Review: 17.7 *Workbook:* Multiple Select: 4
17.8	Describe the stages of the female sexual response.	The female sexual response: pp. 639–641 Figure 17.13		*Text:* Review: 17.8 *Workbook:* Multiple Select: 5
17.9	Explain the effects of aging on the female reproductive system.	Effects of aging on the female reproductive system: pp. 641–642		*Text:* Spot Check: 7, 8 Review: 17.9 *Workbook:* Critical Thinking: 1 Case study: 1–3
17.10	Describe common diagnostic tests for female reproductive system disorders.	Diagnostic tests for female reproductive system disorders: p. 642 Table 17.1		*Text:* Review: 17.10

Learning Outcomes		Text Readings, Figures, and Tables	Workbook Activities	Text and Workbook Assessments
17.11	Describe female reproductive disorders and relate abnormal function to pathology.	Disorders of the female reproductive system: pp. 643–646 Figures 17.14–17.17 Table 17.2		*Text:* Spot Check: 9 Review: 17.11 *Workbook:* Completion: 5
17.12	List the four requirements of pregnancy.	Pregnancy: p. 647 Figure 17.18	*Lab exercises and activities:* Contraceptives Table 17.2	*Text:* Review: 17.12 *Workbook:* Critical Thinking: 2
17.13	Trace the pathway for a sperm to fertilize an egg.	Pathway for sperm to meet an egg: p. 647		*Text:* Review: 17.13 *Workbook:* Multiple Select: 6
17.14	Describe the events necessary for fertilization and implantation.	Fertilization to implantation: pp. 647–649 Figures 17.19–17.20		*Text:* Spot Check: 10 Review: 17.14 *Workbook:* Multiple Select: 7
17.15	Explain the hormonal control of pregnancy.	Hormonal control of pregnancy: pp. 649–650	*Concept maps:* Hormonal control of pregnancy Figure 17.7	*Text:* Review: 17.15 *Workbook:* Matching: 6–10
17.16	Explain the adjustments a woman's body makes to accommodate a pregnancy.	A woman's adjustment to pregnancy: pp. 650–652 Figure 17.21		*Text:* Review: 17.16 *Workbook:* Multiple Select: 8
17.17	Explain the nutritional requirements for a healthy pregnancy.	Nutritional requirements for a pregnancy: p. 652		*Text:* Spot Check: 11 Review: 17.17 *Workbook:* Completion: 3
17.18	Explain what initiates the birth process.	Initiating the birth process: pp. 652–653 Figure 17.22		*Text:* Review: 17.18 *Workbook:* Multiple Select: 9
17.19	Describe the birth process.	The birth process: pp. 653–654 Figures 17.23–17.24	*Concept maps:* The birth process Figure 17.8	*Text:* Review: 17.19 *Workbook:* Multiple Select: 10
17.20	Explain the process of lactation.	Lactation: p. 655 Figure 17.25		*Text:* Review: 17.20 *Workbook:* Completion: 2 Critical Thinking: 3
17.21	Describe disorders of pregnancy and relate abnormal function to pathology.	Disorders of pregnancy: pp. 656–658 Figures 17.26–17.27 Table 17.2		*Text:* Spot Check: 12 Review: 17.21 *Workbook:* Completion: 4

photo credits

Chapter 1
Figure 1.5: Courtesy Deborah Roiger; 1.6: © F. Schussler/PhotoLink/Getty Images RF; 1.11: © The McGraw-Hill Companies, Inc./Rebecca Gray, photographer/Don Kincaid, dissections; 1.12: © The McGraw-Hill Companies, Inc.; 1.13: © The McGraw-Hill Companies, Inc./Rebecca Gray, photographer/Don Kincaid, dissections; p. 16: © PhotoLink/Getty Images RF.

Chapter 2
Figures 2.33-2.35: Courtesy Deborah Roiger; p. 56: © Medicimage/Visuals Unlimited.

Chapter 3
Opener: © The McGraw-Hill Companies, Inc.; p. 72: © image100/Getty Images RF.

Chapter 4
Figures 4.4-4.6: Courtesy Deborah Roiger; p. 93: © Ryan McVay/Getty Images RF.

Chapter 5
Figures 5.5-5.8b: Courtesy Deborah Roiger; 5.9a-b: © The McGraw-Hill Companies, Inc./Photo and Dissection by Christine Eckel; p. 121: © The McGraw-Hill Companies, Inc./Rick Brady, photographer.

Chapter 6
Figure 6.8a-b: Courtesy Deborah Roiger; p. 148: © Stockbyte/PictureQuest RF.

Chapter 7
Page 170: © Royalty-Free/Corbis.

Chapter 8
Page 192: © FogStock/Alamy RF.

Chapter 9
Page 207: © Science Photo Library RF/Getty Images.

Chapter 10
Page 227: © JGI/Jamie Grill/Blend Images LLC RF.

Chapter 11
Page 249: © Steven Peters/Getty Images.

Chapter 12
Page 266: © Phototake, Inc./Phototake.

Chapter 13
Figure 13.4: © Ward's Natural Science; p. 289: © Fuse/Getty Images RF.

Chapter 14
Page 310: © Laurent/Dielundama/Photo Researchers, Inc.

Chapter 15
Figure 15.5: © C. Sherburne/PhotoLink/Getty Images RF; 15.6: © Kim Taylor and Jane Burton/Getty Images; 15.7-15.8: © The McGraw-Hill Companies, Inc./Jill Braaten, photographer; p. 335: © Blend Images/Alamy RF.

Chapter 16
Page 353: © BSIP/Corbis.

Chapter 17
Page 373: © Jose Luis Pelaez Inc./Blend Images LLC RF.

NOTES

NOTES

NOTES

NOTES

NOTES

NOTES

NOTES

NOTES

NOTES

NOTES

NOTES

NOTES

NOTES

NOTES